United States Nuclear Regulatory Commission

Protecting People and the Environment

I0482779

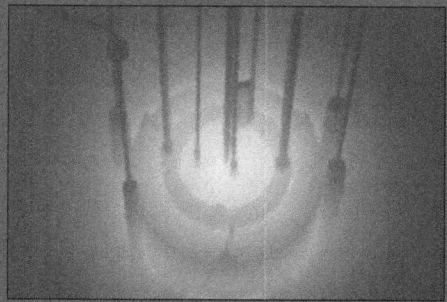

Research Activities

FY 2012–FY 2014

Office of Nuclear Regulatory Research (RES)

AVAILABILITY OF REFERENCE MATERIALS
IN NRC PUBLICATIONS

Research Activities
FY 2012-FY 2014

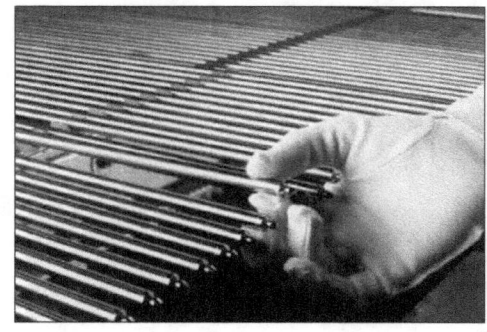
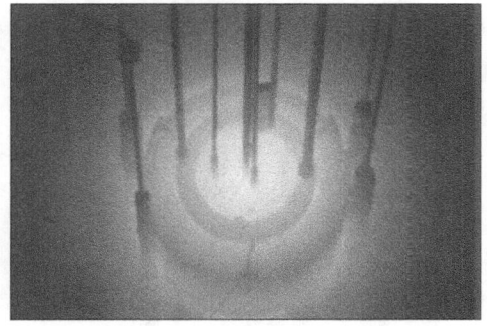

NUREG-1925, Rev. 2
Manuscript Completed: April 2013
Date Published: August 2013

U.S. Nuclear Regulatory Commission
Office of Nuclear Regulatory Research (RES)
Washington, D.C. 20555-0001

www.nrc.gov

Abstract

The Office of Nuclear Regulatory Research (RES) supports the regulatory mission of the U.S. Nuclear Regulatory Commission (NRC) by providing technical advice, tools, and information to identify and resolve safety issues, make regulatory decisions, and issue regulations and guidance. This includes conducting confirmatory experiments and analyses; developing technical bases that support the NRC's safety decisions; and preparing the agency for the future by evaluating the safety aspects of new technologies and designs for nuclear reactors, materials, waste, and security.

The NRC faces challenges as the industry matures, including potential new safety issues, the availability of new technologies, technical issues associated with the deployment of new reactor designs, and knowledge management. The NRC focuses its research primarily on near-term needs related to the oversight of operating reactors, as well as to new and advanced reactor designs. RES develops technical tools, analytical models, and experimental data to allow the agency to assess safety and regulatory issues. The RES staff uses its expertise to develop these tools, models, and data or uses contracts with commercial entities, national laboratories, and universities or in collaboration with international organizations.

This NUREG presents research conducted across a wide variety of disciplines, ranging from fuel behavior under accident conditions to seismology to health physics. This research provides the technical bases for regulatory decisions and confirms licensee analyses. RES works closely with the NRC's licensing offices in the review and analysis of high-risk events and provides its expertise to support licensing. RES has organized this collection of information sheets by topical areas that summarize projects currently in progress. Each sheet provides the names of the RES technical staff who can be contacted for additional information.

Foreword

A Message from the Director

The Office of Nuclear Regulatory Research (RES) is a major U.S. Nuclear Regulatory Commission (NRC) program office, mandated by Congress. The office plans, recommends, and implements a program of nuclear regulatory research, standards development, and resolution of generic safety issues for nuclear power plants and other facilities regulated by the NRC. RES partners with other NRC program offices, Federal agencies, industry research organizations, international organizations, and universities. This NUREG identifies numerous key program projects and their status.

Much of the office's work is available to the public through NUREG and NUREG/CR series reports that describe various research projects and the associated results. In fiscal year (FY) 2012, the RES staff issued 34 NUREG reports on a wide variety of topics including the State-of-the-Art Reactor Consequence Analyses project, seismic source characterization for the Central and Eastern United States, hurricane wind speeds, stress-corrosion cracking, and multiple fire research projects. Some of the highlighted FY 2012–2014 projects include severe accident analysis (Chapter 3), the analysis of cancer risk in populations living near nuclear facilities (Chapter 4), Level 3 Probabilistic Risk Assessment (Chapter 5), human reliability analysis activities (Chapter 6), seismic and structural research (Chapter 8), and international cooperative research (Chapter 12). RES and the regulatory offices also continue to focus on other issues such as dissimilar metal weld cracking inspections and mitigation, cable aging, and other aging-related materials issues, digital instrumentation and control, Fukushima lessons learned, and new and advanced reactors. These are simply a few of the critical research projects contained in this report and expected to continue into the future.

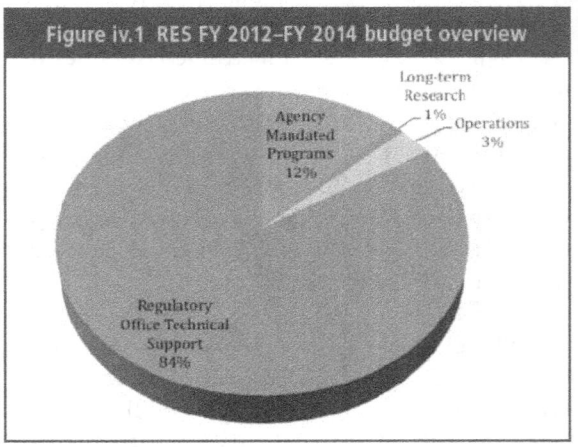

Figure iv.1 RES FY 2012–FY 2014 budget overview

The office accomplishes its regulatory research mission by conducting research both in-house and with the use of contractors. The office's annual budget for contracted work is typically around $50 million; the chart in Figure iv.1 illustrates the funding breakdown, which is described below:

- The needs of NRC's regulatory offices drive over three-fourths of RES activities (through user needs).

- The NRC Commission drives about 10 percent of RES activities (through agency-mandated programs and Commission tasking memoranda).

- A small amount of long-term research supports anticipated future NRC regulatory needs on subjects expected to be critical in 5 to 10 years.

- About 3 percent of the office's budget is spent on operations, which includes staff travel and training, and information technology purchases.

Currently, RES has about 240 staff members. This staff continues to reflect diversity in academic degrees, demographics, and technical disciplines. The wide range of engineering and scientific disciplines includes expertise in nuclear materials, human factors and human reliability, health physics, fire protection, seismology, environmental, and probabilistic risk assessment.

In summary, RES appreciates your interest in these activities and will continue to issue updates of this NUREG for your information. Additional questions or comments on the content should be directed to the technical staff or the division noted on each specific project summary sheet.

Brian W. Sheron, Director
Office of Nuclear Regulatory Research

Table of Contents

Figures

Abbreviations and Acronyms

Numerals

ΔCDP	change in core damage probability
%OLTP	percent of originally licensed thermal power
%RCF	percent of rated core flow
10 CFR	Title 10 of the *Code of Federal Regulations*
4S	Toshiba Super Safe, Small and Simple reactor

A

ABAQUS	Suite of software applications for finite-element analysis and computer-aided engineering
ABWR	advanced boiling-water reactor
ac	alternating current
ACI	American Concrete Institute
ACRS	Advisory Committee on Reactor Safeguards
ADAMS	Agencywide Documents Access and Management System (NRC)
ADS-IDAC	Accident Dynamics Simulator Using Information, Decisions, and Actions in a Crew Context
AEA	Atomic Energy Act
AEC	U.S. Atomic Energy Commission
AECL	Atomic Energy of Canada Ltd.
AERB	Atomic Energy Regulatory Board (India)
AES	Advanced Environmental Solutions, LLC
ALARA	as low as reasonably achievable
AMFL	Advanced Multi-Phase Flow Laboratory
AMP	aging management program
AMPX	Advanced Module for Processing Cross-sections
ANL	Argonne National Laboratory
ANS	American Nuclear Society
ANSI	American National Standards Institute
ANSYS	engineering simulation software developer
AO	abnormal occurrence
AP1000	Advanced Passive 1000 Megawatt
APEX	Advanced Power Extraction
API	application programming interface
APWR	U.S. Advanced pressurized-water Reactor (Mitsubishi)
ARRP	Advanced Reactor Research Program
ARS	Agricultural Research Service
ARTIST	Aerosol Trapping In Steam generator

ASCII	American Standard Code for Information Interchange
ASD	aspirating smoke detectors
ASEP	Accident Sequence Evaluation Program
ASME	American Society of Mechanical Engineers
ASP	accident sequence precursor
ASR	alkali-silica reaction
ASTM	American Society for Testing and Materials
ATHEANA	A Technique for Human Event Analysis
ATWS	anticipated transient without scram

B

BADGER	Boron-10 Areal Density Gauge for Evaluating Racks
BAM	German Federal Institute for Materials Research and Testing
BDD	binary decision diagrams
BETHSY	loop for the study of thermal-hydraulic system
BFBT	BWR full-size fine-mesh bundle test
BFN	Browns Ferry Nuclear Power Plant
BIP	Behavior of Iodine Project (CSNI)
BMI	bottom-mounted instrumentation
BNL	Brookhaven National Laboratory
BRIIE	Baseline Risk Index for Initiating Events
BUC	burnup credit
BWR	boiling-water reactor
B&W	Babcock & Wilcox

C

C	Celsius
C-SGTR	consequential steam generator tube rupture
Cal/g	calorie per gram
CAMP	Code Application and Maintenance Program (NRC)
CANDU	Canada Deuterium Uranium reactor
CAROLFIRE	Cable Response to Live Fire
CBDT	cause-based decision tree
CCDP	conditional core damage probability
CCF	common-cause failure
CCI	core-concrete interaction
CDA	critical digital asset
CDF	core damage frequency
CE	Combustion Engineering
CEUS	Central and Eastern United States
CEUS SSC	Central and Eastern United States Seismic Source Characterization
CFAST	Consolidated Fire Growth and Smoke Transport Model

CFD	computational fluid dynamics	DICB	Digital Instrumentation and Control Branch (NRC)
CFR	*Code of Federal Regulations*	DI&C	digital instrumentation and control
CHF	critical heat flux	DIRS	NRR Division of Inspection and Regional Support
CHRISTI-FIRE	Cable Heat Release, Ignition, and Spread in Tray Installations during Fire	DM	dissimilar metal
COL	combined license	DMW	dissimilar metal weld
CONOPS	concepts of operations	DNA	deoxyribonucleic acid
CONTAIN	Containment Analysis Code	DOE	U.S. Department of Energy
CoP	community of practice	DRA	RES Division of Risk Analysis
CP	computerized procedure		
CPLD	complex programmable logic device		
CR	control room	**E**	
CRAC	Calculation of Reactor Accident Consequences	EAC	environmentally assisted cracking
		EAF	environmentally assisted fatigue
CRDM	control rod drive mechanism	EAL	emergency action level
CRGR	Committee to Review Generic Requirements	EC	economic consequences
CRPPH	International Commission on Radiation Protection	EC	emergency classification
		ECCS	emergency core cooling system
CRT	crew response tree	ECI	exterior communications interface
CRUD	Chalk River Unidentified Deposit	ECL	emergency classification level
CSARP	Cooperative Severe Accident Research Program (NRC)	ECR	equivalent cladding reacted
		EDF	Electricite de France
CSAU	Code Scaling, Applicability, and Uncertainty	EDO	Executive Director for Operations
CSM	conceptual site model	EGOE	Export Group on Occupational Exposure
CSNI	Committee on the Safety of Nuclear Installations	ELECTRA-FIRE	Electrical Cable Test Results and Analysis during Fire Exposure
CTP	crack-tip parameter	ENDF	Evaluated Nuclear Data File
CUF	cumulative usage factor	EP	emergency preparedness
CV	cross vessel	EPA	U.S. Environmental Protection Agency
		EPAct	Energy Policy Act of 2005
		EPICUR	Experimental Program for Iodine Chemistry Under Irradiation
D			
D3	diversity and defense-in-depth	EPIX	Equipment Performance and Information Exchange System
DAKOTA	Design Analysis Kit for Optimization and Terascale Applications		
		EPR	U.S. Evolutionary Power Reactor
DART	Deep-ocean Assessment and Reporting of Tsunamis	EPRI	Electric Power Research Institute, Inc.
		EPU	extended power uprates
DBA	design-basis accident	EQ	environmental qualification
DBT	design-basis threat	ESBWR	Economic Simplified boiling-water Reactor (General Electric)
DC	design certification		
dc	direct current	EST	extended storage and transportation
DCSS	dry cask storage system	ETB	Environmental Transport Branch
DES	discrete element simulation		
DESIREE-FIRE	Direct Current Electrical Shorting in Response to Exposure Fire	**F**	
		FAQ	frequently-asked-questions
DF	decontamination factor	FCF	fuel cycle facility
DFWCS	digital feedwater control system	FCOP	Fuel Cycle Oversight Process
DHS	U.S. Department of Homeland Security	FDS	Fire Dynamics Simulator

FDT	Fire Dynamics Tools	HEP	human error probability
FE	Finite-element	HERA	Human Event Repository and Analysis
Fe	iron	HF	human factors
FEA	finite-element analysis	HFE	human factors engineering
FERC	Federal Energy Regulatory Commission	HFE	human failure event
FIVE	Fire-Induced Vulnerability Evaluation	HGL	hot gas layer
FFD	fitness for duty	HHA	hierarchical hazard assessment
FLASH-CAT	Flame Spread over Horizontal Cable Trays	HMR	hydrometeorology report
FLECHT	Full Length Emergency Cooling Heat Transfer	HPP	human performance profile
FLUENT	computer code used for CFD and FEA	HQ	headquarters
FP	fission product	HRA	human reliability analysis
FPGA	field programmable gate array	HRP	Halden Reactor Project
FPRP	Fire Protection Research Program	HRR	heat release rate
FPT	fission product transport	HRRPUA	heat release rate per unit area
FR	Federal Register	HSI	human-system interface
FPRA	Fir Probabilistic Risk Assessment	HTGR	high-temperature gas-cooled reactor
FRAPCON3	Fuel Rod Analysis Program (CONstant (steady state) version)	HZP	hot zero power
FRAPTRAN	Fuel Rod Analysis Program (TRANsient version)	**I**	
		IA	International Agreement
FRB	Fire Research Branch	IAD	irradiation-assisted degradation
FSME	Office of Federal and State Materials and Environmental Management Programs	IAEA	International Atomic Energy Agency
		IAGE	Integrity and Aging of Components and Structures
FY	fiscal year	IASCC	irradiation-assisted stress-corrosion cracking
G		I&C	instrumentation and control
GCC	graphite core component	ICAP	International Code Assessment and Application Program
GDC	general design criterion	ICRP	International Commission on Radiological Protection
GDP	gross domestic product		
GI	generic issue	IEEE	Institute of Electrical and Electronics Engineers
GIP	Generic Issues Program		
GMC	Ground Motion Characterization	IFE	Institutt for Energiteknikk (Norwegian Institute for Energy Technology)
GMPE's	Ground Motion Prediction Equations		
GSC	Geological Survey of Canada	IFRAM	International Forum for Reactor Aging Management
GSI	generic safety issue		
GTAW	gas tungsten arc welded	IHX	Intermediate Heat Exchanger
GUI	graphical user interface	INL	Idaho National Laboratory
GWd/MTU	gigawatt day per metric ton of uranium	IPEEE	individual plant examination of external events
GWd/t	gigawatt day per ton		
		iPWR	Integral pressurized-water Reactor
H		IRIS	International Reactor Innovative and Secure Light Water Reactor (Westinghouse)
HAMMLAB	Halden Man-Machine Laboratory		
HBWR	Halden boiling-water Reactor	IROFS	items relied on for safety
HDPE	high-density polyethylene	IRSN	Institut de Radioprotection et de Surete Nucleaire (French Institute for Radiological Protection and Nuclear Safety)
HDR	high dose rate		
HEAF	high energy arcing faults		
HEB	Health Effects Branch		
HELB	high-energy line break	ISA	integrated safety analysis

iSALE	impact Simplified Arbitrary Lagragean Eulerian
ISEMIR	Information System on Occupational Exposure in Medicine, Industry, and Research
ISFSI	independent spent fuel storage installation
ISG	interim staff guidance
ISI	inservice inspection
ISL	In-Situ leach
ISOE	Information System on Occupational Exposure
ISP	International Standard Problem
ISP-48	International Standard Problem on containment integrity
IST	Integrated System Test
ISTP	International Source Term Program
IT	information technology
ITP	Industry Trends Program

J

JAEA	Japanese Atomic Energy Agency
JAERI	Japan Atomic Energy Research Institute
JACQUE-FIRE	Joint Assessment of Cable Damage and Quantification of Effects from Fire
JCCRER	Joint Coordinating Committee for Radiation Effects Research
JISAO	Joint Institute for the Study of the Atmosphere and Ocean
JLD	Japan Lessons Learned Directorate
JNES	Japan Nuclear Energy Safety Organization
Joan of Arc	Joint Analyses of Arc Faults

K

KATE-FIRE	Kerite Analysis in Thermal Environment of Fire
KM	knowledge management

L

LANL	Los Alamos National Laboratory
LAR	licensee amendment request
LBB	leak before break
LBDMW	large-bore dissimilar metal weld
LD	leak detection
LER	licensee event report
LERF	large early release frequency
LLW	low-level waste
LOCA	loss-of-coolant accident
LOFW	loss of feedwater
LOOP	loss-of-offsite-power

LPSD	low-power/shutdown
LRA	license renewal application
LRB/CLB	lead rubber and cross linear bearing
LSDYNA	Livermore Software Technology Corporation for dynamic explicit finite-element analysis
LSTF	large-scale test facility
LTRP	Long-Term Research Program
LWR	light-water reactor

M

MACCS	MELCOR Accident Consequence Code System
MAG	modeling application guide
MAGIC	fire modeling tool
MARIAFIRES	Methods for Applying Risk Analysis to Fire Scenarios
MARSAME	Multi-Agency Radiation Survey and Assessment of Materials and Equipment Manual
MARSSIM	Multi-Agency Radiological Survey and Site Investigation Manual
MASLWR	Multi-Application Light Water Reactor
MASS	MELCOR Accident Simulation Using SNAP
MATLAB	MATrix LABoratory
MCAP	MELCOR Code Assessment Program
MCCI	Melt Coolability and Concrete Interaction
MCNP	Monte Carlo N-Particle Transport Code
MD	monitoring device
MD	management directive
MELCOR	computer code for analyzing severe accidents in NPPs
MELLLA+	the maximum extended load line limit Analysis Plus
MgO	magnesium oxide
MIC	microbiologically induced corrosion
MIRD	medical internal radiation dose
MIT	Massachusetts Institute of Technology
MOR	monthly operating report
MOST	Method of Splitting Tsunami
MOU	memorandum of understanding
MOX	mixed oxide
MOX FFF	Mixed Oxide Fuel Fabrication Facility
MP	monitoring point
MRP	Materials Reliability Project
MSIP	Mechanical Stress Improvement Process
MSLB	main steamline break
MSPI	Mitigating Systems Performance Index
MTO	Man-Technology-Organization

MW	megawatt		OGC	Office of the General Counsel
			OIG	Office of the Inspector General
N			OIP	Office of International Programs
NAS	U.S. National Academy of Sciences		ORNL	Oak Ridge National Laboratory
NCI	U.S. National Cancer Institute			
NCRP	National Council of Radiation Protection and Measurements		**P**	
			PA	performance assessment
NDE	nondestructive examination		PAGs	protective action guidelines
NEA	Nuclear Energy Agency		PARENT	Program to Assess Reliability of Emerging Nondestructive Techniques
NEI	Nuclear Energy Institute			
NERC	North American Electric Reliability Corporation		PANDA	Passive Non-Destructive Assay of Nuclear Materials
NEPA	National Environmental Policy Act		PARCS	Purdue's Advanced Reactor Core Simulator
NESCC	Nuclear Energy Standards Coordination Collaborative		PA-UT	phased array ultrasonic
NFPA	National Fire Protection Association		PBMR	pebble bed modular reactor
NGA	next generation attenuation		PBP	paper-based procedure
NGNP	Next Generation Nuclear Plant		PBPM	planning, budgeting, and management (process)
NGO	Non-Governmental Organization			
NIST	National Institute of Standards and Technology		PBR	pebble bed reactor
			PCCV	Prestressed Concrete Containment Vessel
NMSS	Office of Nuclear Material Safety and Safeguards		PCFC	pyrolysis combustion flow calorimeter
NOAA	National Oceanic and Atmospheric Administration (U.S. Department of Commerce)		PEER	Pacific Earthquake Engineering Research (Center)
			PEO	period of extended operation
NPP	nuclear power plant		PFM	probabilistic fracture mechanics
NPP FIRE MAG	Nuclear Power Plant Fire Modeling Application Guide		Phebus-FP	Phebus-Fission Products
			Phebus-ISTP	Phebus-International Source Term Program
NRC	U.S. Nuclear Regulatory Commission		PI	performance indicator
NRO	Office of New Reactors		PIMAL	phantom with moving arms and legs
NRR	Office of Nuclear Reactor Regulation		PINC	Program for the Inspection of Nickel-Alloy Components
NSIR	Office of Nuclear Security and Incident Response			
			PIRT	Phenomena Identification and Ranking Table
NTTF	Near-Term Task Force		PKL	Primarkreislauf-Versuchsanlage (German for primary coolant loop test facility)
NUPEC	Nuclear Power Engineering Corporation (Japan)			
			PM	project manager
NUREG	NRC technical report designation		PMDA	Proactive Materials Degradation Assessment
NUREG/CR	NRC technical report designation/contractor report		PMMD	Proactive Management of Materials Degradation
NUREG/IA	NRC technical report designation/ international agreement		PMP	probable maximum precipitation
			PMR	prismatic modular reactor
NUSSC	Nuclear Safety Standards Committee		PNNL	Pacific Northwest National Laboratory
NWS	National Weather Service		POS	plant operating state
			PPS	Package Performance Study
O			PRA	probabilistic risk assessment or probabilistic risk analysis
ODCM	offsite dose calculation models			
OECD	Organisation for Economic Co-operation and Development		PSA8	Probabilistic Safety Conference 2008
			PSHA	probabilistic seismic hazard assessment
			PSI	Paul Scherrer Institut

PTS	pressurized thermal shock	**S**	
PUMA	Purdue University Multi-Dimensional Integral Test Assembly	S&T LLC	Standards and Technology Limited Liability Company
PWR	pressurized-water reactor	SAIC	Science Applications International Corporation
PWSCC	primary water stress-corrosion cracking	SACADA	Scenario Authoring, Categorization, and Debriefing Application
Q			
QA	quality assurance	SAMG	severe accident mitigation guideline
QHO	quantitative health objective	SAPHIRE	Systems Analysis Programs for Hands-on Integrated Reliability Evaluation
R		SBDMW	small-bore dissimilar metal weld
RACKLIFE	software calculation package used for mapping of degradation	SBO	station blackout
		SCALE	modeling and simulation computer code for nuclear safety analysis
RADS	Reliability and Availability Data System	SCIP	Studsvik Cladding Integrity Project
RADTRAD	Radionuclide Transport, Removal, and Dose code	SC	Office of Science (DOE)
RAMONA	T/H computer code	SCC	stress-corrosion cracking
RASP	Risk Assessment Standardization Project	SDP	Significance Determination Process
RBHT	Rod Bundle Heat Transfer Program	SEASET	Separate Effects and Systems Effects Tests
RCS	reactor coolant system	SECY	Office of the Secretary
R&D	research and development	SERF	small early release frequency
REAcct	Regional Economic Accounting Tool	SFR	sodium-cooled fast reactor
REIRS	Radiation Exposure Information and Reporting System	SFP	spent fuel pool
		SFPS	Spent Fuel Pool Scoping Study
RELAP5	Reactor Excursion and Leak Analysis Program	SG	steam generator
REMIX	Regional Mixing Model	SGAP	Steam Generator Action Plan
RES	Office of Nuclear Regulatory Research	SGTR	steam generator tube rupture
RG	regulatory guide	SKC	susceptibility, knowledge, and confidence
RGDB	Regulatory Guide Development Branch (NRC)	SI Units	International System of Units (abbreviated SI from the French Ie Systeme International)
RIC	Regulatory Information Conference	SIMULIA	engineering simulation software vendor previously known as ABAQUS
RIDM	risk-informed decisionmaking		
RIM	Reliability and Integrity Management	SMAW	shielded metal arc welding
RIS	regulatory issue summary	SME	subject matter expert
RMIEP	Risk Methods Integration and Evaluation Program	SNAP	Symbolic Nuclear Analysis Package
		SNF	spent nuclear fuel
RMTF	Risk Management Task Force	SNFT	spent nuclear fuel transportation
ROE	red oil excursion	SNL	Sandia National Laboratories
ROP	Reactor Oversight Process	SOARCA	State-of-the-art Reactor Consequence Analysis
ROSA	Rig of Safety Assessment	SOKC	state-of-knowledge correlation
RPB	Radiation Protection Branch (NRC)	SPAR	Standardized Plant Analysis Risk
RPV	reactor pressure vessel	SPAR-H	Standardized Plant Analysis Risk—Human Reliability Analysis Method
RR	round robin		
RSICC	Radiation Safety Information Computational Center	SPE	standard problem exercise
		SRM	staff requirements memorandum
RuO4	ruthenium tetroxide	SRP	Standard Review Plan
RV	reactor vessel	SS	stainless steel
		SSC	structure, system, or component

SSHAC	Senior Seismic Hazard Analysis Committee	**W**		
SSU	safety system unavailability	WEP	wired equivalent privacy
SSWICS	small-scale water ingression and crust strength	WGRisk	OECD/NEA/CSNI Working Group on Risk
STAR CCM+	computer code used for CFD	WIR	waste-incidental-to-reprocessing
STCP	Source Term Code Package	wppm	weight parts per million
Std	standard	WRS	weld residual stresses
STEM	Source Term Evaluation and Mitigation	WTC	World Trade Center
STSET	Source Term Separate Effects Test Project			
S/U	sensitivity/uncertainty	**X**		
		xLPR	extremely low probability of rupture
T				
T/H	thermal-hydraulic	**Z**		
TEPCO	Tokyo Electric Power Company	ZIRLO	fuel rod cladding material
THERP	Technique for Human Error Rate Prediction			
THI	Thermal-Hydraulics Institute			
THIEF	Thermally-Induced Electrical Failure			
TID	Technical Information Document			
TIP	tube integrity programm			
TMI	Three Mile Island (Nuclear Power Plant)			
TR-33	Plastic Pipe Institute revised document for fusion procedure			
TRAC	Transient Reactor Analysis Code			
TRACE	TRAC/RELAP Advanced Computational Engine			
TRISO	Tristructual-Isotropic			
TWG	task working group			
U				
U	uranium			
UA	uncertainty analysis			
UCF	University of Central Florida			
UMD	University of Maryland			
UO2	uranium dioxide			
U.S. APWR	U.S. Advanced pressurized-water Reactor (Mitsubishi)			
USEGC	U.S. east and gulf coasts			
USGS	U.S. Geological Survey			
V				
V&V	verification and validation			
VARSKIN	code used to model and calculate skin dose			
VEGA	Verification Experiments of radionuclides Gas/Aerosol release			
VERCORS	French test program			
VEWFD	very early warning fire detection			
VHTR	very-high-temperature gas-cooled reactor			
VSL	value of statistical life			
VTT	Technical Research Center of Finland			

Chapter 1: Regulatory Support

Regulatory Guide Program

Regulatory and Economic Analysis Support

Consensus Codes and Standards

Generic Issues Program

Fuel Cycle Oversight Process

Knowledge Management in the Office of Nuclear Regulatory Research

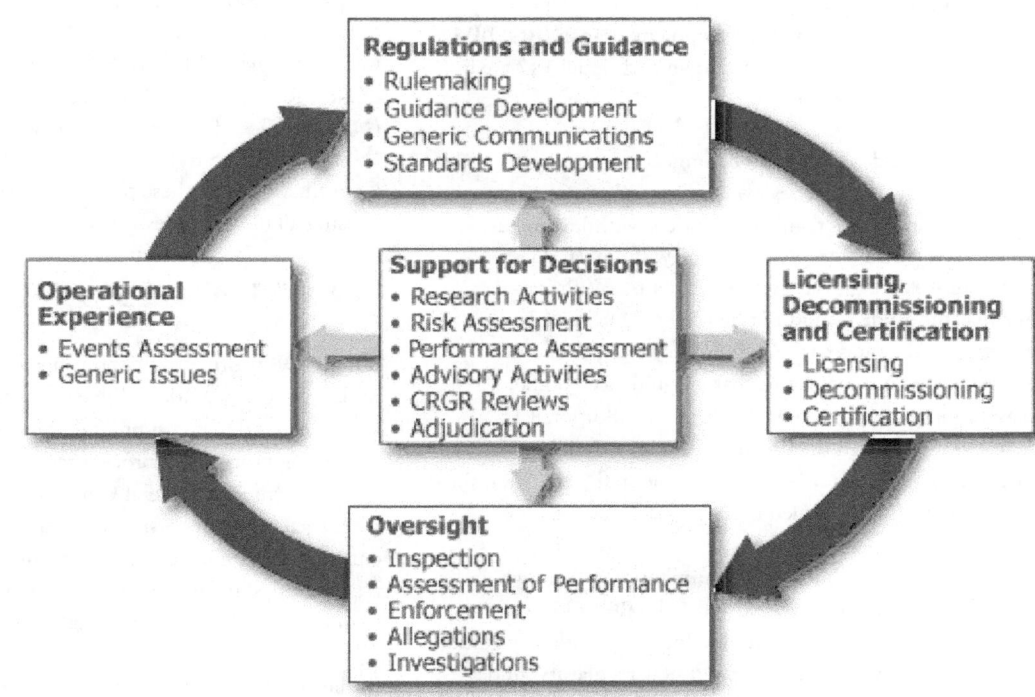

Regulatory Guide Program

Background

The U.S. Nuclear Regulatory Commission (NRC) issues regulatory guides for public use to present approaches that the staff considers acceptable in implementing the agency's regulations. The Office of Nuclear Regulatory Research (RES) provides the tools and methods used by NRC program offices to issue and maintain regulatory guides. The Regulatory Guide Update Project was initiated in 2006 at the direction of the Commission to review, prioritize, and update all regulatory guides. The initial project identified 426 regulatory guides for evaluation, many written in the 1970s that describe methods or approaches for meeting the current regulations. Out of the 426 regulatory guides, those guides necessary to support the design and construction of new nuclear power plants were the first priority. These 29 high priority guides were completed by the end of March 2007.

Since the inception of the 2006 Regulatory Guide Update Project, the Regulatory Guide Development Branch (RGDB) has completed 284 regulatory guides, and an additional 142 are in the process of being reviewed.

Some guides were determined to be unnecessary during the initial regulatory guide review. These guides are being withdrawn. Although a regulatory guide is withdrawn, current licensees may continue to use it and the withdrawal does not affect any existing licenses or agreements. Withdrawal means that the guide should not be used for future NRC licensing activities, and changes to existing licenses should be accomplished using other regulatory guidance. So far, 45 regulatory guides have been withdrawn. An additional 15 regulatory guides are in the process of being withdrawn. In addition, 106 of the guides reviewed by the staff were determined to be acceptable for continued use.

The review process also identified a number of areas in which no formal regulatory guidance existed. The RGDB is working with the NRC program offices to develop new regulatory guides to fill these gaps. As of April 2013, 44 new regulatory guides were identified that needed to be created. To date, 18 of those planned regulatory guides were cancelled, 21 have been issued, and five are currently under review. Since 2006, the RGDB has continued to add regulatory guides to the update project, on an ongoing basis. Figure 1.1 depicts the current status of the Regulatory Guide Update Project.

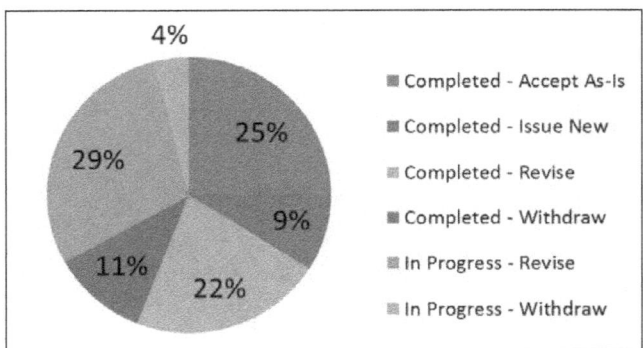

Figure 1.1 NRC Regulatory Guide status

Approach

The RGDB authored Management Directive 6.6, "Regulatory Guides," and Handbook 6.6, to formalize the regulatory guide development and revision process. The process is depicted in Figure 1.2.

The RGDB coordinates with the Office of Federal and State Materials and Environmental Management Programs (FSME), Office of Nuclear Material Safety and Safeguards (NMSS), Office of New Reactors (NRO), Office of Nuclear Reactor Regulation (NRR), Office of Nuclear Security and Incident Response (NSIR), and RES to issue revised and new regulatory guides. Coordination with these program offices led to prioritization for revising all of the guides.

Program Management

The RGDB is primarily responsible for program management of the regulatory guides. The RGDB performs many activities, including working with program offices to revise the guides, explaining process and procedures, problemsolving, working with the Office of General Counsel (OGC) to ensure the guides meet legal requirements, facilitating reviews of backfit requests, ensuring that public comments on draft guides are addressed, establishing standardized RG and draft guide templates, frequently used memos, and staff updates.

In the event that a program office is unable to meet the schedule, regulatory guide project managers have the additional responsibility of authoring guides, when possible, as well as resolving comments from other program offices and the public. Ultimately, the RGDB works with all parties involved to ensure a timely and quality product.

Schedule

To ensure guides continue to be reviewed with reasonable frequency; the RGDB has developed a 5-year maintenance cycle. Using SharePoint, the RGDB tracks each guide and notifies the appropriate NRC program office a year in advance of an upcoming 5-year review. Using the results of the review, the program office decides whether a Regulatory Guide is acceptable as-is, or whether it should be revised or withdrawn.

If a guide is to be revised, the RGDB has established a 65-week schedule that is broken into 21 scheduling activities. This aggressive schedule includes a 15 week drafting period for the technical lead, as well as three different review periods, two for internal stakeholders and one for the public and other external stakeholders.

Tracking Database

The Regulatory Guide SharePoint site was developed to track the schedule. In addition to helping maintain the schedule, the Regulatory Guide SharePoint site allows the RGDB program manager to include important information, including the lead office contact, status of the guide, draft regulatory guide number, delays, dates, and notes that are vital information to developing and issuing a guide.

Contracts

The RGDB manages several contracts to develop the technical basis for new and revised guides. Depending on the topic, the RGDB serves as technical monitors to ensure the lead NRC program office receives necessary quality technical content required to write the regulatory guide.

Checklists

To facilitate the update process for guides, the RGDB has developed checklists with criteria for "Withdrawn" and "Acceptable As Is" decisions. Both checklists are written to ensure that the technical lead critically evaluates the guide and has a verifiable reason for the decision.

Future

The RGDB is working to enhance its support to the program offices by facilitating updates to a broader spectrum of regulatory infrastructure when updates to regulatory guidance are needed, particularly in response to emerging technical issues and stakeholder input. Examples of regulatory infrastructure include Standard Review Plans, inspection procedures, and technical basis documents. The RGDB is also coordinating development and maintenance of regulatory infrastructure with the agency's efforts in the use of consensus codes and standards.

For more information
Contact Carol Moyer, RES/DE, at Carol.Moyer@nrc.gov.

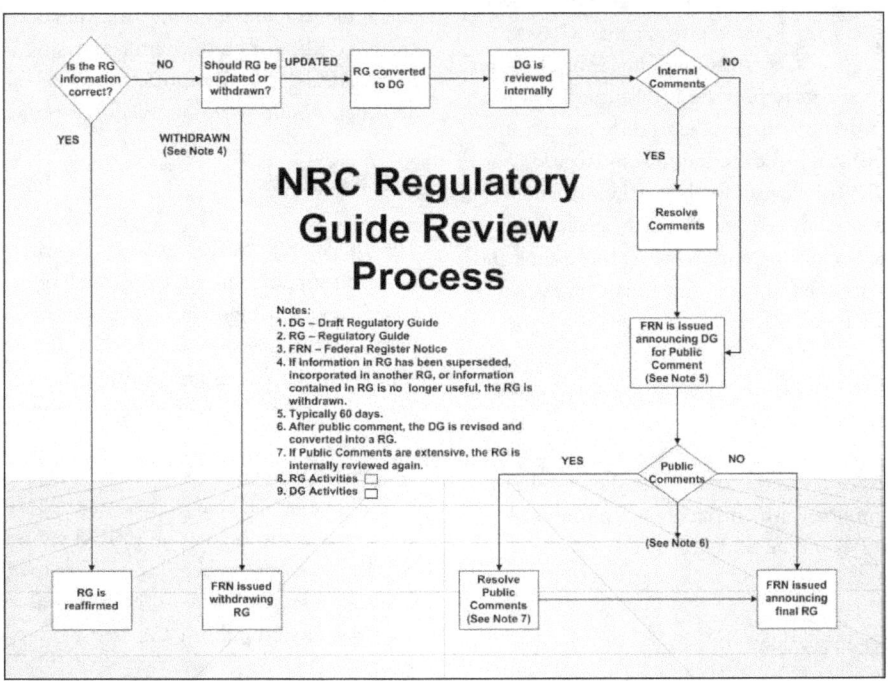

Figure 1.2 NRC Regulatory Guide review process

Regulatory and Economic Analysis Support

Background

A regulatory analysis is an analytical tool that Federal agencies use to anticipate and evaluate the likely consequences of rules. The U.S. Nuclear Regulatory Commission's (NRC's) decisionmakers use regulatory analyses to assist in determining whether a proposed regulatory action is cost beneficial, which means that the benefits of the proposed action are equal to, or exceed, the costs of the proposed action. No legislation or regulation requires a regulatory analysis for NRC-initiated actions. However, multiple Executive Orders have been issued on this topic over the past several years, and the NRC has been voluntarily performing such analyses since 1976, and voluntarily complying with Office of Management and Budget's Circular A-4, "Regulatory Analysis," since 1981. Nonetheless, the regulatory analysis process may be modified or eliminated at the discretion of an NRC office director or higher authority.

Similar to a regulatory analysis, a backfit analysis is an analytical tool the NRC uses to assist in determining whether a proposed regulatory action applicable to nuclear facilities, already licensed when the new requirement is being considered, should be adopted. The requirements set forth in 10 CFR 50.109, "Backfitting," govern backfitting for nuclear power reactors. In addition, 10 CFR Part 70, "Domestic Licensing of Special Nuclear Material," 10 CFR Part 72 "Licensing Requirements for the Independent Storage of Spent Nuclear Fuel, High-Level Radioactive Waste, and Reactor-Related Greater than Class C Waste," and 10 CFR Part 76, "Certification of Gaseous Diffusion Plants," include backfit regulatory provisions for other facilities. Analogous backfitting provisions applicable to early site permits and standard design certifications, differing in some regards from those in 10 CFR 50.109, are set forth in 10 CFR Part 52, "Licenses, Certifications, and Approvals for Nuclear Power Plants." In general, the backfitting requirements for reactor and materials facilities consider the following three main steps (see Enclosure 5 for a more detailed discussion of backfitting):

1. Evaluate whether a backfit analysis exemption for adequate protection or compliance applies.

2. Determine whether a substantial increase in the overall protection of the public health and safety or common defense and security would be achieved by the proposed change.

3. Complete a cost-benefit evaluation.

Although regulatory analyses and backfit analyses are distinct types of evaluations, a regulatory analysis may be sufficient to satisfy the cost-evaluation requirements for a backfit analysis. Furthermore, as part of the implementation of the National Environmental Policy Act (NEPA) requirements, the NRC evaluates the costs and benefits of severe accident mitigation alternative and severe accident mitigation design alternative analyses, for certain nuclear reactor licensing and application reviews.

In performing cost-benefit determinations for regulatory, backfitting, and environmental analyses, the NRC traditionally has considered many different attributes. Such attributes include public health, occupational health, onsite property, and offsite property. Documents that provide descriptions of these attributes and staff guidance for conducting cost-benefit analyses include NUREG/BR-0058 Revision 4, "Regulatory Analysis Guidelines," NUREG/BR-0184, "Regulatory Analysis Technical Evaluation Handbook," and NUREG-1530, "Reassessment of NRC's Dollar per Person-Rem Conversion Factor Policy." For instance, as stated in NUREG/BR-0058, onsite property is generally owned or controlled by the license- or certificate-holder and located within the boundaries of the licensed facility, whereas offsite property is located outside of the site boundaries, and is not owned or controlled by the license- or certificate-holder. However, in a cost-benefit analysis, the distinction between offsite and onsite property goes beyond the location or ownership of the property. Onsite property costs include replacement power, decontamination costs, and costs associated with refurbishment or decommissioning. Offsite property costs include both the direct costs associated with property damage (e.g., diminution of property values) and indirect costs (e.g., tourism, manufacturing, and agriculture disruption). Additionally, the NRC uses its current dollar per person-rem conversion factor, published in NUREG-1530, to capture the dollar value of the health detriment resulting from radiation exposure.

Objective

The Office of Nuclear Regulatory Research (RES) provides technical support for regulatory analysis and economic cost-benefit policy. Ongoing projects include reassessing the dollar per person-rem conversion factor policy and evaluating the NRC's consideration of economic consequences of the unintended release of licensed nuclear materials to the environment.

The NRC last revised its value of a person-rem averted to be $2,000 per person-rem averted in 1995. In 2010, the Office of Nuclear Reactor Regulation (NRR) contracted the services of ICF International to begin to reassess the dollar per person-rem conversion factor. In 2011, NRR requested that RES further this research and publish a revised conversion factor policy in the form of a NUREG.

Furthermore, the accident at the Fukushima Dai-ichi nuclear power plant in Japan initiated discussion of how the NRC's regulatory framework considers offsite property damage caused by a significant radiological release from an NRC-licensed facility and licensed material. In response to this discussion, staff evaluated the analyses in which offsite property damage is considered, held two public meetings to solicit stakeholder input, and submitted SECY-12-0110, "Consideration of Economic Consequences within the U.S. Nuclear Regulatory Commission's Regulatory Framework," dated August 14, 2012, to the Commission. The SECY paper can be found at http://www.nrc. gov/reading-rm/doc-collections/commission/secys/2012/2012-0110scy.pdf. In this SECY and associated enclosures, the staff summarizes a broad spectrum of background information on this subject.

Approach

To reassess the NRC's dollar per person-rem conversion factor policy, RES reviewed the ICF report and the value of statistical life (VSL) other Federal agencies used to determine whether the recommendations of ICF were up-to-date and comparable to that of other agencies. To facilitate information gathering and exchange with other Federal agencies, RES sponsored an interagency regulatory analysis workshop focusing on the VSL, a major component of the dollar per person-rem conversion factor. The workshop was held March 19–20, 2012. It brought together approximately 50 participants from 10 different Federal agencies and included representatives from the U.S. Department of Energy, the U.S. Department of Homeland Security, the U.S. Department of Transportation, the U.S. Environmental Protection Agency, the U.S. Food and Drug Administration, the National Oceanic and Atmospheric Administration, and the U.S. Department of Agriculture. The participants exchanged lessons learned about calculating, updating, applying, and communicating the VSL, and identified potential areas for future interagency collaboration in regulatory analysis. The workshop highlighted similar and unique challenges regarding the VSL that each agency faces and provided useful insights for the NRC's updating efforts.

The staff is continuing work on determining an updated dollar per person-rem conversion factor and researching the feasibility of developing a well-defined process to periodically update this factor. The staff expects to complete research on an updated dollar per person-rem factor and publish a final NUREG documenting the revised value in 2014. The staff will engage external stakeholders and seek approval from the Commission before finalizing this NUREG.

For More Information
Contact Kevin Coyne, RES/DRA, at Kevin.Coyne@nrc.gov.

Consensus Codes and Standards

Background

The U.S. Nuclear Regulatory Commission (NRC) cooperates with professional organizations that develop consensus standards associated with systems, structures, equipment, or materials that the nuclear industry uses. A standard contains technical requirements, safety requirements, guidelines, characteristics, and recommended practices for performance. The consensus standards process is based on openness, balance of interests, due process with written records, and consensus—more than a majority but not necessarily unanimity. Codes are defined as standards or groups of standards that have been incorporated by reference into the regulations of one or more governmental bodies and have the force of law.

For example, the American Society of Mechanical Engineers (ASME) developed the Boiler and Pressure Vessel Code, which is widely acknowledged as an acceptable set of standards used to design, construct, and inspect pressure-retaining components, including nuclear vessels, piping, pumps, and valves. Similarly, the National Fire Protection Association (NFPA) has developed a series of consensus standards to define acceptable methods to design, install, inspect, and maintain fire protection systems. The NRC has incorporated into its regulations parts of the ASME Boiler and Pressure Vessel Code and a key NFPA standard, with some limitations, as well as other consensus standards. The Regulatory Guide Development Branch (RGDB) in the Office of Nuclear Regulatory Research (RES) coordinates the NRC's use of consensus codes and standards.

Objective

The objective of this program is to optimize the NRC's development and use of consensus codes and standards as part of its regulatory framework and in voluntary compliance with Public Law 104-113, the "National Technology Transfer and Advancement Act of 1995."

Approach

The NRC's use of consensus standards is consistent with the requirements of this Act, as further described in the Office of Management and Budget's Circular A-119, "Federal Agency Participation in the Development and Use of Voluntary Consensus Standards and in Conformity Assessment Activities." Participation of NRC staff in consensus standards development is essential because the codes and standards are an integral part of the agency's regulatory framework. The benefits of this active involvement include cost savings, improved efficiency and transparency, and regulatory requirements of high technical quality. The agency acknowledges the broad range of technical expertise and experience of the individuals who belong to the many consensus standards organizations. Thus, participation in standards development minimizes the expenditure of NRC resources that would otherwise be necessary to provide guidance with the technical depth and level of detail of consensus standards.

In 2011, about 190 NRC staff members participated in more than 300 standards activities, such as membership on a standards-writing committee. The organizations governing these committees include ASME, NFPA, the Institute of Electrical and Electronics Engineers, the American Concrete Institute, and many others.

In addition to issuing regulations that incorporate consensus standards, the NRC staff issues guidance on acceptable methods for complying with its regulations, such as regulatory guides. These guidance documents frequently reference consensus standards as acceptable methods for compliance with NRC regulations. A principal reason for using standards is to provide the regulatory stability and predictability that stakeholders desire.

Most codes and standards evolve over time, through a process that includes development of new standards and revision of existing ones. For example, work is underway with standards developing organizations to update voluntary consensus standards that may be applied to license renewal or new nuclear plant construction, including advanced reactor technologies and small modular reactors.

Nuclear Energy Standards Coordination Collaborative

In 2009, in cooperation with other Federal agencies, the NRC helped establish a new information exchange forum called the Nuclear Energy Standards Coordination Collaborative (NESCC). The NESCC is a cross-stakeholder forum to identify and respond to the needs of the U.S. nuclear industry for updates to codes and standards. The NESCC is a joint effort of the NRC, the Department of Energy, the National Institute of Standards and Technology, standards developing organizations, and the nuclear industry. Its goals are to identify standards needs, prioritize standards for development or revision, and initiate or support collaboration in writing or updating standards. The group works to facilitate and coordinate the timely identification, development, and revision of standards for the design, operation, development, licensing, and deployment of nuclear power plants and other nuclear technologies. Central to the mission of the NESCC is developing a standards database that will provide government agencies, standards developers, and nuclear industry users with clear information about available consensus standards and how the industry can use those standards to meet regulatory requirements.

Review of International Atomic Energy Agency Safety Standards

The International Atomic Energy Agency (IAEA) Commission on Safety Standards is a body of senior government officials from member nations that oversees the development of international safety standards. IAEA has four Safety Standards Committees that participate in the development, review, and update of standards and guidance documents related to nuclear safety, radiation protection, waste management safety, and transport safety. The Director of RES' Division of Engineering serves as the U.S. delegate to one of these four committees, the Nuclear Safety Standards Committee (NUSSC). This participation helps harmonize NRC standards and guidance with international standards and guidance. A member of the RGDB provides technical support to the division director for this effort. This staff member compiles information needed to support attendance at the NUSSC meetings, collects and resolves U.S. stakeholder comments on the draft standards, disseminates documents to other NRC offices for input, and promotes awareness of such safety standards.

In addition to safety standards and guides issued by the IAEA, the NRC staff is also evaluating other international standards, such as documents published by the International Standards Organization and the International Electrotechnical Commission. Where applicable, these documents are referenced for information or guidance. The NRC staff is exploring the possibility of future endorsement of international standards within the agency's regulatory framework.

For more information
Contact Carol E. Moyer, RES/DE, at Carol.Moyer@nrc.gov.

Generic Issues Program

Background

The U.S. Nuclear Regulatory Commission (NRC) staff developed the Generic Issues Program in response to Commission and congressional directives in 1976 and 1977, respectively. The Generic Issues Program enables the public and NRC staff to raise issues with potential significant generic safety or security implications. The program ensures that those safety and security issues are considered through an effective, collaborative, and open process and that any needed actions are taken to ensure safety at licensees. The Generic Issues Program also disseminates information pertinent to these issues.

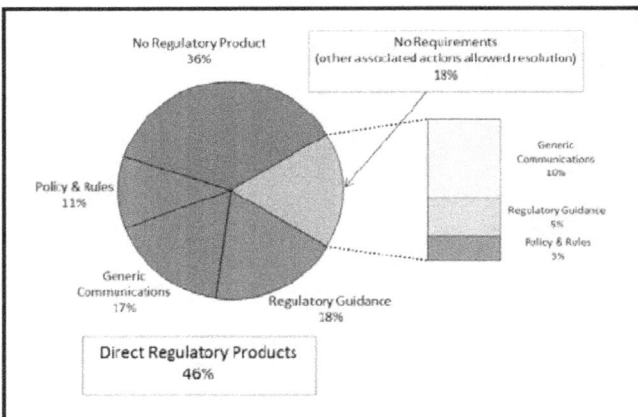

Figure 1.3 Breakdown of resolution products for GIs

Originally, the program included identifying Generic Issues (GIs), assigning priorities, developing detailed action plans, projecting costs, providing continuous high-level management oversight of progress, and disseminating information to the public about the issues as they progressed. The Generic Issues Program process for resolution of issues has evolved gradually since 1977. The latest process improvements were initiated by SECY-07-0022, "Status Report on Proposed Improvements to the Generic Issues Program," dated January 30, 2007, and later implemented in Management Directive (MD) 6.4, "Generic Issues Program," dated November 17, 2009. The program has identified more than 850 GIs to date, resulting in important safety improvements at NRC licensees and a variety of regulatory products, such as generic communications and regulatory guidance.

Approach

MD 6.4 describes the process used to resolve the GIs. Its guidance provides a consistent framework for handling, tracking, and defining the minimum documentation associated with processing GIs. The Generic Issues Program uses a five-stage process to resolve issues that meet the MD 6.4 screening criteria. The NRC staff applies the screening criteria to include only those issues in the Generic Issues Program that the program could most effectively resolve. The five stages of the program include: (1) identification, (2) acceptance review, (3) screening, (4) safety or risk assessment, and (5) regulatory assessment. The Generic Issues Program staff apply a graded approach (i.e., as an issue proceeds through the program, it is analyzed with more rigor, and more resources are devoted to it). Similarly, issues with potentially greater safety significance receive more resources and priority. Because of the varying technical disciplines and levels of difficulty, flexibility is built into the program.

Generic Issues Program Products

The Generic Issues Program has contributed significantly to the NRC's mission. Because of the diverse nature of the GI topics, the NRC has developed a variety of products to resolve them. GIs that have satisfied the screening criteria or historically were prioritized with significant rankings could lead to a regulatory product. About 300 issues reached the resolution stage and could have resulted in a regulatory product. Roughly two thirds of these issues were resolved by a regulatory product from four categories (shown in blue): (1) new policies and rules, (2) generic communications, (3) regulatory guidance, and (4) no direct regulatory products but associated actions allowed resolution. Roughly a third of the issues did not require any new regulatory products for resolution (shown in red). Figure 1.3 shows a breakdown of the resolution products for GIs processed under the Generic Issues Program from 1983 to 2012. Approximately two-thirds of the issues prioritized from 1983 to 1999 or screened after 1999 were not pursued further for resolution. These issues either were integrated with other issues, their safety concerns were addressed by other issues, or their prospect of safety improvements was not necessary. For issues that did not need to be pursued to the resolution stage and, consequently, their disposition did not result in a formal regulatory product, completing the prioritization or screening stages for these issues provided an in-depth insight as to their risk and safety significance. Figure 1.3 does not include these issues because the NRC did not pursue them to the resolution stage.

List of Active Generic Issues

More information for open and recently closed GIs, and past GIs is available online at:
http://www.nrc.gov/reading-rm/doc-collections/generic-issues/

Also, the annual summary of the Generic Issues Program activities provided to the Commission can be found online at:
http://www.nrc.gov/reading-rm/doc-collections/generic-issues/annual/index.html

For More Information
Contact John Kauffman, RES/DRA, at John.Kauffman@nrc.gov.

Fuel Cycle Oversight Process

Background

In the last decade, the U.S. Nuclear Regulatory Commission (NRC) has been evaluating how to include risk insight information and decisionmaking into its fuel cycle facility regulatory framework. Since 2007, specific staff initiatives have intended to improve the Fuel Cycle Oversight Process (FCOP) objectivity, predictability, transparency, and consistency and to incorporate Risk-Informed and Performance-Based tools. As part of this effort, the Office of Nuclear Materials Safety (NMSS) and Safeguards asked the Office of Nuclear Regulatory Research (RES) to support the exploration of the use of probabilistic risk assessment (PRA) tools and guidance for use in fuel cycle facility applications. Figure 1.4 provides an overview of the activities involved in the nuclear fuel cycle.

In SECY 010 0031, "Revising the Fuel Cycle Oversight Process," dated March 19, 2010, the staff proposed a qualitative and a quantitative option for continuing to revise the FCOP. However, the Commission directed the staff to compare integrated safety analysis (ISA) used in fuel facilities with PRA methods used in nuclear power plants. Fuel cycle facilities (FCF) are required in Title 10 of the *Code of Federal Regulations* (10 CFR) 70.62, "Safety Program and Integrated Safety Analysis," to perform an ISA to include radiological, chemical and facility hazards, potential accident sequences, consequence and likelihood of occurrence of each accident sequence, and the identification of items relied on for safety (IROFS). In addition, the Commission directed the staff to continue making modest adjustments to the existing FCOP to enhance its efficiency and efficacy.

In December 2010, NRC staff published a report entitled, "A Comparison of Integrated Safety Analysis and Probabilistic Risk Assessment" (Agencywide Documents Access and Management System Accession No. ML103330478). This report explored and compared major differences from an ISA used in FCFs versus PRA used in nuclear power plants and how PRA could be used in FCFs.

Objective

The objective of this project is to assist the Office of Nuclear Materials Safety and Safeguards (NMSS) in evaluating differences between an ISA and a PRA and in developing tools and guidance for the FCOP.

Approach

RES contracted with Brookhaven National Laboratory to support NMSS in improving the FCOP with PRA insights and tools. The comparison report comparing ISA and PRA was part of this effort.

The comparison report concluded that fuel cycle ISAs and reactor PRAs are performed for different purposes. Some ISAs have used several PRA methods extensively, whereas other ISAs have used them selectively. ISAs were performed to identify potential accident sequences, designate IROFS to prevent or mitigate them, and describe management measures to be applied to ensure the reliability and availability of IROFS. Because ISAs are not performed to support risk significance, the staff expects that modifications would be needed in some cases to obtain reasonable and consistent evaluations on risk to be used in the FCOP.

Longer term tasks that NMSS could request include developing tools and guidance for chemical safety, criticality safety, and human reliability, and undertaking a pilot project to test tools and methods that the NRC could apply to the FCOP.

Future Work

The NRC will continue to improve its FCOP with risk insights and PRA as it continues to mature. Future work will be determined based on the results of the current work, available resources, and future needs as NMSS determines.

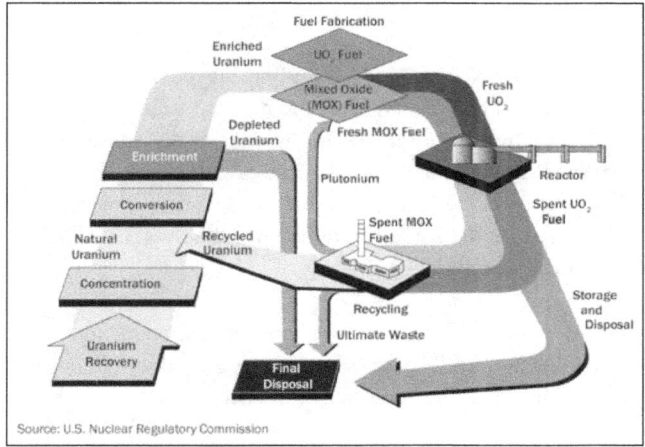

Figure 1.4 The nuclear fuel cycle

For More Information:
Contact:
Felix E. Gonzalez, RES/DRA, at Felix.Gonzalez@nrc.gov.
Michelle Gonzalez, RES/DRA, at Michelle.Gonzalez@nrc.gov.
Brian Wagner, RES/DRA, at Brian.Wagner@nrc.gov.

Knowledge Management in the Office of Nuclear Regulatory Research

Background

The mission of the Office of Nuclear Regulatory Research (RES) is to support the regulatory mission of the U.S. Nuclear Regulatory Commission (NRC) by providing technical advice, technical tools, and information for identifying and resolving safety issues, performing the research necessary to support regulatory decisions, and issuing regulations and guidance. RES's principal product is knowledge; therefore, knowledge management (KM) is an integral part of the RES mission.

Objective

RES's objective is to capture, preserve, and transfer key knowledge among employees and stakeholders. The body of knowledge can be used when making regulatory and policy decisions and ensures that issues are viewed and analyzed within a historical context.

Approach

RES KM activities fall into several categories as described below.

Agency-Level KM Steering Committee and KM Staff Leads

The NRC has a KM Steering Committee in which senior-level managers listen to new KM ideas and discuss future plans. The meetings cultivate an awareness of the value of KM initiatives agencywide and support staff with KM-oriented projects and goals. It also provides an opportunity for senior level managers to participate in the agency's various KM initiatives, such as Marketing and Standardization.

RES is a member of the committee and sends a representative to the meetings, which occur several times annually. The office presents KM ideas and concepts for discussion.

As referenced previously, the KM Steering Committee also participates in agency KM initiatives. In 2011 and 2012 these included: Marketing, Standardization, and IT. These three groups assist the agency in both highlighting the agency's KM program (Marketing subgroup) as well as identifying better information management and access systems (Standardization and IT subgroups). In addition, the KM Staff Leads meet several times a year and provide assistance to the KM Steering Committee and their individual office staff.

RES KM Focus Area Group Inclusion

RES identifies "focus areas" each year to pool additional attention and resources on high priority issues. RES' knowledge management area is part of the "Promote Employee Well-Being and Self-Development" focus area for 2012–2013. The office's KM working group supports this focus area and focuses on such initiatives as:

- Expertise Exchange Program.

- Communities of Practice (CoPs).

- NUREG/KM series development.

- "Inside the Regulator's Studio:" Interactive Knowledge Transfer Sessions.

RES Seminars

For several years, RES has sponsored monthly seminars on technical topics of broad agency interest. RES also sponsors special in-depth technical symposia on topics such as the Three Mile Island (TMI) accident, Chernobyl, and the September 11, 2001, attack on the World Trade Center (WTC) Twin Towers and Building 7. These events include staff presentations and also may feature special guests who have unique knowledge of the topic. For example, for the TMI seminar in 2009, speakers included Governor Richard Thornburgh of Pennsylvania (see Figure 1.5) and Ed Frederick, who was an operator on shift at the time of the accident in 1979. The two September 11 seminars (WTC Twin Towers and WTC Building 7) were presented by the scientists and researchers from the National Institute of Standards and Technology as mandated by Congress to determine why the structures collapsed.

Figure 1.5 Governor Dick Thornburgh (PA) at a RES seminar on the 1979 accident at Three Mile Island

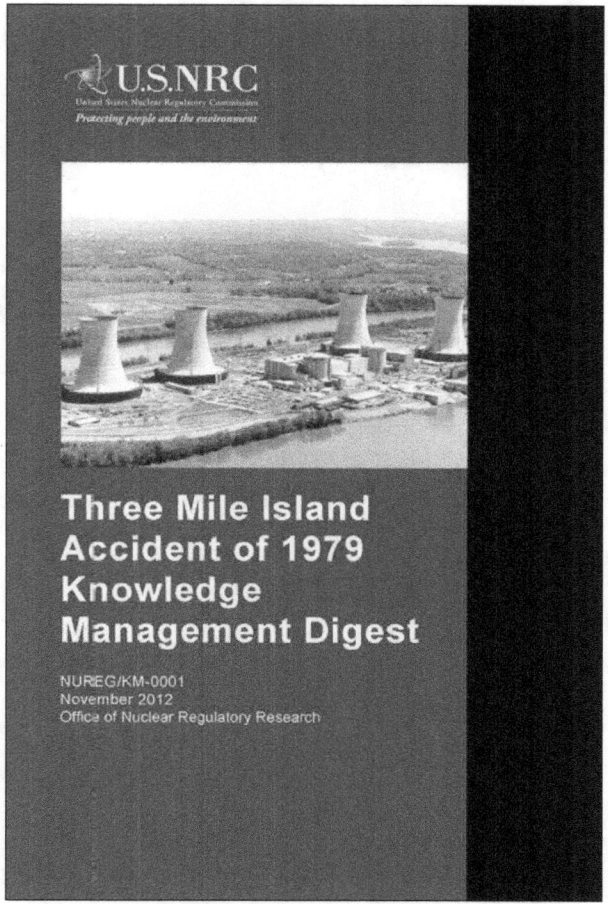

Figure 1.6 NUREG/KM-0001, A History of the Three Mile Island Nuclear Power Plant Accident

Communities of Practice

To be successful, the NRC staff must have access to existing sources of technical information. A key aspect of the RES KM Program is the development of virtual CoPs in which RES staff members can share and collect information in their area of interest. RES now has several CoPs on topics such as human factors; HTGRs; liquid metal cooled reactors; fire protection; health effects; and structural, geotechnical, and seismic engineering.

Publications—NUREGS

Official NRC publications are called NUREGs. RES is the agency leader for publishing KM focused NUREGs that compile historic information, video, and references. The Fire Research Branch, in particular, has contributed much to the office, agency, and industry through its KM efforts. The following NUREGs, which include some from the Fire Research Branch, are publicly available:

- NUREG/BR-0465, Revision 1, "Fire Protection and Fire Research Knowledge Management Digest" (Note: NUREG/KM-0003 is currently in draft development and will supersede NUREG/BR-0465 Revision 1).

- NUREG/BR-0175, "A Short History of Nuclear Regulation, 1946–1990"

- NUREG/BR-0364, "A Short History of Fire Safety Research Sponsored by the U.S. NRC, 1975–2008"

- NUREG/KM-0002, "The Browns Ferry Nuclear Plant Fire of 1975 Knowledge Management Digest," supersedes NUREG/BR-0361, "The Browns Ferry Nuclear Plant Fire of 1975 and the History of NRC Fire Regulations"

In 2010, RES proposed a new publication series focused exclusively on collecting and interpreting historical information on technical topics for the benefit of future generations of NRC professionals. A publication in the proposed NUREG series would be called a NUREG/KM. The series has been approved and the first item is: "KM-0001, A History of the Three Mile Island Nuclear Power Plant Accident." (See Figure 1.6)

Expertise Exchange Program

The Expertise Exchange Program matches seasoned professionals with newer employees or those who want to learn more to facilitate information sharing. The program provides a way to preserve institutional knowledge, expertise, and opinions gained through on-the-job experiences. It also has been a goal for the office's KM Focus Area Group.

The general approach to the program includes exposing employees to key topic areas, which are recorded in a formal knowledge transfer plan. Employees gain exposure to key topic areas and become known to management through attendance at selected internal meetings and management briefings. Knowledge is also gained through attendance at select conferences when possible. Finally, in addition to well-defined short-term and long-term tasks, employees may be asked to provide support to other offices to build their skill set and familiarity with their subject area.

"Inside the Regulator's Studio" Interactive Knowledge Transfer Sessions

The agency recognized the need to develop a standardized process to document subject matter expert (SME) knowledge before staff changes positions or departs from the agency. Inside the Regulator's Studio was proposed to blend knowledge capture and knowledge transfer activities into one interactive event. To date, pilot sessions have been conducted within the agency, with future plans including a panel presentation consisting of the RES Office Director and several other NRC senior staff on thermal-hydraulics.

For More Information
Contact Felix Gonzalez, RES/DRA, at Felix.Gonzalez@nrc.gov.

Chapter 2: Reactor Safety Codes and Analysis

Code Application and Maintenance Program (CAMP)

TRAC/RELAP Advanced Computational Engine (TRACE)
Thermal-Hydraulics Code

Symbol c Nuclear Analysis Package (SNAP) Computer Code
Applications

Thermal-Hydraulic Experimental Programs

Thermal-Hydraulic Simulations of Operating Reactors

Simulat on of Anticipated Transients Without SCRAM with
Core Instability for Maximum Extended Load Line Limit
Analysis Plus-Boiling Water Reactors

Therma -Hydraulic Analyses of New Reactors

Therma -Hydraulic Analysis of Integral Pressurized-Water
Reactors (iPWR)

Compu:ational Fluid Dynamics in Regulatory Applications

Nuclear Analysis and the SCALE Code

High-Burnup Light-Water Reactor Fuel

Spent Nuclear Fuel Burnup Credit

Burnup Credit Methodology –
Pressurized-Water Reactors

Fuel Rod Thermal and Mechanical
Modeling and Analyses

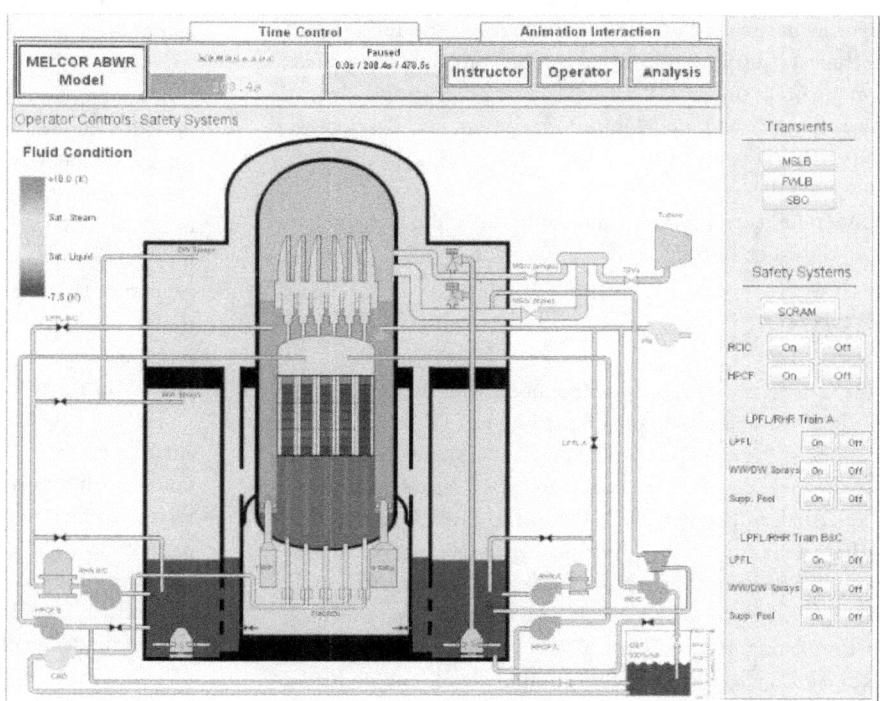

Animating analysis results using SNAP

Code Application and Maintenance Program (CAMP)

Background

In 1985, the U.S. Nuclear Regulatory Commission (NRC) developed the International Code Assessment and Application Program (ICAP) to assess and improve its thermal-hydraulic (T/H) transient computer codes. Approximately 14 nations signed bilateral cooperative agreements with the United States, providing contributions in the form of model development, code assessment, and information generated from applying the codes to operating nuclear power plants. ICAP members held 14 management and specialist meetings between 1985 and 1991. During this time, the NRC published approximately 130 NUREG/IA reports on ICAP work in a number of areas, including core reflood, stratification in horizontal pipes, vertical stratification, postcritical heat flux, and blowdown and quench. ICAP used a variety of test facilities to assess the codes independently. The information generated from this cooperative international work helped the NRC to improve the accuracy, reliability, and speed of its T/H codes. Input from the program also supported the development and application of the Code Scaling, Applicability, and Uncertainty code evaluation methodology in the late 1980s.

In the early 1990s, ICAP developed into CAMP. The CAMP agreement involved monetary contributions and in-kind technical contributions. The technical contributions include, among other things, (1) sharing code experience and identifying areas for code and model improvements, and (2) developing expertise in the use of the codes.

CAMP holds two meetings annually, one in the United States and the other abroad.

Approach

The CAMP program provides members with the TRAC/RELAP Advanced Computational Engine (TRACE), Purdue's Advanced Reactor Core Simulator (PARCS), Symbolic Nuclear Analysis Package (SNAP) codes, and the Reactor Excursion and Leak Analysis Program (RELAP5). The TRACE code is the NRC's primary T/H reactor system analysis code. PARCS is a multidimensional reactor kinetics code coupled to TRACE. SNAP is a graphical user interface to the codes that provides preprocessing, runtime control, and postprocessing capabilities. RELAP5 is a legacy NRC T/H computer code and no further development is being done; however, bugs are patched when found. These codes are used to analyze accidents and transients

in operating reactors, support the resolution of generic issues, evaluate emergency procedures and accident management strategies, confirm licensees' analyses, test the fidelity of NRC simulators, provide training exercises for NRC staff, and support the certification of advanced reactor designs. During the biannual CAMP meetings, members have an opportunity to present their technical findings to the NRC. More specifically, the members (1) share experience with NRC T/H computer codes to identify errors, perform assessments, and identify areas for additional experiments, model development, and improvement, (2) maintain and improve user expertise, (3) develop and improve user application guidelines, (4) develop a well-documented T/H code assessment database, and (5) share experience in the use of the codes to resolve safety and other technical issues (e.g., scalability and uncertainty).

Accomplishments

The CAMP program has provided more than 70 NUREG/IAs that have contributed to the development, assessment, and application of the NRC T/H analysis codes. The NUREG/IAs are listed on NRC's public Web site at http://www.nrc.gov/reading-rm/doc-collections/nuregs/agreement/. Technical areas span the entire range of accident and transient analysis. These include low-pressure, low-power transients; advanced reactor design applications; coupling between the primary system and containment; operation of passive core cooling systems during accidents; boron dilution transients; neutronics coupling; reflood; and condensation with noncondensables. The reports document the contributions made to assessment, plant analysis, and physical model development.

In several recent cases, contributions to the CAMP program provided important code improvements and saved the NRC time and money. For example, analyses of proposed supercritical water reactor designs by CAMP members identified problems in the RELAP5 water properties near the critical point, an area now being improved. (TRACE also uses the RELAP5 water properties.) Although the NRC is not currently analyzing supercritical water reactors, water properties near the critical point are important in calculations for pressurized-water reactor (PWR) anticipated transients without scram.

Another example of efficiency is the Republic of Korea's in-kind contributions on Canada Deuterium Uranium (CANDU) reactors, which were used during ACR-700 T/H code development. This in-kind contribution allowed the NRC to start analyzing the ACR-700 during the pre-application review, sooner than it could have without the Korean contributions. Korean modeling of the advanced accumulator in the proposed AP1400 reactor design has helped guide NRC efforts to model the advanced accumulator of the U.S. Advanced pressurized-water Reactor (APWR), which has similar design features.

Further, the NRC's Office of New Reactors is currently reviewing this model for design certification. Another recent Korean in-kind contribution was it sharing of a multi-energy-group solver for NRC's PARCS code. This addition to PARCS removes the present limitation of two neutron energy groups and allows PARCS to more accurately model situations in which a multigroup approach is desirable (e.g., mixed oxide (MOX) fueled cores).

Future Work

When CAMP began, the NRC was using four primary T/H and reactor kinetics codes specifically designed for modeling transient and accident behavior in PWRs and boiling-water reactors (BWRs). The codes used 1980-era computer languages and T/H modeling. In the late 1990s, the NRC began a code consolidation effort to merge the features of these codes into a new code, using a modern software architecture that would more easily support the addition of modern T/H models and be easily portable to new computer hardware and operating systems. That new code is TRACE, which has reduced the personnel resources and funding needed to maintain and improve multiple codes and the associated training costs.

TRACE is the primary T/H code the NRC uses to review and audit license amendments for operating reactors, advanced reactor license applications, generic safety issues, and power uprate requests.

Several CAMP members have built large, detailed TRACE models to facilitate their in-kind technical contributions. For example, CAMP members have shown good results in TRACE assessments of the Rig of Safety Assessment (ROSA) and Primarkreislauf-Versuchsanlage (PKL) integral test facilities, in separate effects condensation tests, and in the BWR full-size fine-mesh bundle test (BFBT) single channel steady-state and transient tests. Members also have demonstrated coupling TRACE to computational fluid dynamics (CFD) codes. As TRACE matures, CAMP will continue to be an important contributor to its future development and assessment. CAMP contributions will provide information to the NRC code development staff to improve the speed, accuracy, robustness, and usability of TRACE, thereby improving the NRC's reviews, analyses, and audits of licensee products and its protection of public health and safety.

For More Information
Contact Antony Calvo, RES/DSA, at Antony.Calvo@nrc.gov.

TRAC/RELAP Advanced Computational Engine (TRACE) Thermal–Hydraulics Code

Background

The NRC uses thermal-hydraulic (T/H) codes to perform operational and accident transient analyses. Before the late 1990s, the NRC developed and used four system computer codes—Reactor Excursion and Leak Analysis Program (RELAP5), Transient Reactor Analysis Code for pressurized-water reactors (TRAC PWR), Transient Reactor Analysis Code for boiling water reactors (TRAC BWR), and RAMONA—to perform independent safety analyses of PWR and BWR nuclear power plants. These computer codes used architecture and modeling methods developed in the 1970s. The NRC decided that it would be more cost effective to maintain a single modernized computer code that could be used to analyze all the reactor designs and operational conditions that the four older computer codes addressed.

In an effort to meet this goal, over the last 10 years the NRC decided to consolidate the four analysis codes into a single modernized computational platform. The code consolidation project began with the vision "to have the capability to perform thermal-hydraulic safety analysis in the future that allows for solutions to the full spectrum of important nuclear safety problems in an efficient and effective manner, taking complete advantage of state-of-the-art modeling, hardware, and software capabilities." The NRC has successfully consolidated resources needed for maintaining and improving its T/H analysis capability while creating a single code that has improved ease of use, speed, robustness, flexibility, maintainability, and upgradability.

Version 5.0 of TRACE is the culmination of that effort. It can analyze operational and safety transients, such as small and large break loss-of-coolant accidents (LOCA) in PWRs and BWRs, as well as the interactions between the related neutronic and T/H systems.

The T/H and neutronic capabilities of TRACE V5.0 enable the NRC to make independent evaluations of transients for existing and new reactor designs. The NRC uses these capabilities to perform sensitivity assessments of system hardware and phenomena, which can be modeled using different analytical or modeling approaches.

Approach

Development and assessment is an ongoing process. Recently, the NRC addressed modeling issues identified during (1) an independent peer review, completed in 2008, (2) the development of input models used to support the licensing of new and operating reactors, and (3) code assessment activities leading up to the release of Version 5.0. These efforts ultimately led to the release of TRACE V5.0 Patch 3 in June 2012.

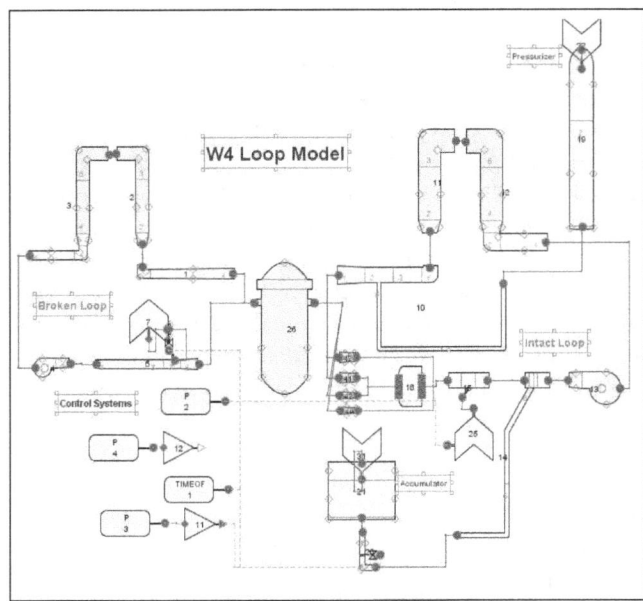

Figure 2.1 Simplified plant model nodalization

Modeling Capabilities

The code features a two fluid, compressible, nonequilibrium hydrodynamics model that can be solved across a one, two, or three-dimensional mesh topology. It also features a three-dimensional reactor kinetics capability through coupling with Purdue's Advanced Reactor Core Simulator (PARCS). The code is capable of performing any type of reactor analysis previously performed by each of the predecessor codes and has component models and mesh connectivity that allow a full reactor and containment system to be easily modeled. Figure 2.1 shows an example of a simplified reactor system nodalization for TRACE.

The NRC added a significant number of new features to the code as a result of the consolidation project. The most notable achievements include the addition of many BWR specific component types; a single junction component (to capture RELAP5 style mesh connectivity); 3D kinetics (through coupling with PARCS); a new heat structure component; an improved set of constitutive models for reflood, condensation, and other basic phenomena; an improved level tracking model; numerous usability enhancements; and countless bug fixes.

A significant advance in the modeling capability of TRACE is the addition of a parallel processing capability that allows the code to communicate with itself or other codes. This feature is known as the exterior communications interface (ECI). ECI is a request driven interface that allows TRACE to communicate with any code that implements the ECI, without actually having to make any modifications to TRACE. ECI has allowed TRACE to be easily coupled to codes such as Symbolic Nuclear Analysis Package (SNAP), Containment Analysis Code (CONTAIN), Regional Mixing Model (REMIX), and Matrix Laboratory (MATLAB). The interface should allow TRACE to be coupled to computational fluid dynamics (CFD) or other special purpose codes in the future.

TRACE Development

TRACE uses a modern code architecture that is portable, easy to maintain, and easy to extend with new models to address future safety issues (a graphical representation of this is shown below in Figure 2.2). TRACE has run successfully on multiple operating systems, including Windows NT/2000/XP/7, Linux, and Mac OSX.

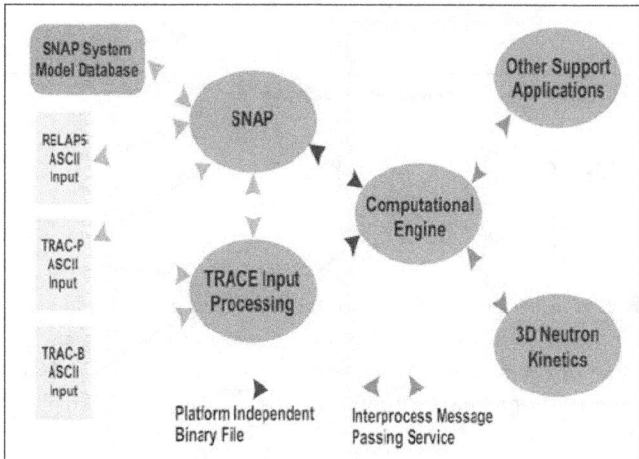

Figure 2.2 TRACE architecture

Code quality is the goal of a stringent development process. Some of the principal elements of this process include:

- Configuration control.
- Establishment and strict enforcement of coding guidelines and development standards.
- Documented development process.
- Software requirements document.
- Software design and implementation test plan.
- Completion report.

- Three tiered testing process.
- Comprehensive regression set.
- Automated robustness testing.
- Automated code assessments.
- Multiplatform testing.
- Automated bug tracking system.

The final stage before any periodic official release of TRACE involves a thorough developmental assessment to identify any deficiencies in its physical models and correlations. The NRC may develop new physical models when it identifies a need for them.

The current assessment test matrix for TRACE contains more than 500 cases. The TRACE assessment test matrix contains a comprehensive set of fundamental, separate effects, and integral tests. These tests range from 1/1,000th scale to full scale and include new and advanced plant specific experiments for both BWRs and PWRs. In addition to data from NRC funded experiments, the assessment matrix includes experimental data obtained through international collaboration. Among these are experiments at the loop for the study of thermal-hydraulic systems (BETHSY), Rig of Safety Assessment (ROSA), and passive decay heat removal and depressurization test (PANDA) facilities. The set of experimental data against which TRACE has been validated is more comprehensive than any other NRC T/H code in terms of scope, quantity, and quality.

Improvements underway for future versions of TRACE are focused on enhancing capabilities related to the simulation of advanced reactor designs, such as the U.S. Advanced pressurized-water Reactor, the U.S. Evolutionary Power Reactor, and the Advanced Passive 1000 Megawatt, as well as small-scale integral reactors. Significant efforts are also directed towards fixing bugs, addressing peer review findings, and improving code robustness and run time performance. The TRACE development team recently released V5.0 Patch 3 to address some of the issues identified to date, and additional patch releases are planned. TRACE will provide a robust and extensible platform for safety analyses well into the future.

For More Information
Contact Chris Hoxie, RES/DSA, at Chris.Hoxie@nrc.gov.

Symbolic Nuclear Analysis Package (SNAP) Computer Code Applications

Background

The U.S. Nuclear Regulatory Commission (NRC) recognizes that analytical capability and expertise are essential to ensure design adequacy and safe operation of nuclear power plants. This mission is, in part, accomplished by analyzing operational and postulated accident transients using analytic modeling software. The NRC has developed and uses several computer codes to perform safety analyses of pressurized-water reactors (PWR) and boiling water reactors (BWR). The input models for most of these codes are text based, requiring the user to write an input file (or deck) in a text editor and then run the analysis program. These input files are often very complex, difficult to read, and time consuming to prepare. Additionally, each computer code uses different input formats and variable names. This adds to the burden on the analysts, who usually use more than one of these modeling programs to perform a review. To lessen this model development burden, the NRC decided that it would be cost effective to develop a single, standardized graphical user interface (GUI) that could be extended for use with any analytical code. This code is the Symbolic Nuclear Analysis Package (SNAP).

and remember several different interfaces and, therefore, is less likely to make an error based on differences in input formats. Currently, SNAP has interfaces for the Reactor Excursion and Leak Analysis Program (RELAP5), TRAC/RELAP Advanced Computational Engine (TRACE), Scale, Containment Analysis Code (CONTAIN), MELCOR, Radionuclide Transport, Removal, and Dose code (RADTRAD), and FRAPCON3 (see Figures 2.3, 2.4, 2.5, and 2.6).

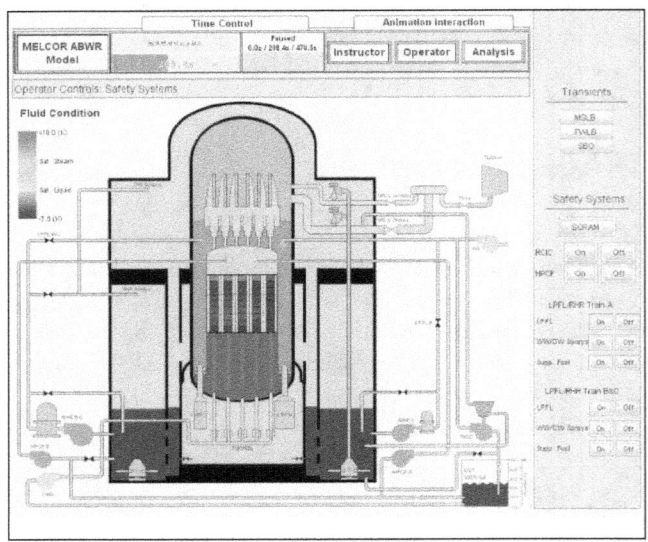

Figure 2.4 Animating analysis results using SNAP

Approach

Development

Recent development focused on implementing a SNAP plug in for the SCALE TRITON depletion sequence, developing an interface for nonmodel based uncertainty parameters in TRACE, as well as improving the existing uncertainty quantification plug in's abilities.

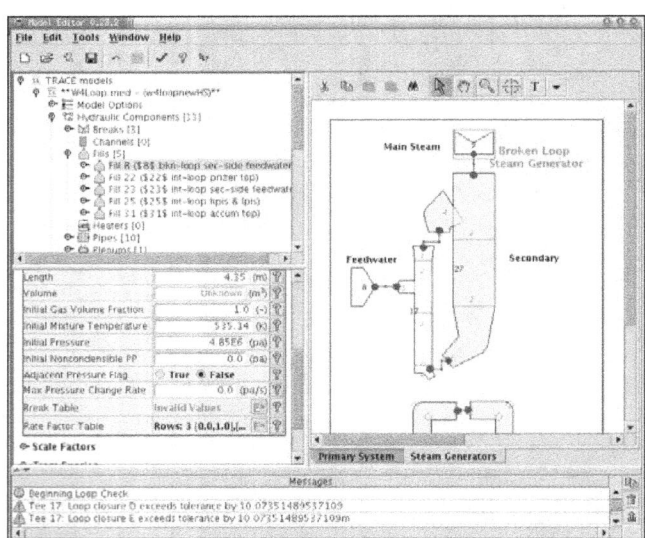

Figure 2.3 Creating input models using SNAP

SNAP removes the need for analysts to use the text based entry methods and to transfer or replicate data among several different packages. It does this by providing a powerful, flexible, and easy to use GUI, both to prepare analytical models and to interpret results. Since the core look and feel of SNAP is the same for different programs, the analyst does not have to learn

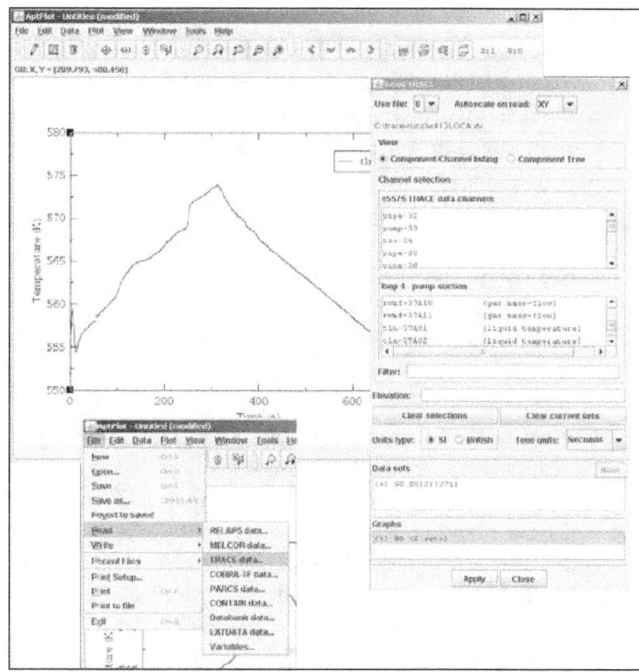

Figure 2.5 Plotting analysis results using SNAP

In more detail, the changes to SNAP during 2012 were as follows:

- The RADTRAD code, a code that the NRC's Office of Nuclear Reactor Regulation uses to review licensees' offsite dose calculations and which was previously integrated with SNAP was moved out of SNAP to allow for 3rd party development of the RADTRAD code. Also, improvements to the SNAP-RADTRAD user interface were introduced based on user feedback and stakeholder comments. A major portion of these improvements included a new source term editor. The new source term editor features built in source terms referenced in the commonly used Regulatory Guide 1.183, "Alternative Radiological Source Terms for Evaluating Design-basis Accidents at Nuclear Power Reactors," for offsite dose estimation, as well as access to the International Commission on Radiological Protection (ICRP) Publication 38, 838 nuclide database.

- The SNAP-TRACE plug in maintained compatibility with the current versions of TRACE during the recent TRACE development efforts. New nonmodel based uncertainty quantification inputs were added as well as improvements to the uncertainty quantification user interface and generated reports.

- A SNAP-SCALE plug in was developed that provides a user interface for the current SCALE 6.1 code. Specifically, this new SNAP-SCALE interface currently only supports the TRITON depletion sequence in SCALE. The user interface previously used for the TRITON sequence

was re-engineered and implemented in SNAP to further consolidate user interface functionality for the analytical codes that the NRC uses. Development plans for SCALE 7 will be tightly coupled with the SNAP–SCALE interface to ensure that the SNAP user interface fully supports SCALE users.

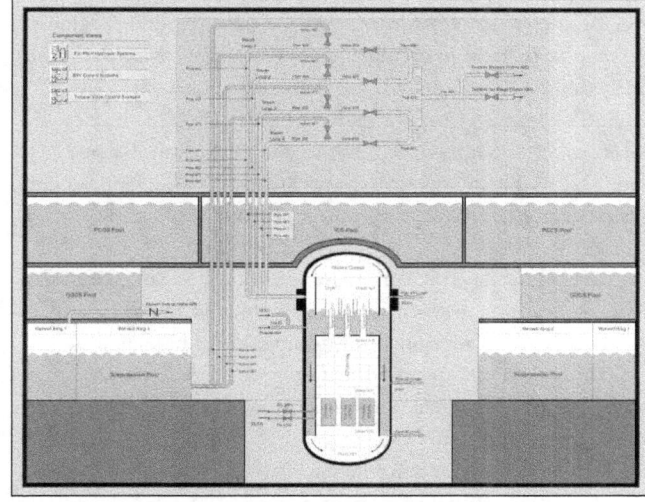

Figure 2.6 Updated model editor display capabilities

Application

SNAP has now been adopted by a large number of analysts using TRACE, MELCOR, and to a lesser extent, by analysts using RELAP5, CONTAIN, RADTRAD, SCALE, and FRAPCON3. A SNAP plug in has been developed for MELCOR, and the integration of MELCOR and SNAP provides a more user friendly system for input deck preparation and accident simulation. SNAP continues to gain greater acceptance and use throughout the agency, as well as in other organizations involved with nuclear analysis.

For More Information
Contact Chester Gingrich, RES/DSA, at
Chester.Gingrich@nrc.gov.

Thermal-Hydraulic Experimental Programs

Background

The U.S. Nuclear Regulatory Commission (NRC) maintains several experimental research programs that directly support reactor safety code development. These experimental programs investigate thermal-hydraulic phenomena and provide data and analysis used to improve the predictive capability of the TRAC/RELAP Advanced Computational Engine (TRACE) reactor safety code, which the NRC uses to analyze operational and safety transients in pressurized-water reactor (PWR) and boiling water reactor (BWR) nuclear power plants.

The TRACE code is currently assessed against a matrix of more than 500 cases. However, when a new phenomena or design is identified that falls outside of the assessment base, new experimental programs must be developed to collect relevant data to support further TRACE development. The data collected in these programs is used to develop TRACE models as well as the validation of those models as assessment cases that are added to the already substantial assessment matrix.

Facilities and Activities

Three primary experimental research programs have played a fundamental role in providing necessary thermal-hydraulic data for improving TRACE code predictive capability.

- Thermal-Hydraulics Institute (THI): The Thermal-Hydraulics Institute is a consortium of universities that has been performing separate effects experiments for the NRC since 1997. Several unique test facilities are used to perform a wide variety of thermal-hydraulic experiments. The emphasis of these tests has been interfacial area transport in pipes, annuli, and rod bundles. In addition, work has been conducted to investigate post critical heat flux (CHF) heat transfer and to support the Generic Issue and Advanced Reactor Programs.

- Rod Bundle Heat Transfer (RBHT) Program: The RBHT program has been performing separate effect experiments using a full length rod bundle designed to simulate a light water reactor rod bundle. The facility is capable of high temperatures and is heavily instrumented. Additionally, the RBHT facility has the capability for advanced droplet visualization techniques. The tests focus on steam cooling and reflood thermal-hydraulics, including the influence of spacer grids and the behavior of droplets because these items are important in determining key regulatory figures of merit, such as peak clad temperature.

- Advanced Multi-Phase Flow Laboratory (AMFL): The AMFL performs two-phase flow experiments in a highly instrumented flow loop facility that is used to design and perform scaled experiments as well as pursue theoretical and computational treatment of multiphase flows. Researchers have used the AMFL to enhance the database for Interfacial Area Transport Models. The objective of the research is to generate data for vertical and co-current downward, horizontal two-phase flow configurations, as well as the phenomena due to transition from horizontal to vertical downward or a vertical to horizontal flow through a 90 degree elbow. The experimental data are acquired by state-of-the-art two-phase flow instrumentation, including the four sensor conductivity probe, high speed camera, and laser Doppler anemometer. The obtained data will be used for developing the two group interfacial area transport model, which has been implemented in test versions of the TRACE code. This new interfacial area transport model will improve TRACE code capabilities in predicting two-phase flow characteristics and heat transfer phenomena. The use of this new model will effectively avoid the shortcomings of the traditional experimental correlations that are based on flow regimes and regime transition criteria.

Accomplishments

The thermal-hydraulic experimental programs conducted in support of TRACE have successfully completed a large number of tasks that provide data for an assessment basis and validation framework for the reactor safety code. For example, the THI program has delivered experimental data on void fraction, pressure drop, and interfacial area transport. Among other things, this data has been used to develop assessment cases for several geometric configurations and in the development and validation of interfacial area transport models for a future version of TRACE. Likewise, the RBHT and AMFL programs have provided valuable data that is being applied to two-phase flow, spacer grid, and droplet behavior models.

In addition to data gained from NRC's participation in international test programs—such as the Rig of Safety Assessment (ROSA) and Primärkreislauf Versuchsanlage [primary coolant loop test facility] (PKL) programs—these experimental programs have proven to be valuable in providing experimental verification and validation of NRC's TRACE code as well as providing the data upon which many of TRACE's correlations depend. See the TRACE assessment manuals available in NRC's public Agencywide Documents Access and Management System at www.nrc.gov at Accession No. ML120060403.

Summary

TRACE is used at the NRC for confirmatory analyses and to gain regulatory insights. Licensees and applicants are using increasingly sophisticated best-estimate methods to meet the requirements of Title 10 of the *Code of Federal Regulations* (10 CFR) 50.46, "Acceptance Criteria for Emergency Core Cooling Systems for Light Water Nuclear Power Reactors." The thermal-hydraulic experimental research programs provide an assessment basis and validation framework that help to ensure that the NRC will continue to have audit tools available that are sufficiently sophisticated to confirm industry calculations submitted to the NRC.

Contact: Chris Hoxie, RES/DSA, at Chris.Hoxie@nrc.gov.

Thermal-Hydraulic Simulations of Operating Reactors

Background

The Office of Nuclear Regulatory Research (RES) provides the tools and methods that the U.S. Nuclear Regulatory Commission (NRC) program offices use to review licensee submittals and evaluate and resolve safety issues. For thermal-hydraulic (T/H) analyses, the NRC uses the TRAC/RELAP Advanced Computational Engine (TRACE) computer code to perform the following:

- Confirmatory calculation reviews of licensee submissions, such as those for extended power uprates.

- Exploratory calculations to establish the technical bases for rule changes, such as the proposed revisions to the emergency core cooling system rule in Title 10 of the *Code of Federal Regulations* (10 CFR) 50.46, "Acceptance Criteria for Emergency Core Cooling Systems for Light Water Nuclear Power Reactors."

- Exploratory calculations for the resolution of generic issues, such as Generic Issue 191, "Assessment of Debris Accumulation on PWR [pressurized-water reactor] Sump Performance."

RES is developing a library of TRACE input decks for simulating currently operating PWRs and boiling water reactors (BWRs).

Approach

TRACE plant input decks are developed for specific simulations. These can be design-basis loss-of-coolant accidents (LOCAs), anticipated operational occurrences, anticipated transient without scram (ATWS), and other transients. Depending on the simulations to be performed, the size and complexity of plant input decks can range from single system components to the entire nuclear steam supply system. TRACE is able to simulate the multifaceted evolution of these events, capturing all of the major system operations and T/H processes that unfold (see Figure 2.7).

Each physical piece of equipment in a plant can be represented as some type of TRACE component, and each component can be further nodalized into a number of physical volumes—also called cells—over which the fluid, conduction, and kinetics equations are averaged. TRACE input decks representing entire plants consist of an array of one dimensional and three-dimensional TRACE components arranged and sized to match plant specifications.

Because of the modeling flexibility available to the user, the "TRACE User's Manual" (Ref. 1) contains the best practice modeling guidelines. The User's Manual shows modelers the most effective methods to arrange generic one dimensional components to depict particular systems and to employ function specific components, such as the PWR accumulator and pressurizer and the BWR jetpump and channel components, to achieve the desired results.

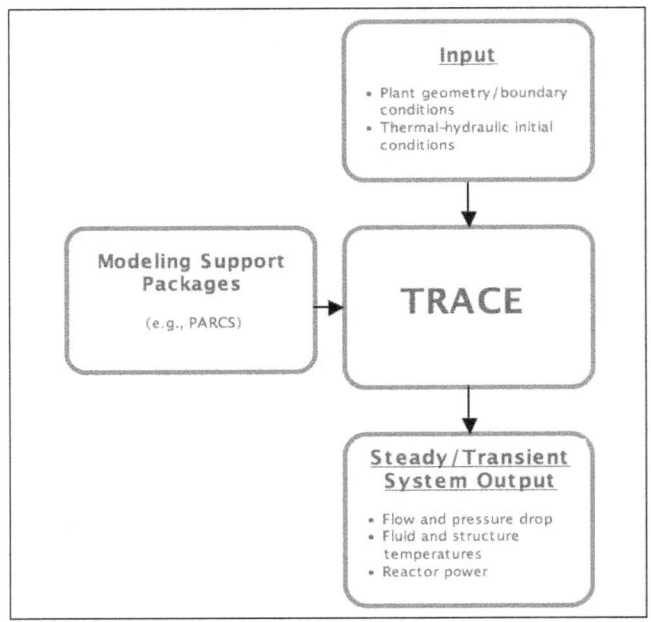

Figure 2.7 TRACE, an advanced, best-estimate reactor system code used to model the T/H performance of nuclear power plants

User input includes plant geometry and process conditions (e.g., temperature, flow). The code supports integration with detailed modeling packages (e.g., the three-dimensional kinetics code, PARCS), used to model specific performance issues, including neutronics.

Once the arrangement of the plant deck is complete and each component is set with initial values for normal operating pressures, temperatures, and flow conditions, TRACE is run in steady-state mode for a period of time to test the model and to develop appropriate steady-state initial conditions for the specified operating state and boundary conditions. TRACE models transients and accidents by simulating an initiating event after steady initial conditions have been reached. Developmental assessments support the applicability of TRACE in modeling these events (Ref. 2).

The NRC updated plant input decks developed for other system codes and converted them into TRACE format to support the licensing reviews of extended power uprate applications. It uses these models to assess the effects of increased power on system behavior and safety margins.

BWR Models

The NRC has developed representative LOCA and design-basis accident input decks for most General Electric type BWRs, including the BWR3, BWR4, and BWR5 plants (see Figure 2.8). TRACE significantly enhanced component specific features to improve the modeling of containment pressurization and feedback during design-basis events.

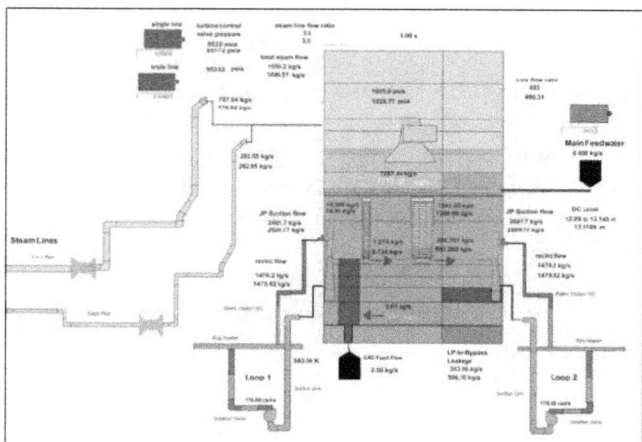

Figure 2.8 Steady-state conditions in a BWR

PWR Models

Representative LOCA and design-basis accident models have been developed for Westinghouse PWRs with two, three, and four loops, several Combustion Engineering plants, and two Babcock and Wilcox plants (see Figure 2.9).

Building a comprehensive library of plant input decks will enhance the ability of the NRC staff to perform timely confirmatory analyses in support of regulatory decisions.

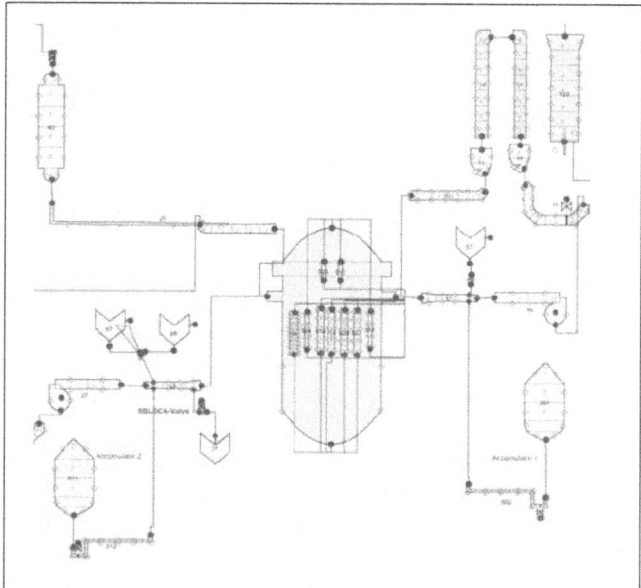

Figure 2.9 Key primary coolant T/H components, including reactor vessel, pumps, and steam generator, for a two-loop PWR

References

1. TRACE 5.0 User's Manual, Volume 2: Modeling Guidelines (Agencywide Document Access and Management System [ADAMS] Accession No. ML071720510).

2. TRACE 5.0 Assessment Manual—Main Report (ADAMS Accession No. ML071200505).

For More Information
Contact Scott Elkins, RES/DSA, at Scott.Elkins@nrc.gov.

Simulation of Anticipated Transients Without SCRAM with Core Instability for Maximum Extended Load Line Limit Analysis Plus-Boiling Water Reactors

Background

The industry has proposed the maximum extended load line limit analysis plus (MELLLA+) domain for boiling water reactors (BWRs) that have extended power uprates (EPUs). The MELLLA+ domain would allow operation at high reactor thermal power (up to 120 percent of originally licensed thermal power [%OLTP]) at reduced reactor core flow (as low as 80 percent of rated core flow [%RCF]). The high power-to-flow operating point (120 %OLTP / 80 %RCF) introduces new concerns related to the consequences of anticipated transient without SCRAM (ATWS) events initiated from this point [Ref. 1]. In particular, the plant will evolve to a condition of high power to flow ratio during an ATWS, in which large amplitude power oscillations are expected to occur. Figure 2.10 illustrates the transient evolution of postulated ATWS events for a plant operating at the low flow corner of the MELLLA+ upper boundary.

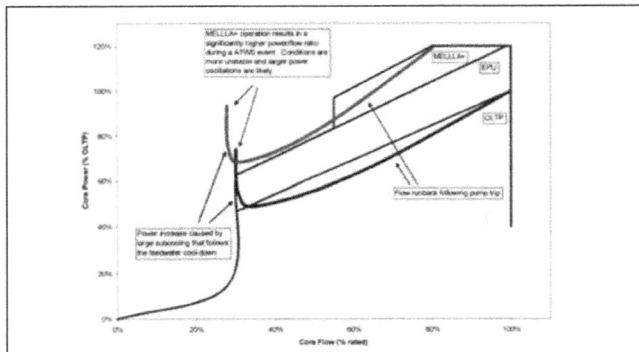

Figure 2.10 Operating state evolution during ATWS for different operating domains

The TRAC-RELAP Advanced Computational Engine (TRACE) and Purdue Advanced Reactor Core Simulator (PARCS) codes have previously been applied to analyze and study complex BWR transients [Ref. 2 and Ref. 3]. The U.S. Nuclear Regulatory Commission's Office of Nuclear Regulatory Research (RES) has performed studies of ATWS with instability (ATWS-I) events using TRACE/PARCS to understand better the dynamic coupling during ATWS-I and the safety implications associated with the MELLLA+ operating domain.

Approach

Simulation of ATWS-I requires using several codes and a defined methodology for the use of these codes and interfaces. RES developed a methodology for generating large core models in TRACE comprised of many channels to represent the thermal-hydraulic and fuel thermal-mechanical response of the core. The model uses FRAPCON calculations to generate dynamic gap conductance properties for the fuel. Figure 2.11 illustrates the process for generating detailed multichannel BWR core models.

Once the core model has been generated and incorporated into the TRACE model, calculations are performed using TRACE and PARCS in a coupled manner. Figure 2.12 illustrates the process for performing these coupled calculations. One key feature of the TRACE/PARCS method is the use of flux harmonic calculations to excite in-phase and out-of-phase core oscillations.

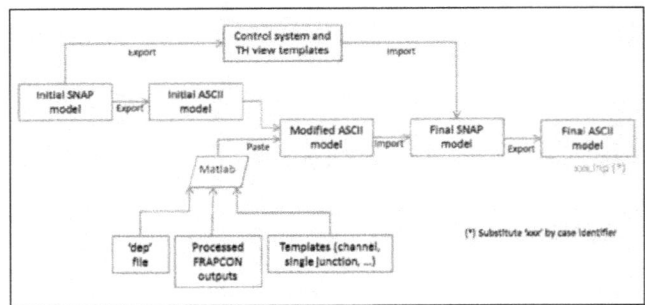

Figure 2.11 Generation of the BWR core model

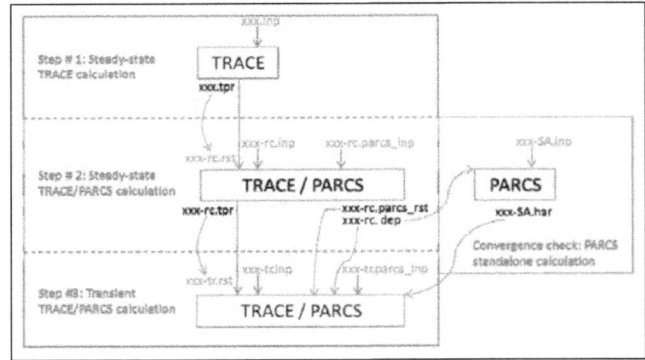

Figure 2.12 TRACE/PARCS coupled methodology

Results

Using the TRACE/PARCS methodology, complex transients such as ATWS-I can be simulated. Visualization tools have been developed to analyze the evolution of the power oscillations during ATWS-I and to study the oscillation contour. Figure 2.13 illustrates the result of an ATWS-I simulation. TRACE/PARCS predicts the onset of large amplitude power oscillations and the evolution of an out of phase oscillation contour in this example. Results from this analysis are still under investigation.

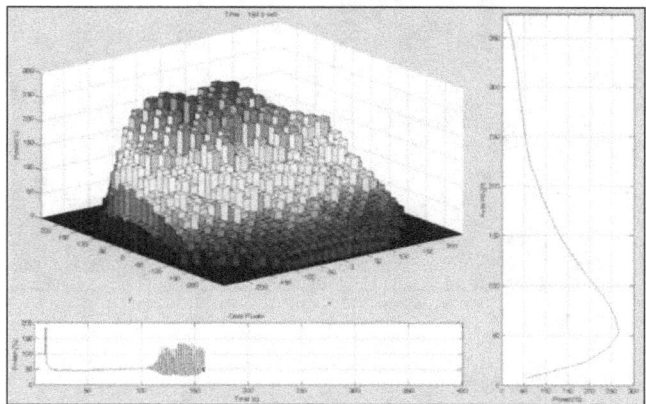

Figure 2.13 Power oscillation visualization during simulated ATWS-I

References

1. "General Electric Boiling Water Reactor Maximum Extended Load Line Limit Analysis Plus," NEDC-33006P-A, Rev. 3, General Electric Nuclear Energy, June 2009.

2. P. Yarsky, "Applicability of TRACE/PARCS to MELLLA+ BWR ATWS Analyses," U.S. Nuclear Regulatory Commission, Office of Nuclear Regulatory Research, Rev. 1, November 18, 2011.

3. "TRACE V5.0 Assessment Manual," U.S. Nuclear Regulatory Commission, 2007.

For More Information
Contact Scott Elkins, RES/DSA, at Scott.Elkins@nrc.gov.

Thermal-Hydraulic Analyses of New Reactors

Background

The U.S. Nuclear Regulatory Commission (NRC) uses the Transient Reactor Analysis Code/Reactor Excursion and Leak Analysis Program (TRAC/RELAP) Advanced Computational Engine (TRACE) code to perform confirmatory calculations in support of design certification and combined operating license reviews for all new reactors—the Advanced Passive 1000 Megawatt (AP1000), U.S. Advanced Pressurized-Water Reactor (U.S. APWR), the U.S. Evolutionary Power Reactor (EPR), the Economic Simplified Boiling-Water Reactor (ESBWR), and the Advanced Boiling-Water Reactor (ABWR). The modeling of various integral pressurized-water reactor (iPWR) designs has been undertaken to assess the applicability of NRC codes in anticipation of confirmatory analyses.

New reactor designs include evolutionary advances in light water reactor technology and thus pose unique modeling challenges as a result of novel systems and operating conditions. Many of these modeling challenges are associated with passive systems that rely on phenomena such as gravity, pressure differentials, natural convection, or the inherent response of certain materials to temperature changes. Most developmental assessments conducted for currently operating light water reactors cover the phenomenology necessary in thermal-hydraulic simulations for new reactor designs. However, the modeling of some of the novel systems and operating conditions of new reactors requires further code development and additional assessments against specific experimental data.

New Reactor Designs

AP1000

The AP1000 (see Figure 2.14) relies extensively on passive safety systems. Passive systems are used for core cooling, containment cooling, main control room emergency habitability, and containment isolation. These systems challenge system codes in predicting fluid flow induced by small driving heads. The applicability of TRACE to simulate AP1000 transients was demonstrated through comparisons with data from relevant integral and separate effects test facilities.

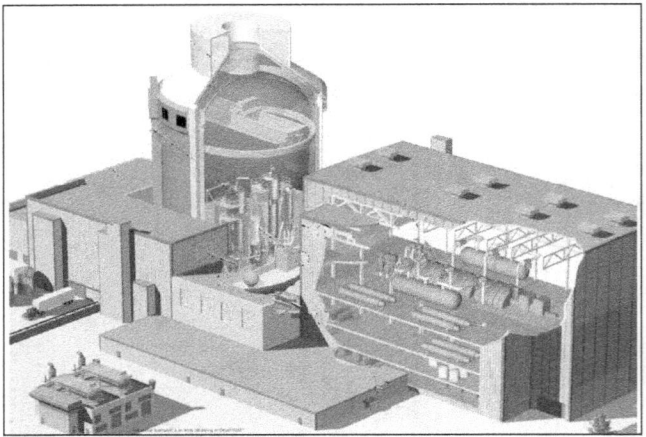

Figure 2.14 AP1000

U.S. APWR

Most of the major components of the U.S. APWR (see Figure 2.15) are very similar to those of existing pressurized-water reactors (PWRs). The major exception is the advanced accumulator that eliminates the need for pumped low pressure safety injection. The ability of TRACE to predict the behavior of advanced accumulators has been demonstrated with separate effects data. Furthermore, detailed three-dimensional phenomena, such as cavitation, nitrogen ingress, and mass flow rate, have been modeled using computational fluid dynamics tools, and the results were coupled as needed with system code simulations.

Figure 2.15 U.S. APWR

EPR

The EPR (see Figure 2.16) is an evolutionary PWR design that uses rapid secondary side depressurization for mitigation of loss-of-coolant accidents (LOCAs). This increases the emphasis on the ability of TRACE to predict reflux condensation in steam generator tubes. To demonstrate the applicability of TRACE to the EPR, code predictions were assessed against data acquired from separate and integral test facilities, such as Advanced Power Extraction (APEX), Full Length Emergency Cooling Heat

Transfer Separate Effects and Systems Effects Tests (FLECHT SEASET), Rig of Safety Assessment (ROSA) IV, and ROSA V.

Figure 2.16 U.S. EPR

ESBWR

The ESBWR (see Figure 2.17) has a passively driven containment cooling system and a gravity driven cooling system. Both systems rely entirely on natural phenomena for the convection of mass and energy. The prediction of void distributions and two-phase natural circulation is very important for the ESBWR. Integral test data from the Purdue University Multi-Dimensional Integral Test Assembly (PUMA) and Passive Non Destructive Assay of Nuclear Materials (PANDA) facilities were used to assess the code for this application. In addition, proper modeling of film condensation in the presence of non condensable gases at low power levels posed a significant challenge in the ESBWR analysis. Improved models in TRACE predicted these phenomena very well.

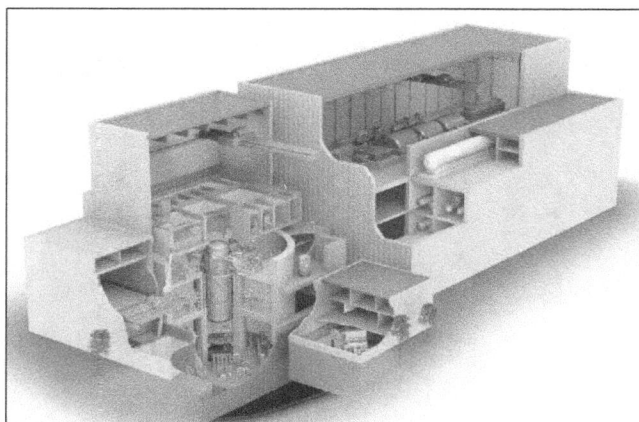

Figure 2.17 ESBWR

ABWR

The ABWR (see Figure 2.18) is an evolutionary boiling water reactor that includes such design enhancements as recirculation pumps internal to the reactor vessel and digital controls. TRACE will be used to simulate the plant response to LOCAs, as well as to anticipated operational occurrences and other transients. Modeling internal pumps and incorporating the logic needed for digital controls will pose potential challenges to the code.

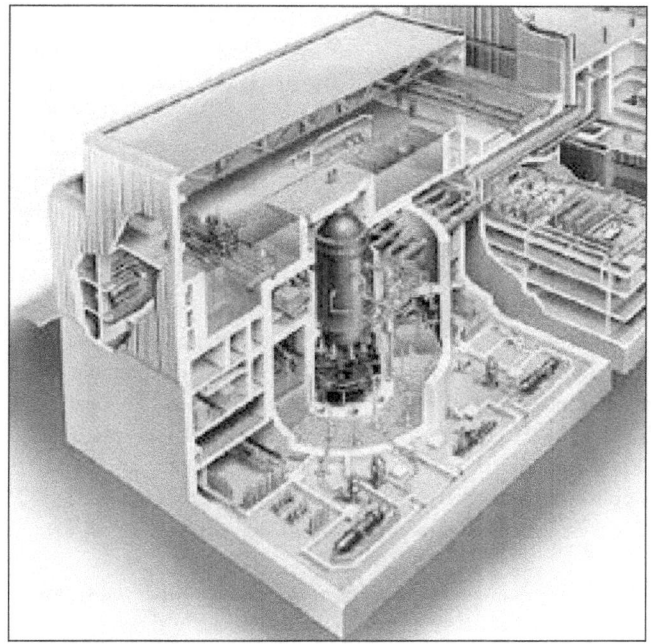

Figure 2.18 ABWR

For more information

Contact Scott Elkins, RES/DSA, at Scott.Elkins@nrc.gov.

Thermal-Hydraulic Analysis of Integral Pressurized-Water Reactors (iPWR)

Background

The U.S. Nuclear Regulatory Commission (NRC) has been informed that design certification applications for iPWRs are planned by Babcock & Wilcox (B&W), NuScale, and Westinghouse. Applications are expected in the near future for B&W's mPower Reactor and NuScale's Power Reactor. As part of NRC's pre-application review activities for these designs, efforts have been initiated to identify and examine key technical and policy issues potentially shared by various iPWRs designs. Activities including the identification of unique and important phenomena, development of phenomena identification and ranking tables (PIRTs), and assessment of thermal-hydraulic code capabilities are important since they can inform policy and technical issues related to iPWR licensing activities. They are also anticipated to assist with the selection and preparation of methods and computer codes that will be used to review potential iPWR applications submitted for design certification. Early modeling of iPWR designs performed as part of the pre-application activities assists in assessing the applicability of NRC codes in anticipation of confirmatory analyses.

iPWRs are small and medium sized PWRs in which the steam generators and reactor core are integrated within the vessel. Current iPWR designs (e.g., Figure 2.19) eliminate external reactor coolant system piping and integrate the steam generator and pressurizer within the reactor vessel as one integral primary system. The NuScale iPWR design uses helical tube steam generators, and the mPower iPWR design uses once through tube steam generators to produce superheated steam in the secondary system. Test data from the Multi Application Light Water Reactor (MASLWR) and Integrated System Test (IST) facilities are being used to assess NRC thermal-hydraulic codes for applicability for use in system analysis of these designs.

Approach

Recent iPWR pre-application studies and preparations consisted of six tasks:

1. Identify event scenarios for pre-application studies.

2. Obtain "best available design information" for analysis and assessment.

3. Develop pre-application PIRT results.

4. Conduct a pre-application gap analysis for the TRAC/RELAP Advanced Computational Engine (TRACE) and Purdue Advanced Reactor Core Simulator (PARCS) computer codes.

5. Determine pre-application code and model modifications.

6. Complete a pre-application code applicability assessment.

To date, the two iPWR designs that have been advanced for pre-application studies have been the mPower and NuScale designs. Therefore, two NuScale-like scenarios and two mPower-like scenarios were used to build plant simulation models. These chosen scenarios were used to exercise both the thermal-hydraulic and neutronic capabilities of the TRACE/PARCS computer codes. For the NuScale-like design, the thermal-hydraulic event analyzed was an inadvertent opening of a reactor recirculation valve, and the neutronic event analyzed was an inadvertent control rod bank withdrawal from hot zero power (HZP). For the mPower-like design, a double-ended break of a condensate return line was the thermal-hydraulic scenario considered, while an inadvertent control rod bank withdrawal from HZP was the simulated neutronic event. TRACE (with PARCS) is capable of modeling these scenarios. However, modeling of some iPWR-related phenomena will require further assessment using experimental data or development of iPWR-specific correlations. The pre-application studies have been based on design and technical information that vendors and the U.S. Department of Energy (DOE) have provided. Applicability of the pre-application results to subsequent iPWR application reviews depends on the degree of similarity of the pre-application design information used for this work to that of the final design submitted through the license application process.

The pre-applicability capabilities assessment indicates that the TRACE and PARCS codes can be effectively used to perform thermal-hydraulic confirmatory analyses on iPWR reactor designs. Results from these activities are currently being used to prepare for review of design certification applications.

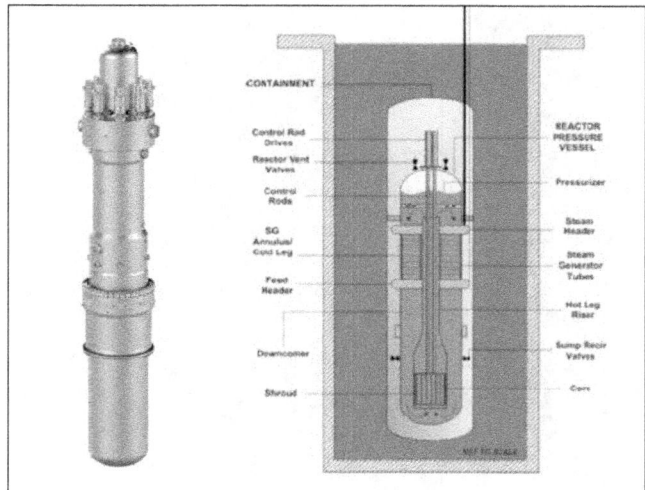

Figure 2.19 Example of the mPower IPWR design

For More Information

Contact Scott Elkins, RES/DSA, at Scott.Elkins@nrc.gov.

Computational Fluid Dynamics in Regulatory Applications

Background

Computational fluid dynamics (CFD) has reached the maturity necessary to play an increased role in the nuclear power generation industry. CFD provides detailed three-dimensional fluid flow information not available from system code thermal-hydraulic simulations. These multidimensional details can enhance the understanding of certain phenomena and thus play a role in reducing the uncertainty in the technical bases for licensing decisions.

The U.S. Nuclear Regulatory Commission's Office of Nuclear Regulatory Research (RES) has developed a state-of-the-art CFD capability that supports multiple offices within the agency. RES uses the commercial CFD codes from ANSYS Inc. (FLUENT) and CD adapco (STAR CCM+) and has supported the development of multiphase modeling capabilities in research codes. The office maintains a Linux cluster with more than 200 processors to provide the capability needed to solve the large scale problems that are characteristic in the nuclear industry. RES staff is actively involved in national and international CFD programs and maintains a high level of expertise in the field. This state-of-the-art capability provides a robust infrastructure for both confirmatory and exploratory CFD computations.

Applications

Spent Fuel Transportation and Storage

RES works closely with the Office of Nuclear Material Safety and Safeguards in areas concerning the analysis of spent fuel storage cask designs.

The CFD approach has been used to study cask designs under a variety of external conditions, such as fires, reduced ventilation, and hotter fuels. This work supports dry cask certification efforts by further informing the agency's technical bases for licensing decisions (see Figure 2.20).

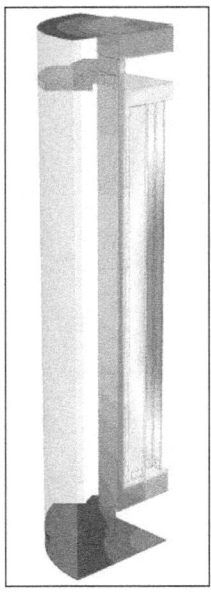

Figure 2.20 Temperature contours of a ventilated dry cask that uses ambient air to passively cool the spent fuel stored inside the canister surrounded by a concrete overpack

Operating Reactors

CFD predictions have also aided in understanding detailed fluid behavior for broad scope analyses, such as pressurized thermal shock, induced steam generator tube failures, boron dilution and transport, and spent fuel pool analyses. In most cases, CFD results are used iteratively with system code predictions, or they provide boundary or initial conditions for other simulations (see Figure 2.21).

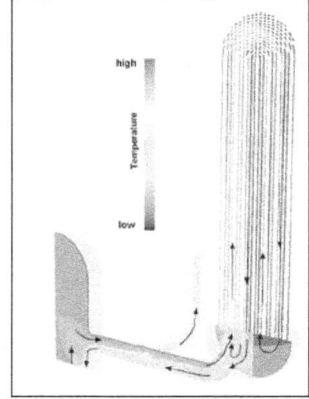

Figure 2.21 During a particular severe accident scenario, hot gases from the core circulate through the hot legs and steam generator in a counter-current flow pattern. The risk of induced failures is considered

New and Advanced Reactors

The agency has used CFD to confirm the distribution of injected boron in the ESBWR. In the design certification of the U.S. APWR, CFD was used to investigate the performance of an advanced accumulator (see Figure 2.22). The phenomena of interest are cavitation and nitrogen ingress, which exceed typical system code capabilities. CFD results also were used to examine possible scale effects.

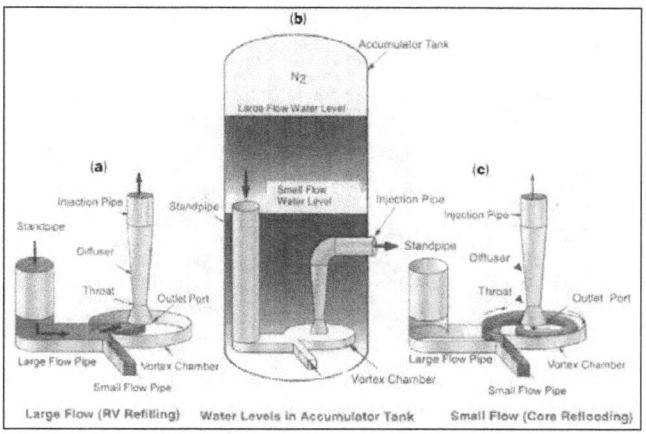

Figure 2.22 The advanced accumulator (b) is a water storage tank with a flow damper in it that switches the flow rate of cooling water injected into a reactor vessel from a large (a) to small (c) flow rate

For More Information

Contact Scott Elkins, RES/DSA, at Scott.Elkins@nrc.gov.

Nuclear Analysis and the SCALE Code

Background

As used here, the term nuclear analysis describes the use of analytical tools and experimental data to predict and understand the interactions of nuclear radiation and matter within various nuclear systems. Nuclear analysis thus encompasses the analyses of (1) fission reactor neutronics, both steady-state and dynamic, (2) nuclide generation and depletion, as applied to predicting reactor and spent-fuel decay heat power, fixed radiation sources, and radionuclide inventories potentially available for release, (3) radiation transport and attenuation, as applied to the evaluation of fluence leading to material damage, material dosimetry, material activation, radiation detection, and radiation protection, and (4) nuclear criticality safety (i.e., the prevention and mitigation of self-sustaining fission chain reactions outside reactors).

Objective

An objective of the U.S. Nuclear Regulatory Commission's (NRC's) Office of Nuclear Regulatory Research (RES) is to perform independent neutronics and criticality analyses for existing and new nuclear reactor designs, spent fuel pools (SFPs), and spent fuel storage and transportation casks.

Approach

Overview

Nuclear analysis efforts support the staff's ongoing and anticipated nuclear safety evaluation activities for the licensing and oversight of (1) existing reactors, front-end fuel cycle activities, and spent fuel storage, transport, and disposal systems, and (2) proposed new and advanced reactors (see Figures 2.23 and 2.24) and their associated front-end and back-end fuel cycle activities. The primary nuclear analysis tools used for these activities are (1) the Purdue Advanced Reactor Core Simulator (PARCS) core neutronics simulator code, (2) the SCALE 6.1 modular code system, and (3) the Advanced Module for Processing Cross Sections (AMPX) code for processing fundamental nuclear data in the Evaluated Nuclear Data File (ENDF) into code-usable libraries of continuous energy or fine-group nuclear Cross-sections and related nuclear data. When appropriate, RES integrates planned nuclear analysis activities into larger NRC research plans for the respective applications.

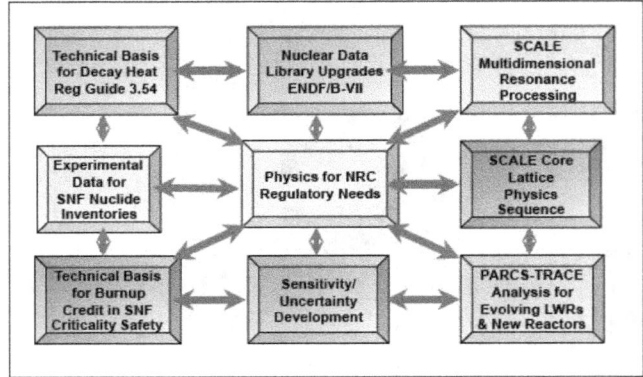

Figure 2.23 Coupled reactor and fuel cycle nuclear analyses

Identification of Issues and Needs

An example of the need for additional data for current and near-term activities is in the area of burnup credit for the criticality safety analysis of spent fuel casks. Operating and new reactors need experimental data to validate codes and reduce uncertainties. Such validation currently relies on limited data or code-to-code comparisons. The NRC has validated nuclear codes for partial mixed-oxide fueling in pressurized-water reactors (PWR) and is validating codes against plant operating and test data for use in steady-state and transient analyses of modern boiling-water reactor (BWR) cores, including the Economic Simplified BWR.

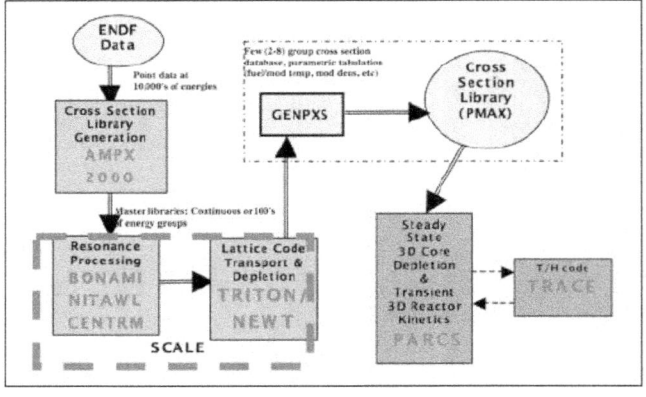

Figure 2.24 NRC nuclear analysis codes for reactor physics

The NRC is currently modifying and extending codes to accommodate different fuel, core, and control configurations and operating features of high-temperature gas-cooled reactors and small modular reactors. A new SCALE automated calculation sequence is being developed to allow quicker lattice Cross-section generation execution times and engineering evaluations. In addition, the NRC is updating the radiation shielding codes for application to high-capacity spent fuel cask systems and advanced reactor systems.

For More Information
Contact Dr. Mourad Aissa, RES/DSA, at Mourad.Aissa@nrc.gov

High-Burnup Light-Water Reactor Fuel

Background

Fuel rod cladding is the first barrier for retention of fission products, and the structural integrity of the cladding ensures a coolable geometry during hypothetical reactor transients and accidents. Ensuring cladding integrity also allows simplifying assumptions to be made in spent fuel cask criticality calculations. Regulations and regulatory guidance documents contain fuel and cladding damage criteria. Licensees compare the criteria to predictions of fuel rod behavior during reactor operation and following discharge during spent fuel transportation and storage.

The fuel damage criteria were originally developed from a data base of unirradiated and low-burnup fuel with Zircaloy cladding. From more recent test data, it became clear that extrapolation from a low-burnup data base was not satisfactory for regulatory purposes, and the NRC initiated a high-burnup fuel research program to address this issue.

Objective

The current research program is designed to provide information in the following areas:

- Embrittlement Criteria and Oxidation Correlations for Loss-of-Coolant Accidents (LOCA) Title 10 of the *Code of Federal Regulations* (10 CFR) 50.46, "Acceptance Criteria for Emergency Core Cooling Systems for Light-Water Nuclear Power Reactors"

- Coolability Criteria and Threshold Failure Correlations for Reactivity-Initiated Accidents (Regulatory Guide 1.77, "Assumptions Used for Evaluating a Control Rod Ejection Accident for Pressurized-Water Reactors"; NUREG-0800 section 4.2, "Fuel System Design").

- Fuel Rod Properties for Transportation and Storage Analysis (10 CFR 71.55, "General Requirements for Fissile Material Packages"; 10 CFR 72.122, "Overall Requirements").

- Fuel Rod Computer Codes, used to audit licensees' evaluation models that demonstrate compliance with criteria and to analyze test data (10 CFR Part 50, "Domestic Licensing of Production and Utilization Facilities," Appendix K, "ECCS Evaluation Models").

Approach

NRC sponsors experimental programs at Oak Ridge National Laboratory and Studsvik Nuclear AB hot cell laboratory (Sweden) to provide data for the various objectives of the current research program. NRC also contracts with the Pacific Northwest National Laboratory for support for NRC's fuel rod computer codes, along with support for a code users' group consisting of 24 U.S. and international participants.

The NRC is actively engaged with international research programs conducted by the Halden Reactor Project (Norway), the Institute for Radiological and Nuclear Safety (France), the Japan Atomic Energy Agency, and Studsvik Nuclear AB, which are also providing valuable data related to the research program objectives.

Loss-of-Coolant Accidents

During a postulated LOCA, the fuel rod cladding would experience very high temperatures and severe oxidation. The NRC's regulations specify limits for temperature and oxidation to preserve ductility and thereby ensure a coolable geometry following this postulated accident. However, additional phenomena occur with high-burnup fuel that the original embrittlement criteria do not address. Nevertheless, current plant operations provide adequate assurance of safety, largely through the use of conservative methods.

Based on its research (Research Information Letter 0801, "Technical Basis for Revision of Embrittlement Criteria in 10 CFR 50.46," dated May 30, 2008, Agencywide Documents Access and Management System [ADAMS] Accession No. ML081350225, and "Update to Research Information on Cladding Embrittlement Criteria in 10 CFR 50.46" dated December 29, 2011, ADAMS Accession No. ML113050484), the NRC is developing new performance-based criteria to account for high-burnup phenomena and to permit the use of new cladding materials without requiring license exemptions.

With the loss of reactor pressure and high temperatures experienced during a LOCA, some fuel rods will deform outwards and burst or rupture. The NRC conducted a confirmatory research program aimed particularly at the behavior of the ballooned and burst region of high-burnup fuel under LOCA conditions. Figure 2.25 illustrates extensive bending of a ballooned and ruptured sample which did not experience high temperature oxidation. The sample demonstrated significant ductility, demonstrating a large capacity for plastic deformation without fracture. This result, in combination with bending tests of heavily oxidized samples that fractured in a brittle manner, supported the NRC's treatment of embrittlement behavior for ballooned and ruptured fuel rods in LOCA analysis within the proposed rulemaking for 10 CFR 50.46.

Figure 2.25 Illustration of the total displacement of a ballooned and ruptured sample, which did not experience high temperature oxidation, by superimposing the initial and final images of the bent rod. This rod demonstrated significant ductility, demonstrating a large capacity for plastic deformation without fracture

Reactivity-Initiated Accidents

Following an accidental control rod ejection in a pressurized-water reactor (PWR) or control blade drop in a boiling-water reactor (BWR), the fuel rod cladding would experience very large stresses at relatively low temperatures. The NRC requirements specify limits on the energy deposited in these events to avoid cladding failure or dispersal of hot fuel particles with the potential for energetic fuel coolant interactions, core damage, and loss of coolability. However, additional phenomena occur with high-burnup fuel that lower the cladding's ductility and substantially reduce the amount of deposited energy that can be tolerated. Although additional work is planned in France, using the CABRI reactor, and continuing in Japan, using the Nuclear Safety Research Reactor, enough research has been completed to revise regulatory criteria based on these results. These revisions to NRC requirements apply to all new reactor designs and will apply to existing plants when they become final. Meanwhile, current plant operations provide adequate assurance of safety, largely as the result of the voluntary use of conservative methods.

Transportation and Storage

During transportation and storage of spent fuel, the fuel rod cladding experiences higher temperatures and pressure differences than during full-power operation, and the fuel rods experience large impact loads in postulated accidents. Because of the fuel rod cladding's reduced ductility at high burnup, its mechanical properties and failure conditions are substantially altered.

Testing on high-burnup specimens of most commercial cladding types will provide the mechanical properties that are needed for safety analyses.

Storage conditions can also lead to changes in the morphology of hydrogen precipitates, leading to changes in the fracture properties of cladding material. Hydrogen is absorbed in the cladding during the burnup-related corrosion process under normal operation and typically precipitates in hydrides oriented in the cladding circumferential direction. Under cask loading conditions the hydride precipitates may reorient in the cladding radial direction, resulting in a reduction in cladding ductility. The NRC has recently completed a research program aimed at identifying the conditions under which this ductility loss takes place and the results have been published as an original research article, *"Ductile-to-brittle transition temperature for high-burnup cladding alloys exposed to simulated drying-storage conditions,"* in the Journal of Nuclear Materials in February 2013. A full technical report is available in the Agencywide Documents Access and Management System (ADAMS), ML12181A238.

Fuel Rod Computer Codes

The NRC maintains computer codes for the analysis of both steady-state and transient conditions. The agency uses these codes to evaluate experimental data and to audit licensees' safety analyses. As new cladding alloys are introduced (e.g., Areva's M5 and Westinghouse's Optimized ZIRLO), burnable poisons are changed (e.g., high concentrations of gadolinia), and higher burnups are sought (beyond 62 gigawatt day per ton), the materials' properties and models in the codes must be revised. In-reactor tests are often used to obtain data for these changes. Halden results are particularly valuable. The ability to perform quantitative analyses of fuel rod behavior is an essential part of the NRC's assessment of safety in reactor operations and spent fuel transportation and storage.

For More Information
Contact Harold Scott, RES/DSA, at Harold.Scott@nrc.gov. Michelle Flanagan, RES/DSA, at Michelle.Flanagan@nrc.gov.

Spent Nuclear Fuel Burnup Credit

Background

Spent nuclear fuel (SNF) refers to uranium-bearing fuel elements that have been used at commercial nuclear reactors and are no longer producing enough energy to sustain full power reactor operation. The fission process has stopped once the spent fuel is removed from the reactor, but the spent fuel assemblies still generate significant amounts of radiation and heat. Because of the residual hazard, spent fuel must be stored or shipped in containers or casks that shield and contain the radioactivity and dissipate the heat. Further, the SNF storage or shipping system needs to ensure subcriticality, thereby preventing criticality accidents.

The United States stores SNF at a variety of sites (e.g., in reactor spent fuel pools (SFPs) or in dry cask storage at reactor sites). Over the last 30 years, thousands of shipments of commercially generated SNF have been made over highways, through towns, and along railroads in the United States without causing any radiological releases to the environment or harm to the public. It is also crucial to have no criticality accidents during storage and transportation.

Most of these spent fuel shipments occur between reactors owned by the same utility to share storage space. In addition, spent fuel may be shipped to a research facility to perform tests on the spent fuel. To minimize the number of such shipments, as much nuclear material as possible is loaded into each shipment without violating criticality safety. Based on work performed under this research project, the U.S. Nuclear Regulatory Commission's Office of Nuclear Material Safety and Safeguards (NMSS) recently revised its Interim Staff Guidance 8 (ISG8, Revision 3) to allow full burnup credit for pressurized-water reactor (PWR) spent fuel in transportation and storage casks. Burnup credit is the concept for taking credit for the reduction in reactivity due to fuel burnup as neutron-absorbing fission products and built-in actinides (see Figure 2.26).

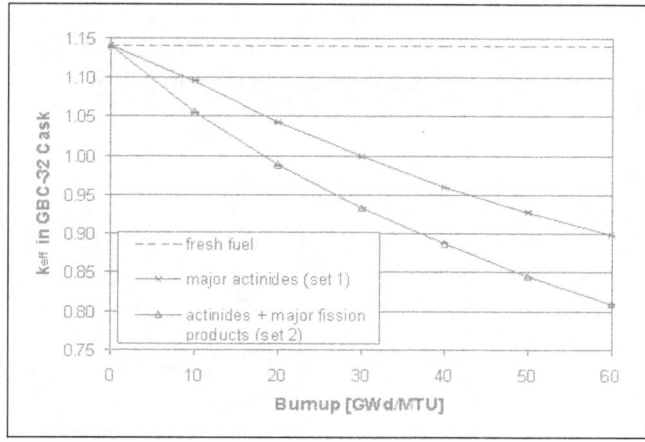

Figure 2.26 Comparison of typical reactivity decrements associated with actinides only and with a combination of actinides and fission products

Currently, the Office of Nuclear Reactor Regulation (NRR) allows licensees full burnup credit for PWR spent fuel stored in their SFPs; however, in the case of boiling-water reactors (BWRs), it is limited to the assembly peak reactivity accompanying burnable poison depletion.

Objective

The purpose of this research is to develop a technical basis to support the allowance of full (fission product and actinides) burnup credit for spent fuel located in SFPs and transportation and storage casks.

The research will provide the technical basis to support an agencywide, integrated approach to further expand the application of burnup credit in spent nuclear fuel storage and transportation systems to BWR fuel.

Approach

In contrast to PWR fuel burnup credit, which this project has evaluated for several years, relatively few studies have been performed to establish the detailed understanding of the relevant issues and phenomena associated with BWR fuel burnup credit. A detailed understanding is required to provide a technical basis for regulatory guidance on the use of burnup credit for BWR fuel in SFPs and dry cask storage and transport.

The following approach for BWR fuel is based on the work performed in implementing the PWR burnup credit methodology:

1. **Sensitivity studies to determine and document the reactivity influence of variations in input parameters required for BWR burnup credit criticality safety evaluations.**

   Studies are currently being performed to address the directional sensitivities to reactor operating conditions, including fuel temperature, moderator temperature and density, power, and control blade usage; the reactivity behavior of different BWR fuel assembly designs; the influence of spatial burnup variations; the influence of cooling time; and the influence of isotopic validation. The studies are being performed with relevant SFP storage racks and dry casks for storage and transportation design and will be documented in NUREG/CR reports.

2. **Evaluation of available measured isotopic composition data from BWR spent fuel to support isotopic validation.**

   Under this activity, two-dimensional (SCALE/TRITON) depletion calculations will be performed for comparison to the available measured data with the goals of developing a basis for isotopic validation, determining a representative bias and bias uncertainty for the SCALE/TRITON code, and determining the range of applicability associated with the bias and bias uncertainty. Much of the existing and recently available measured BWR data has not been previously modeled, thus considerable effort is required in this activity to first model and then evaluate these data.

3. **Evaluation of available critical experimental data to support criticality validation for spent BWR fuel.**

   Under this activity, the sensitivity/uncertainty tools (TSUNAMI) in SCALE will be used to evaluate relevant critical experiments and identify those that are applicable for validation of BWR SFP racks and dry cask storage and transportation designs. The evaluation will consider experiments from the International Criticality Safety Benchmark Experiment Project Handbook, as well as other proprietary and nonproprietary experiments, with the goals of developing a basis for criticality validation, determining a representative bias and bias uncertainty for the SCALE/KENO code, and determining the range of applicability associated with the bias and bias uncertainty.

For More Information
Contact Dr. Mourad Aissa, RES/DSA, at Mourad.Aissa@nrc.gov.

Burnup Credit Methodology— Pressurized-Water Reactors

Background

Burnup credit (BUC) is the term used when accounting for the reduced reactivity in spent nuclear fuel (SNF) in criticality analyses. This includes the calculation of a depletion code bias and bias uncertainty. BUC is used in both spent fuel pool (SFP) and transportation and storage system criticality analyses. This work focuses on the application of BUC to pressurized-water reactor (PWR) fuel.

BUC takes credit for the fact that during irradiation, the amounts of uranium-235 and embedded neutron absorbers in the uranium dioxide fuel are reduced. At the same time, the content of other fissile isotopes increases because of nonfission neutron capture, and irradiation of the fuel produces neutron-absorbing fission product nuclides. This results in a net reduction in reactivity of the fuel relative to burnup.

For the appropriate use of BUC, the criticality analysis requires that the computer code used be validated over the range of interested nuclides and burnup. This is a challenging requirement because critical experiments with actinide and fission product nuclides similar to SNF are not generally available.

Currently, the regulatory environment has the Office of Nuclear Regulation (NRR) and the Office of Nuclear Materials and Safeguards (NMSS) using different methodologies for the application of BUC. To address this issue, the U.S. Nuclear Regulatory Commission (NRC) initiated a project to establish a consistent, agencywide, technical basis for the application of BUC to criticality analyses.

Objective

The objective of this work is to establish a consistent, agencywide, technical basis for SFP and dry cask criticality safety evaluations that NRC staff can confidently use to assess licensee applications. The scope will be for PWRs.

Approach

Currently, the offices involved with SNF criticality-related applications are NRR for the SFP and NMSS for dry storage and transportation systems. These offices are using different methods and allow for different isotopes to be credited. NRR reviewers allow credit to be taken for all the calculated isotopes of interest to BUC. Of particular note, NRR uses guidance from an internal memorandum called the Larry Kopp Letter (ML003728001) that accounts for the calculation of depletion uncertainty. This memorandum recommends the use of an uncertainty equal to 5 percent of the reactivity decrement to the burnup of interest. The technical basis for this depletion uncertainty is engineering judgment and is not well documented. NMSS uses Interim Staff Guidance (ISG) 8 Revision 3 (ISG 8 Revision 2 at the time of this document's development) for BUC application to dry storage and transportation systems for a specified subset of validated nuclides. Thus, significant differences exist between NRR's and NMSS's approach.

This work will (1) develop and establish a validation approach for commercial SNF criticality safety evaluations based on best-available data and methods and (2) demonstrate its application via a representative SNF storage and transport system.

Research Activities

This research has now concluded with the development of two NUREG/CRs: NUREG/CR-7108, entitled "An Approach for Validating Actinide and Fission Product Burnup Credit Criticality Safety Analyses—Isotopic Composition Predictions," and NUREG/CR-7109, entitled "An Approach for Validating Actinide and Fission Product Burnup Credit Criticality Safety Analyses—Criticality (k_{eff}) Predictions." The NUREG/CRs cover two different areas of consideration for BUC analysis: (1) isotopic composition predictions and (2) criticality predictions. These NUREG/CRs form the technical bases for new guidance from their respective program offices on BUC application to SFPs and dry storage and transportation systems.

For More Information
Contact Don Algama, RES/DSA, at Don.Algama@nrc.gov.

Fuel Rod Thermal and Mechanical Modeling and Analyses

Background

To comply with safety regulations, licensees must demonstrate the acceptable thermal and mechanical performance of nuclear fuel during steady-state operation and anticipated transients and accidents.

The U.S. Nuclear Regulatory Commission (NRC) maintains the FRAPCON and FRAPTRAN computer codes to predict fuel rod thermal and mechanical behavior under steady-state and transient conditions, respectively. The ability to perform quantitative analysis of fuel rod behavior is an essential part of the NRC's assessment of safety in reactor operations and spent fuel transportation and storage.

Objective

FRAPCON and FRAPTRAN must reliably provide a reasonable prediction of fuel rod behavior, to independently verify licensee safety analyses, and to analyze fuel behavior under hypothetical steady-state and transient conditions. The NRC fuel behavior codes must be able to model current fuel designs deployed in the United States.

Approach

Code Development

Early versions of FRAPCON and FRAPTRAN date back to the 1970s, and both codes have evolved to incorporate new modeling capabilities and new fuel and cladding materials to follow industry trends (Figure 2.27 and 2.28). In recent years, the focus of code development has leaned more toward the improvement of material property models including corrosion and hydriding, uncertainty analysis, the addition of new fuel and cladding materials, and the benchmarking of the code's models at high burnup levels. Publicly available data as well as in-pile and out-of-pile testing sponsored or cosponsored by the NRC have been used to validate the code material property models, and the latest code versions FRAPCON3.4 and FRAPTRAN 1.4 have been extensively validated. Details of this validation are documented in NUREG/CR-7022, "A Computer Code for the Calculation of Steady-State, Thermal-Mechanical Behavior of Oxide Fuel Rods for High Burnup," and NUREG/CR-7023, "A Computer Code for the Transient Analysis of Oxide Fuel Rods."

In examining the predictive bias and sensitivity in the fuel performance codes (NUREG/CR-7001, "Predictive Bias and Sensitivity in NRC Fuel Performance Codes"), several broad sources of uncertainty have been identified. These include uncertainty in the models used in the code, in the manufacturing parameters used as code input, and in the power history assumed in the analysis. In addition, some uncertainty arises from options and choices that the user makes. The use of input preprocessing and formal guidelines for selecting model options has significantly reduced this final source of uncertainty.

Currently, the NRC fuel behavior codes model uranium dioxide (UO_2) pellets, as well as mixed-oxide pellets (MOX), gadolinia (Gd_2O_3) doped pellets, and zirconium diboride (ZrB_2) coated pellets (Integral Fuel Burnable Absorber–IFBA fuel). Furthermore, new pressurized-water reactor (PWR) cladding alloy models were added to the code as these new alloys were introduced in the U.S. fleet of reactors. Examples include AREVA M5™ and Westinghouse ZIRLO™. Finally, the codes have been validated up to the current licensed U.S. burnup limit of 62 GWd/MTU peak rod average.

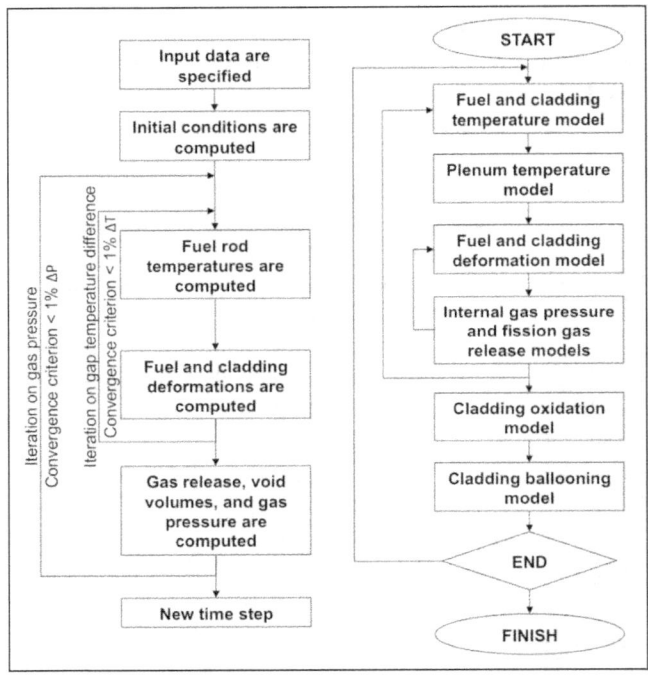

Figure 2.27 Simplified resolution scheme for FRAPCON (left) and FRAPTRAN (right)

Although a number of material property correlations are common to both FRAPCON and FRAPTRAN, the codes can operate separately, or in tandem—with FRAPCON output used as input to FRAPTRAN.

Ongoing development efforts aim to better integrate FRAPCON and FRAPTRAN within the NRC's suite of safety codes, including TRAC/RELAP Advanced Computational Engine (TRACE) and Symbolic Nuclear Analysis Package (SNAP). The code benchmarking database is also continuously being expanded, and the material and failure models are constantly being adjusted to incorporate the latest available data. Sources of material and failure data include international experimental programs in which the NRC is participating, such as the Studsvik Cladding Integrity Program (SCIP) and the Halden Reactor Project. Finally, in the ARM-FRAPCON package (ARM: Adaptive Response Modeling), statistical capabilities have been added to FRAPCON to better estimate upper tolerance levels on the fuel rod thermal and mechanical response. ARM-FRAPCON takes a nominal FRAPCON input and runs a chosen number of cases with random biases applied to manufacturing parameters, code models, and power histories to determine the impact of uncertainties on the outputs of regulatory interest.

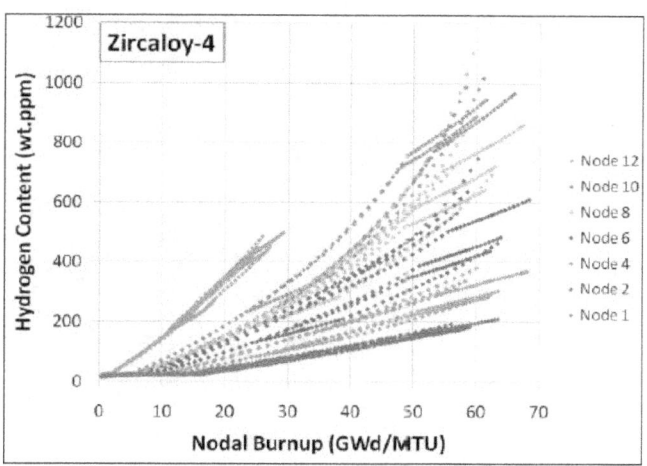

Figure 2.29 FRAPCON prediction of cladding hydrogen content as a function of burnup and axial elevation

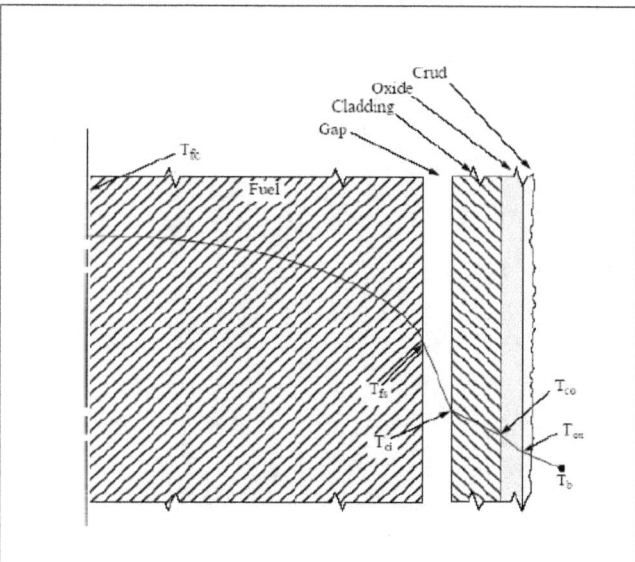

Figure 2.28 Schematic of the fuel rod temperature distribution calculated by FRAPCON

Code Application

The NRC primarily uses FRAPCON and FRAPTRAN to evaluate the effect of experimental data and to audit licensee safety analyses.

In 2011, FRAPCON was used to predict the uptake of hydrogen for the different PWR cladding alloys Zircaloy4, ZIRLO™, and M5™, as well as the boiling-water reactor (BWR) cladding alloy Zircaloy2, for typical power histories (Figure 2.29). The hydrogen predictions were then used to compare these alloys to the new proposed hydrogen-dependent criteria for allowable equivalent cladding reacted (ECR) during a loss-of-coolant accident (LOCA), as well as for allowable fuel enthalpy deposition during reactivity initiated accidents (RIA).

Since 2010, NRC has participated in several code benchmarking exercises. First, as part of the CABRI International Program, FRAPCON and FRAPTRAN were used to predict fuel rod behavior during a simulated RIA. Second, as part of the SCIP2 modeling workshops, 12 different ramp test cases were modeled with FRAPCON and four of those were also modeled with FRAPTRAN to compare code predictions. Best-practice modeling methods have been derived from these exercises.

In addition to RIA and power ramp modeling, FRAPTRAN can be used to predict LOCA fuel behavior, including cladding ballooning, cladding rupture predictions, and ECR calculations (Figure 2.30). Finally, FRAPCON and FRAPTRAN are currently being used with TRACE boundary conditions to produce best-estimate core-wide large-break LOCA fuel rod behavior predictions.

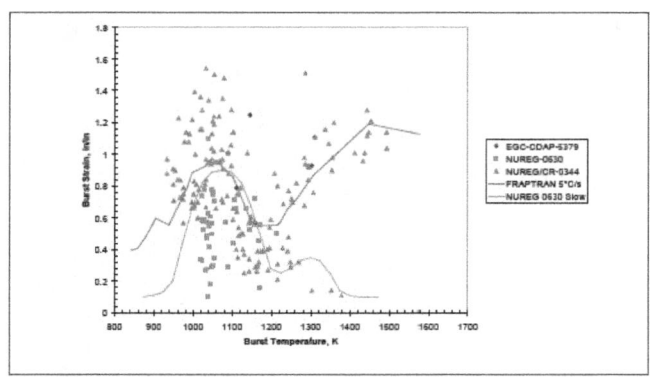

Figure 2.30 Permanent burst strain data and FRAPTRAN predictions for low temperature ramp rates (between 2 and 10°C/s)

For More Information
Contact Patrick Raynaud, RES/DSA, at
Patrick.Raynaud@nrc.gov

Chapter 3: Severe Accident Research and Consequence Analysis

Containment Analyses

Containment Iodine Behavior

Source Term Analysis

Phébus-Fission Product Program and Phébus-International Source Term Program

Melt Coolability and Concrete Interaction Follow-on Program

Severe Accidents and the MELCOR Code

MELCOR Accident Simulation Using SNAP (MASS)

MELCOR Accident Consequence Code System (MACCS2)

State-of-the-Art Reactor Consequence Analyses

Severe Accident Analyses of Integral Pressurized-Water Reactors

Environmental Transport Research Program

Integrated Ground-Water Monitoring and Modeling

In-Situ Bioremediation of Uranium in Ground Water

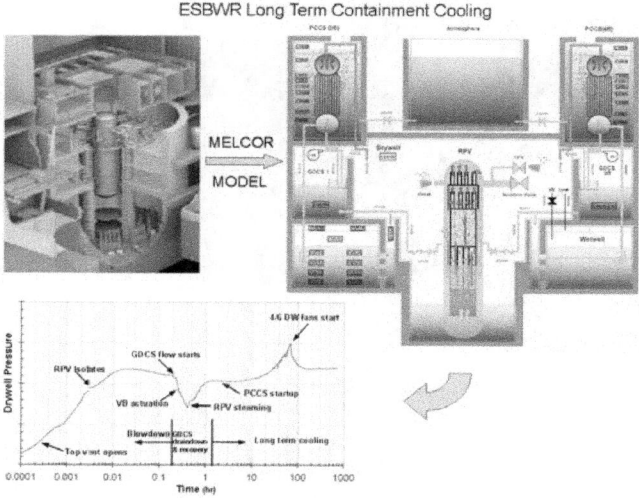

ESBWR long-term containment cooling model

Containment Analyses

Background

The containment encloses the reactor system and is the final barrier against the release of radioactive fission products in the event of a breach of either the primary or secondary coolant system. Evaluations entail a variety of postulated design-basis and beyond-design-basis (including core melt) events that involve accident progression and radiological source term calculations. Computer codes, such as CONTAIN and MELCOR, are used in licensing reviews (including new reactor designs) to address regulatory safety issues (e.g., generic safety issues and risk-informing regulations) and to respond to changes in containment safety margins. These computer codes serve as a repository of accumulated knowledge in the area of containment and severe accident research and will be improved as new information is collected and disseminated.

Objective

The objective of this research is to maintain U.S. Nuclear Regulatory Commission staff expertise and analytical tools on design-basis and beyond-design-basis containment analysis for current light-water reactors and new reactor designs.

Approach

CONTAIN and MELCOR are state-of-the-art lumped parameter codes that offer a greater robustness in analyzing a broader array of reactor containment designs (See Figure 3.1).

For More Information
Contact Allen Notafrancesco, RES/DSA, at
Allen.Notafrancesco@nrc.gov.

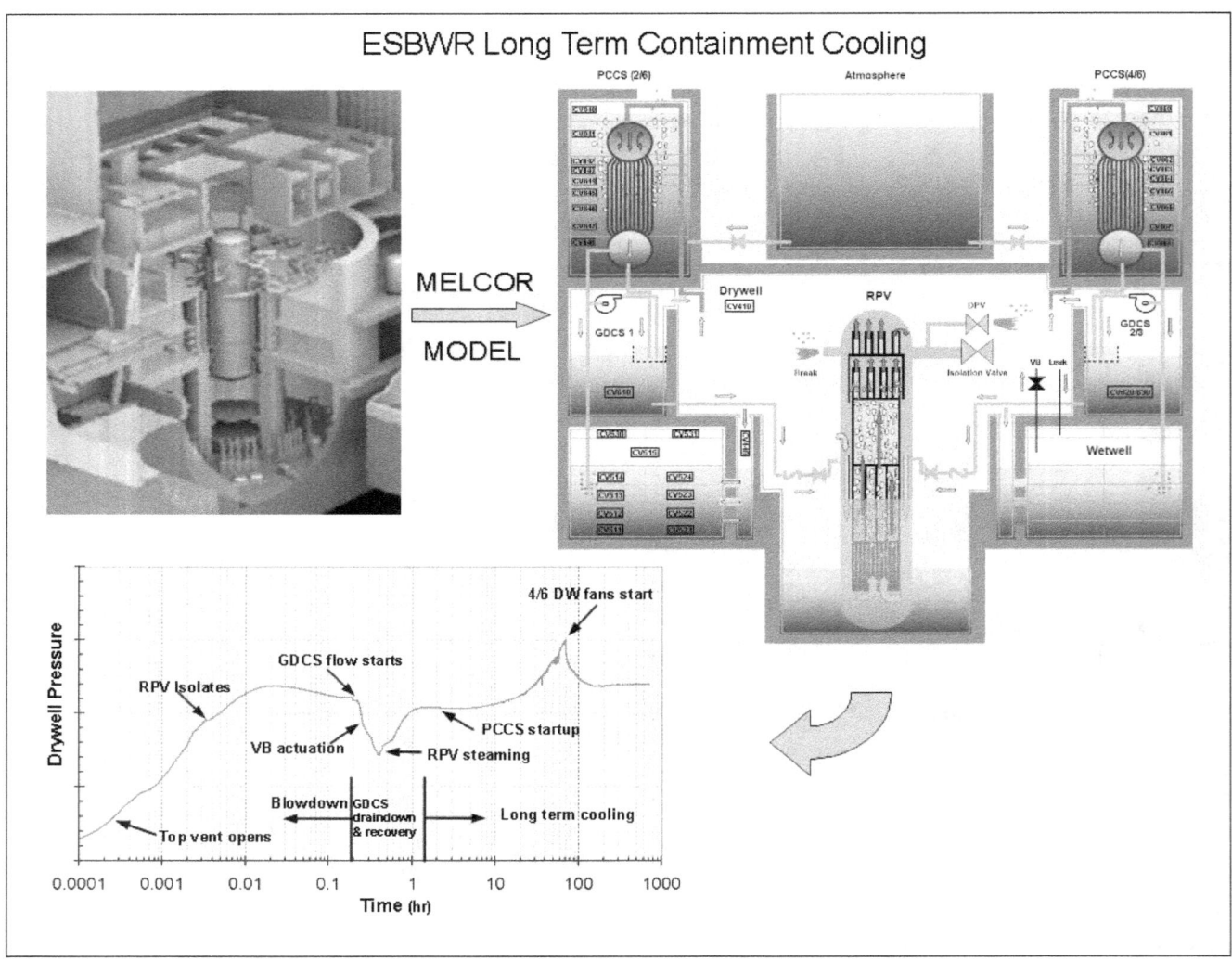

Figure 3.1 ESBWR long-term containment cooling

Containment Iodine Behavior

Background

The integral Phébus-Fission Product (FP) experiments provided an opportunity to test code predictions of overall severe accident behavior. One aspect that could be improved is the prediction of containment iodine behavior. Previously conducted pure water benchtop experiments suggested that preventing pressurized-water reactor (PWR) sump water from becoming acidic is necessary and sufficient to prevent significant gaseous iodine from evolving in a reactor containment following an accident involving core damage. However, the observations of the Phébus-FP experiments, the complexity of which more closely matches prototypic severe accident behavior, show that this may not necessarily be the case for power reactors.

Iodine is one of the major contributors to dose in analyses of postulated reactor accidents and, therefore, merits more attention than less dose-important elements do. Because iodine's dose contribution results from gaseous and particulate fission products contained in gas leaking from the reactor and containment, reducing the amount of airborne fission products reduces the contribution to dose. To minimize the iodine dose, PWR sumps are buffered to keep the sump water alkaline, thus preventing the iodine that reaches the sump from converting to volatile forms that can then be released to the containment atmosphere.

The results of the Phébus-FP tests indicate that controlling the sump pH may not significantly affect the development of a gaseous iodine concentration in the reactor containment in the immediate aftermath of an accident involving core degradation. Two aspects of the Phébus-FP experiments that influenced this iodine behavior were the presence of condensing surfaces and the presence of additional materials in the sump. The buffer in the sump does not affect the liquid films that develop on surfaces; therefore, these films do not remain alkaline. Consequently, the buffer in the sump does not prevent the iodine in these films from converting to volatile forms that may subsequently be released to the containment atmosphere.

In addition to iodine, other fission products and structural materials released from the degrading test bundle reached the sump in the Phébus-FP experiments. Consequently, the Phébus-FP sump chemistry far better represents that of a prototypic reactor than previous experiments did. In the Phébus-FP experiments that used silver-indium-cadmium control rods, the silver that reached the sump reacted with iodine to form a precipitate that effectively prevented iodine in the sump from being released to the containment atmosphere, even when the sump water was acidic. The iodine concentrations observed

in the Phébus-FP experiments cannot be directly applied to reactor containments because the facility was not scaled for gaseous iodine behavior. Accounting for the differences between the Phébus-FP facility and power reactors is necessary to scale this iodine behavior to power reactor containments. Relevant differences include higher dose rates, different containment surface-to-volume ratio and fractional sump settling area, different surface materials, different airborne materials (e.g., from radiolytic destruction of cables), different materials present in sumps, and paint aging. Properly accounting for these differences will require mechanistic models of iodine behavior.

Objective

The objective of this research is to develop mechanistic models of the phenomena that govern the containment iodine behavior observed in the Phébus-FP experiments in order to scale this observed behavior to operating power reactors.

Approach

The Office of Nuclear Regulatory Research (RES) is using the following approach to resolve the iodine issue:

- Test hypotheses against experiments.
- Develop models and validate models with further experiments.
- Simulate the Phébus-FP experiment.
- Simulate power plants.
- Evaluate sensitivities and uncertainty.
- Conduct peer review models and analyses.
- Make recommendations related to gaseous iodine behavior.

The approach for developing models to scale the iodine behavior of the Phébus-FP experiments has been to systematically test various working hypotheses that describe the persistent gaseous iodine behavior.

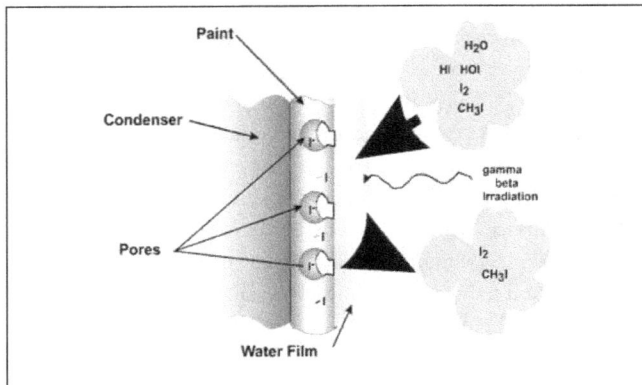

Figure 3.2 Hypothesized mechanism for gaseous iodine source in the Phébus-FP tests

To obtain data to test hypotheses for gaseous behavior and to validate the developed models, RES is participating in international separate effects research programs. These programs are the:

- Behavior of Iodine Project (BIP) and BIP2 (see Figure 3.2 and Figure 3.3).

- PHEBUS-International Source Term Project (ISTP) (see Figure 3.4).

- Source Term Evaluation and Mitigation (STEM).

The Organisation for Economic Co-Operation and Development (OECD) organized BIP, BIP2, and STEM. The Atomic Energy of Canada, Ltd (AECL) conducted the BIP and is conducting BIP2 experiments. Institut de Radioprotection et de Sureté Nucleaire (IRSN) organized and conducted the PHEBUS-ISTP experiments.

Figure 3.3 BIP irradiation vessel with sample coupons

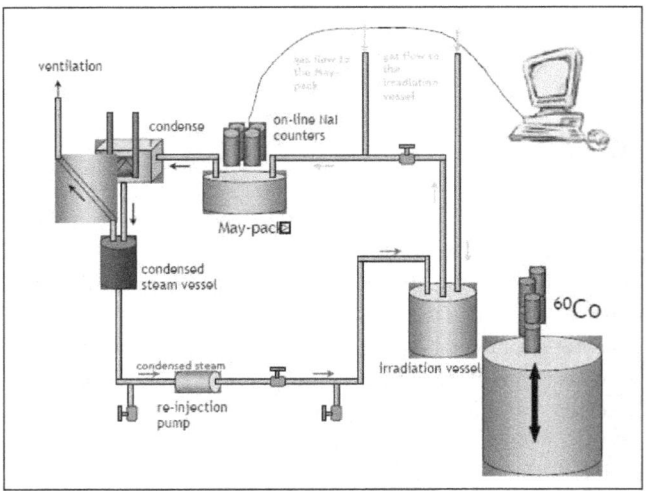

Figure 3.4 EPICUR experimental setup (one of the experiments under the Phébus-ISTP program)

For a steady-state concentration of gaseous iodine to exist, sources of gaseous iodine must balance the sinks of gaseous iodine. The experimental work and modeling is directed toward identifying and characterizing the sources and sinks of gaseous iodine. Based on observations of the Phébus-FP experiments, the results of additional separate-effects experiments, and analyses, the source of the persistent gaseous iodine in the Phébus-FP experiments is believed to be the containment surfaces upon which iodine deposited. Figure 3.2 shows a schematic of the hypothesized mechanism for this source. The general mechanism can be described as follows:

- Particulate and gaseous iodine is released to the containment from the reactor coolant system.

- Particles deposit and gases adsorb on surfaces in the containment.

- Particles decompose and gases absorb into paint.

- Irradiation releases iodine vapors.

- Vapors react in the atmosphere to form iodine oxide particles.

- The particles and vapors can redeposit on paint, thus continuing the cycle.

Development of standalone models to predict the gaseous iodine behavior is currently underway. This modeling is being used to guide further experimental testing.

Applications

The MELCOR severe accident code will include a simplified subset of the developed models. The MELCOR code is used for safety analysis and risk-informed decisionmaking. The results of the iodine modeling and analyses conducted with these models will provide the technical basis for a recommendation on the need for buffering in PWR sumps. The modeling should affect assumptions made in future dose calculations.

For More Information
Contact Michael Salay, RES/DSA, at Michael.Salay@nrc.gov.

Source Term Analysis

Background

The use of postulated accidental releases of radioactive materials is an integral part of defining the U.S. Nuclear Regulatory Commission's (NRC's) regulatory policy and practices. The regulations at Title 10 of the *Code of Federal Regulations* (10 CFR) Part 100, "Reactor Site Criteria," require licensees to postulate, for licensing purposes, the occurrence of an accidental fission product release resulting from "substantial meltdown" of the core into the containment. The regulations also require licensees to evaluate the potential radiological consequences of such a release under the assumption that the containment remains intact but leaks at its maximum allowable leak rate. Radioactive material escaping from the containment is often referred to as the "radiological release to the environment." The radiological release is obtained from the containment leak rate and knowledge of the airborne radioactive inventory in the containment atmosphere. The radioactive inventory within containment is referred to as the "in-containment accident source term."

Regulatory source terms provide a prescription of fission product release magnitude and timings that represents a broad range of accident scenarios. In addition to site suitability, the regulatory applications of this source term (in conjunction with the dose calculation methodology) affect the design of a wide range of plant systems.

Most of the currently operating power reactors in the United States were designed and licensed based on the source term described in Technical Information Document (TID)-14844, "Calculation of Distance Factors for Power and Test Reactors," issued by the U.S. Atomic Energy Commission in 1962. This source term, based on the results of experiments involving the heatup of irradiated fuel fragments in a furnace, was assumed to be instantly available in the containment. Half of the iodine was assumed to deposit in route to the containment. The source to the environment was found by assuming the design-basis leakage rate for the containment and attenuation of the radioactive material available for release by the plant's engineered safety features (e.g., sprays, suppression pools, and ice beds).

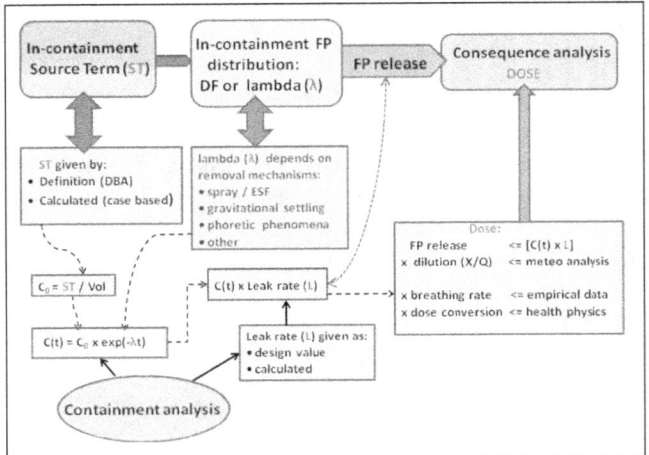

Figure 3.5 Use of source term and relation to other factors in dose calculations

Following the accident at Three Mile Island, radionuclide release evidently did not closely follow the pattern that might be expected based on the TID-14844 source term. Pressure arose from the nuclear industry for a more realistic source term. Because insufficient data were available to define a new source term, research was undertaken to obtain necessary data and to define a new source term.

The route to a more realistic accident source term, as defined by the Office of Nuclear Regulatory Research (RES), was to develop a mechanistic linkage of radionuclide behavior, including release from fuel and transport to the containment to reactor accident phenomena (Figure 3.5). This effort led to the development of the Source Term Code Package , a suite of standalone computer codes linked together to mechanistically predict, for a variety of accidents, the source term to the reactor containment and the attenuation of the inventory of radionuclides in the containment as a result of natural and engineered processes. This first phase of the NRC's study of mechanistic reactor accident source terms culminated in the publication of improved source terms for use in regulatory processes (NUREG-1465, "Accident Source Terms for Light-Water Nuclear Power Plants," issued February 1995) and the publication of a Level III analysis of accident risks at representative U.S. nuclear power plants (NUREG-1150, "Severe Accident Risks: An Assessment for Five U.S. Nuclear Power Plants," issued December 1990).

Objective

The objective of this research is to extend the source term described in NUREG-1465 to cover both light-water reactors with high-burnup cores and light-water reactors with mixed-oxide (MOX) fuel.

Approach

In 2001–2002, the NRC convened an expert panel to develop revisions to the reactor accident source term described in NUREG-1465. The undertaking was prompted by interest in having source terms applicable to conventional reactor fuel taken to high burnups (55 to 75 gigawatt days per ton) and to MOX made with weapons-grade plutonium dioxide. In formulating the revisions, the peer reviewers drew attention to the changes in understanding that have come about because of major experimental investigations of fission product behavior under reactor accident conditions, such as the Phébus-Fission Product program, the VERCORS tests, and the Verification Experiments of Radionuclides Gas/Aerosol Release tests. However, the assessments for that effort were performed without the benefit of accident sequence analyses and without mathematical models validated against the pertinent experiments with high-burnup or MOX fuel. Members of the expert panel developing the revisions to the NUREG-1465 source term attempted to mentally integrate the results of the applicable tests to predict source terms during accidents at nuclear power plants. They extrapolated the phenomenology of source terms from fuel burned to levels in excess of about 60 gigawatt days per ton. The panel members also extrapolated the behaviors of conventional fuels with conventional Zircaloy cladding to estimate the behavior of MOX fuel with zirconium-niobium (M5) cladding. The limitations of the analysis and databases available to the expert panel made research to confirm the panel's estimates necessary.

Confirmatory research has been conducted for both high-burnup and MOX fuels in both boiling-water reactors and pressurized-water reactors. This research was performed by analyzing risk-significant sequences with a version of the MELCOR severe accident code modified for, and validated against, recent fission product release and transport experiments, including experiments involving high-burnup and MOX fuels (VERDON and VERCORS). The integrated systems-level analysis MELCOR code replaced the Source Term Code Package code suite.

This research analyzed source terms for boiling-water reactors with high-burnup fuel and pressurized-water reactors with high-burnup and MOX fuel. Tables generated by the expert panels were updated using the results of this research. It became apparent during the course of the research that advances in modeling of severe accident progression would result in changes from the low-enrichment source terms provided in NUREG-1465. In addition, changes between the NUREG-1465 source terms and those generated for high-burnup and MOX fuels during this research resulted predominantly from the advances in modeling, not from differences between the different fuel types. The most notable change is the reduction of the rate of degradation phenomena resulting from improvements to heat transfer modeling. A draft synthesis report of this research and its findings entitled, "Accident Source Terms for Light-Water Nuclear Power Plants Using High-Burnup or MOX Fuel," has been completed and was peer reviewed in 2011. The synthesis report is being updated based on the reviewers' recommendations.

Applications

The reactor accident source term arises in two distinct ways in the U.S. regulatory process. The first is the release of radioactive material to the environment during a hypothetical reactor accident. This source term is an input to models of radionuclide dispersal and accident consequences. It drives measures taken for emergency preparedness and accident response. It is a crucial element of Level III probabilistic risk assessments and is an important consideration in the cost-benefit analyses of safety improvements that go beyond regulatory requirements to provide adequate protection of public health and safety.

The second source term used in the regulatory process is the release of radioactive material to the reactor containment. This source term is used in the analysis of plant sites. It is a defense-in-depth measure to assess the adequacy of reactor containments and engineered safety systems. This source term also figures into the environmental qualification of equipment within the containment that must function following a design-basis accident.

TID-14844, NUREG-1465, and any updates to this source term are examples of the latter source term. Improvements made to the MELCOR code in the areas of core degradation, fission product release, and fission product transport benefit the former source term whenever the updated code is used to calculate fission product releases.

For More Information
Contact Michael Salay, RES/DSA, at Michael.Salay@nrc.gov.

Phébus-Fission Product Program and Phébus-International Source Term Program

Background

The U.S. Nuclear Regulatory Commission (NRC) has developed computer codes for the analysis of postulated severe accident phenomena and progression. The NRC maintains its analytical tool to evaluate severe accident risk in the transition to a more risk-informed regulatory framework and for use in the study of vulnerabilities of nuclear power plants.

Future needs include developing insights into the severe accident behavior of advanced reactor designs and extending the expertise acquired on current reactor designs to address future design-specific considerations.

The improved understanding of phenomenological behavior and modeling in severe accidents has had direct implications for the analytical methods and criteria adopted for design-basis accidents (e.g., source term research and the revised source term described in NUREG-1465, "Accident Source Terms for Light-Water Nuclear Power Plants," issued February 1995). In addition, the development of improved severe accident, best-estimate models in the future will likely influence the improvement of licensing evaluation models because the development of best-estimate modeling reveals, quantitatively, margins in existing modeling.

Objective

The purpose of the Phébus-Fission Product (FP) program is to conduct integral tests to study the processes governing the transport, retention, and chemistry of fission products under severe accident conditions in light-water reactors and to provide data of integral severe accident behavior to validate severe accident computer codes.

The aim of the follow-on program, Phébus-International Source Term Program (ISTP), is to conduct separate-effects experiments in various experimental facilities to resolve the findings from Phébus-FP and to continue the investigation done in Phébus-FP (e.g., research into air ingress and fission product chemistry, fission product release from high-burnup fuel and mixed-oxide fuel, iodine chemistry, and control rod oxidation and degradation). The follow-on program will also provide data for the improvement of physical models of phenomena in the Phébus-FP experiments that codes did not well predict.

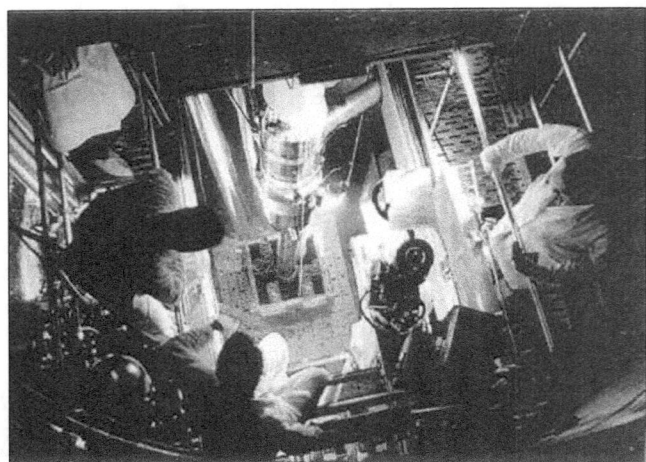

Figure 3.6 Phébus reactor and the test loop (top view)

Approach

The key features of the Phébus-FP program include the following:

- The program uses a loop-type test reactor with a low-enrichment driver core of 20 to 40-megawatt power, using fuel rod elements (Figure 3.6).
- Core cooling and moderation are achieved by demineralized light water.
- Light water and graphite are used as reflectors.
- Tests (four out of five) primarily involve a cluster of 20 fuel rods (about 10 kilograms), 1 meter in length, located in the central hole of the driver core of the Phébus-FP reactor. One test (FPT4) consists of a rubble bed instead of fuel rods.
- The facility is instrumented to measure fission product release, deposition in the primary circuit, and release to the containment.
- The facility includes a representative primary circuit, including a steam generator tube, containment, and sump.

The Phébus-ISTP includes several experimental series, each with its own facility. The experiments cover fission product release, air-cladding oxidation, oxidation of boron carbide steel mixtures by steam, and behavior of iodine both in the reactor coolant system and in containment.

Applications

The Phébus-FP integral experimental data support the assessment and development of new MELCOR models (e.g., iodine chemistry, iodine behavior in containment, and fuel degradation).

The improved MELCOR code is used for safety analysis and risk-informed decisionmaking. The data were also used to confirm many of the important features of the NRC's revised/alternative source term specified in NUREG-1465, such as the finding that iodine release is predominantly in aerosol form, with allowance for small fractions (5 percent) in gaseous form.

The results of the Phébus-FP tests indicate that controlling the sump pH may not significantly affect the development of a gaseous iodine concentration in the reactor containment in the immediate aftermath of an accident involving core degradation. Moreover, interactions between the chemicals used to control sump pH and some insulation materials dispersed to the sump can increase viscosity of the solution rendering it more difficult to pump.

The Phébus-ISTP provides prototypical experimental data on air ingress, fission product chemistry, and fission product release from high-burnup fuel and mixed-oxide fuel for MELCOR code assessment and development (Figure 3.7). The data will enable the NRC to address the issue of ruthenium behavior under accident conditions in an air environment. Situations in which this could occur include a spent fuel pool accident or a reactor accident involving fuel damage in which air enters the reactor vessel. Unlike fuel degradation in steam, which produces relatively nonvolatile ruthenium dioxide, fuel degradation in air can result in the production of the more volatile ruthenium tetroxide, resulting in a greater overall release of ruthenium. If the ruthenium release is significant, it will affect the evaluation of early and latent health effects under Title 10 of the *Code of Federal Regulations* (10 CFR) Part 100, "Reactor Site Criteria." In addition, assessments will be made of the separate-effects results on NUREG-1465 (the NRC's revised/alternative source term). NUREG-1465 is used for design-basis accident analysis in operating plants and in new reactor design certification reviews under 10 CFR Part 100.

The final seminar for Phébus-FP project was held in June 2012, marking the end of this 24-year project, which involved 20 countries and 50 organizations. Much data must still be analyzed. Largely complete, the Phébus-ISTP is still providing data from its small-scale fission-product-release experimental project, VERCORS. The Source Term Evaluation and Mitigation project continues to provide data on iodine and ruthenium.

For More Information
Contact Michael Salay, RES/DSA, at <u>Michael.Salay@nrc.gov.</u>

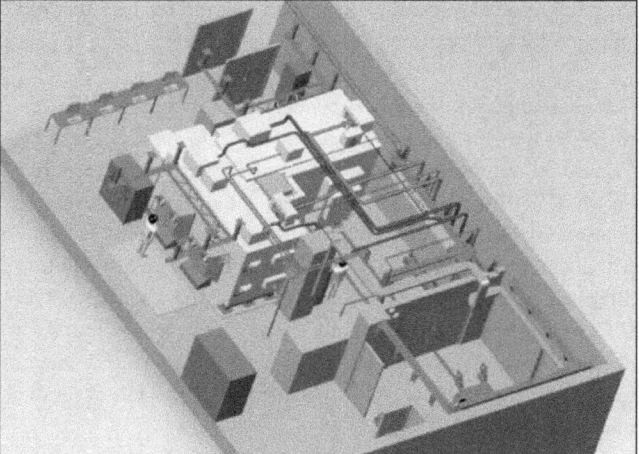

Figure 3.7 VERDON 2-cell FP release experimental facility (one of the experimental facilities constructed under the Phébus-ISTP program)

Melt Coolability and Concrete Interaction Follow-on Program

Background

The goal of the Melt Coolability and Concrete Interaction (MCCI) research program is to conduct reactor material experiments and associated analysis to achieve two technical objectives: (1) to resolve the ex-vessel debris coolability issue through an experimental program that focuses on providing both confirmatory evidence and test data for ex-vessel debris coolability mechanisms and (2) to address the remaining uncertainties related to long-term, two-dimensional core-concrete interactions (CCIs) under both wet and dry cavity conditions through complementary analytical activities. Achievement of these objectives will demonstrate the efficacy of severe accident management guidelines for existing plants and will provide the technical basis for better containment designs. The program described here is a follow-on to the previously completed MCCI program (see Figure 3.8 for the completed MCCI experimental setup) sponsored by the Organisation for Economic Co-operation and Development (OECD) and involves large-scale CCI tests with early water addition to determine its effectiveness as an severe accident management strategy. The experimental program is a joint collaboration among the U.S. Nuclear Regulatory Commission (NRC), Institut de Radioprotection et de Sûreté Nucléaire in France, and Electricite de France and is conducted at Argonne National Laboratory.

Figure 3.8 MCCI Experimental Setup

Approach

The risk to the public from nuclear power generation arises if an unlikely severe accident event progresses to the point at which fuel degradation occurs. In such a situation, molten fuel could hypothetically fail the reactor vessel, leading to melt discharge into the containment. The NRC has analytical tools (i.e., models) and computer codes that simulate the progression of severe accidents. The agency uses these codes to evaluate the consequences of beyond-design-basis accidents; therefore, they are an important tool in the transition to a more risk-informed regulatory framework. The improved models for debris coolability and molten CCI gained from the MCCI program will reduce uncertainties when they are applied to risk assessments of the current fleet and new plant designs.

The approach for the large-scale tests in the follow-on program is to conduct integral effect tests that replicate, as closely as possible, the conditions at plant scale, thereby providing data that can be used to verify and validate the codes directly. The experiments will investigate the effect of top flooding on ex-vessel debris coolability immediately after vessel breach and relocation of the debris in the cavity. Therefore, the results will provide debris coolability data under early flooding conditions and will complement the data obtained previously under late flooding conditions. The input power levels for the tests are selected so that the heat fluxes from the melt to concrete surfaces and the upper atmosphere were initially in the range of the heat flux expected early in the accident sequence.

The tests will provide information that contributes to the database for reducing modeling uncertainties related to two-dimensional molten CCI under both wet and dry cavity conditions. Data from these and other previously conducted test series form a technical basis for developing and validating models of the various cavity erosion and debris cooling mechanisms. These models can then be deployed in integral codes like MELCOR that simulate severe accident phenomena in a reactor, thereby providing a technical basis for extrapolating the experimental results to plant conditions. Furthermore, current experiments are designed to address mitigation features that can enhance coolability in new reactor designs.

Complementary to the experimental program described above, the NRC is participating in a related OECD-sponsored activity focused on the preparation of a state-of-the-art seminal report on MCCI. This report will document the progress made in debris coolability research in the last two decades. The report will document information on experimental research, including international Melt Attack and Coolability Experiments, OECD-MCCI, and other internationally sponsored activities; status of severe accident code development and assessment activities focused on the debris coolability issue; and discussion of residual uncertainties.

For More Information
Contact Sudhamay Basu, RES/DSA, at <u>Sudhamay.Basu@nrc.gov.</u>

Severe Accidents and the MELCOR Code

Background

The risk to the public from nuclear power generation arises if an accident progresses to the point at which fuel degradation occurs, and large quantities of radioactive materials are released into the environment. The U.S. Nuclear Regulatory Commission (NRC) continues to maintain and develop its expertise in severe accident phenomena and has developed computer codes for the analysis of severe accident progression. Expertise on severe accident phenomenological behavior and a quantitative predictive capability for simulating the response of nuclear power systems to severe accidents are essential to the NRC's mission. The role of such expertise and analytical capability is potentially wide ranging in the regulatory environment, which includes the transition to a more risk-informed regulatory framework and the study of vulnerabilities of nuclear power plants. The MELCOR code represents the current state-of-the-art in severe accident analysis, which has developed through the conduct of NRC and international research since the accident at Three Mile Island in 1979.

Objective

The objective of this research is to maintain the NRC staff's expertise on severe accident phenomenological behavior and a computer code for analysis of nuclear power plants' response to severe accidents.

Approach

The MELCOR code is a fully integrated, engineering-level computer code whose primary purpose is to model the progression of postulated accidents in light-water reactors and in non-reactor systems (e.g., spent fuel pool and dry cask). MELCOR is a modular code consisting of three general types of packages: (1) basic physical phenomena (i.e., hydrodynamics—control volume and flowpaths, heat and mass transfer to structures, gas combustion, and aerosol and vapor physics), (2) reactor-specific phenomena (i.e., decay heat generation, core degradation and relocation, ex-vessel phenomena, and engineering safety systems), and (3) support functions (i.e., thermodynamics, equations of state, material properties, data-handling utilities, and equation solvers). These packages model the major systems of a nuclear power plant and their associated interactions (see Figures 3.9 and 3.10). MELCOR 1.8.6 (Fortran 77) was released in September 2005; the code modernization effort resulted in the release of MELCOR 2.0 (Fortran 95) in September 2006. The latest

version (MELCOR 2.1) was released in September 2008. Current activities will include development and implementation of new and improved models to predict the severe accident behavior of various reactor designs.

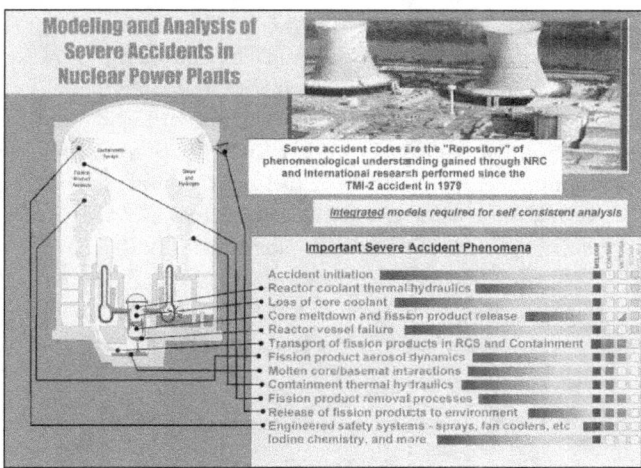

Figure 3.9 MELCOR modeling capabilities

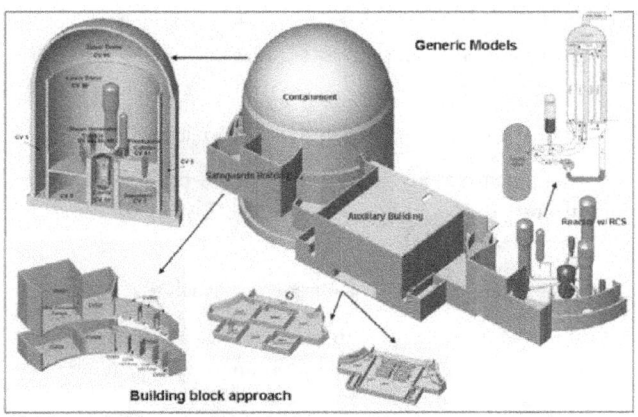

Figure 3.10 MELCOR plant modeling approach

Severe accident competency is needed to evaluate new generic severe accident issues and to address risk-informed regulatory initiatives and operating reactor issues associated with plant changes, such as in the case of steam generator tube integrity. Licensees will continue to pursue plant modifications that require assessment of incremental risk impacts that will necessitate analysis of phenomena related to severe accidents.

Applications

The improved understanding of phenomenological behavior and modeling in severe accidents and their implementation in MELCOR directly resulted in changes to methods used in evaluating design-basis accidents (e.g., revised source term). The NRC expects the development of best-estimate severe accident models in the future to improve the licensing evaluation models.

The development of best-estimate models reveals, quantitatively, margins in existing models.

Activities associated with the development, assessment, and applications of MELCOR include the following:

- Safety analysis and risk decisionmaking, including (1) a revision to the NRC's alternative source term (NUREG-1465, "Accident Source Terms for Light-Water Nuclear Power Plants," issued February 1995) for high-burnup fuel and mixed-oxide fuel and (2) new reactor certification (Advanced Passive 1000 Megawatt (AP1000), Economic Simplified boiling-water Reactor (ESBWR), U.S. Evolutionary Power Reactor (EPR), U.S. Advanced pressurized-water Reactor (U.S. APWR), Advanced boiling-water Reactor (ABWR), and Small Modular Reactors (e.g., mPower).

- Experimental analyses and code validation activities.

- Nuclear power plant beyond-design-basis accidents.

- Aerosol transport and deposition in steam generators during bypass accidents.

- Risk of steam generator tube rupture induced by a severe accident.

- Effects of air ingress on fission product release.

- Consequence Study of a Beyond-Design-Basis Earthquake Affecting the Spent Fuel Pool for a U.S. Mark I Boiling-Water Reactor.

- State-of-the-art Reactor Consequence Analysis.

- Support for site Level 3 probabilistic risk assessments (success criteria thermal-hydraulic and severe accident progression analysis).

- U.S. Department of Energy/NRC Fukushima forensic analysis.

- Support for the NRC's Japan Lessons Learned Directorate and Near-Term Task Force recommendations.

National laboratories, universities, and international organizations (e.g., Paul Scherrer Institute in Switzerland) are involved in the MELCOR code development effort.

A Symbolic Nuclear Analysis Package (SNAP) plugin has been developed for MELCOR, and the integration of MELCOR and SNAP provides a more user-friendly system for input deck preparation and accident simulation. The accident simulation models for new reactor designs, including the EPR, ABWR, U.S. APWR, AP1000, and ESBWR, have been completed, and models for two existing plants (Peach Bottom Atomic Power Station and Byron Station) were completed in 2013. The

models run in severe accident and design-basis accident modes. Figures 3.11, 3.12, and 3.13 illustrate examples of the simulation models for the EPR and ABWR, including core degradation and available system interfaces. The user interface model in the safety system logic provides the capability to initiate, fail, or adjust system functionality by introducing system malfunctions (e.g., loss-of-coolant accident [LOCA]) and controls (e.g., emergency core cooling system [ECCS]) to mitigate the consequences of the accident. The user can visually see the progression of an accident (e.g., core heatup and degradation) as the calculation is progressing. The models are useful for the evaluation of system success criteria.

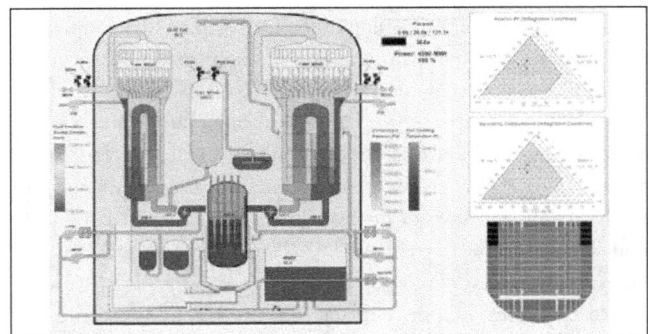

Figure 3.11 EPR simulation model

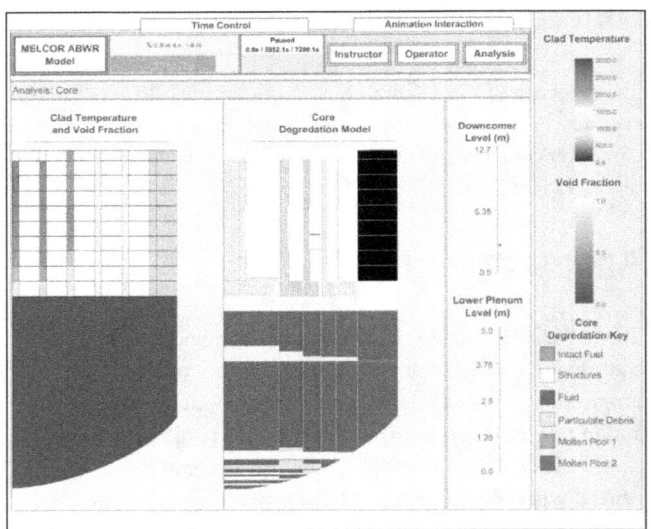

Figure 3.12 ABWR core heatup and degradation

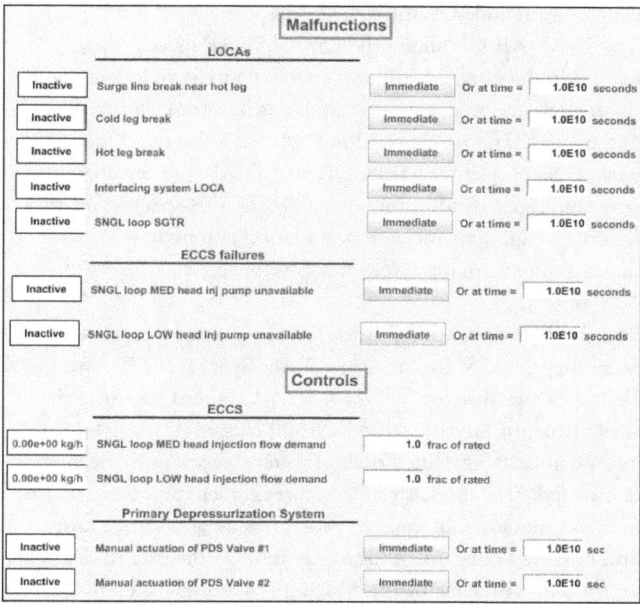

Figure 3.13 EPR user interface model

International Collaborations

The following examples of international collaborations resulted in MELCOR improvements:

- The NRC's Cooperative Severe Accident Research Program

- MELCOR Code Assessment Program and European MELCOR User Group meetings

- Phébus-Fission Products (Phébus-FP), VERCORS (a French test program), and follow-on program (Phébus-Source Term Separate Effects Test Project [STSET]), French Institute for Radiological Protection and Nuclear Safety (IRSN) (This collaboration investigates fission product releases and degradation of uranium dioxide fuel, including burnup greater than 40 gigawatt days per metric ton, and mixed-oxide fuel under severe accident conditions and the effects of air ingress on core degradation and fission product release. The results are used to validate the NUREG-1465 source term and MELCOR code.)

- The German QUENCH experimental program to investigate overheated fuel.

- ARTIST, Paul Scherrer Institute (Switzerland) (This project investigates experimentally the potential mitigation of radioactive material releases through the secondary side of a steam generator. The results from this research would allow the NRC to decide whether improved source term bypass models are needed.)

- Molten Core Concrete Interaction Program, Organisation for Economic Co-operation and Development and Argonne National Laboratory (United States) (This project consists

of separate-effects experiments to further address the ex-vessel debris coolability issue. The results will be used to develop coolability models.)

- Behavior of Iodine Project (BIP), Nuclear Energy Agency, Committee on the Safety of Nuclear Installations (France) (This project involves experimental investigations of iodine behavior in containment during conditions following a severe accident for computer code model development and validation. BIP addresses the uncertainties related to iodine behavior, especially with respect to iodine interactions with paints. Together with complementary testing at Atomic Energy of Canada, Ltd., and IRSN, this project advances and quantifies the state-of-the-art on modeling of iodine behavior in the containment. Adequate modeling of iodine behavior is crucial in determining the need for pH control in containment sump. The proposed research will complement the ongoing IRSN projects of France Phébus-FP and follow-on program, Phébus-STSET.)

For More Information
Contact Hossein Esmaili, RES/DSA, at Hossein.Esmaili@nrc.gov.

MELCOR Accident Simulation Using SNAP (MASS)

Background

MELCOR is a fully integrated computer code that is capable of modeling the progression of severe accidents in light-water reactors. MELCOR is being used for targeted applications, including (1) technical support for the U.S. Nuclear Regulatory Commission's (NRC's) full-scope site Level 3 probabilistic risk assessment, (2) the Spent Fuel Pool Scoping Study, (3) State-of-the-Art Reactor Consequence Analyses, (4) analysis of new and advanced reactors and support of new reactor design certification, including small modular reactors, and (5) analysis of the event at Fukushima and support of the Japan Lessons Learned Directorate and Near-Term Task Force recommendations to more effectively meet the NRC's mission to protect the safety of the public. A Symbolic Nuclear Analysis Package (SNAP) plugin has been developed for MELCOR 2.1, and the integration of MELCOR and SNAP and the development of a graphical user interface for the plant models provide a user-friendly system for accident simulation.

Objective

Simulation models should provide the users with the capability to define accident sequences, alter the system availabilities, and provide a visual progression of the accident using MELCOR for the prediction of the accident outcome and the SNAP animation capabilities.

Approach

The design concept requires minimal user training in both MELCOR and SNAP. The objective is to provide users with an easy to use tool to analyze accident scenarios. The end user access boundary is highlighted in Figure 3.14. Here the end user controls the type of accident (e.g., size and location of a loss-of-coolant accident) and the availability of plant safety systems and any operator actions. For containment design-basis analysis, the mass and energy and fission product sources into the containment can be provided as an external table. The end user can then view the results and perform sensitivity calculations. One of the advantages of the visualization is to provide an overview of the accident progression in terms of interpretation of results, input model checking, and user training.

Because of the desire to make MELCOR more user friendly through the SNAP graphical user interface, an additional program was added to the SNAP suite, the SNAP-KIOSK. The SNAP-KIOSK allows the normal SNAP model editing features to be disabled while still allowing users to interact with the models and to control the simulation. A socket interface and new MELCOR control functions were also developed as part of the project for MELCOR and SNAP to more effectively communicate. In addition, several MELCOR-specific SNAP modules (e.g., dynamic core degradation and hydrogen flammability diagrams) were also developed.

The accident simulation models for new reactor designs, including the U.S. Evolutionary Power Reactor (EPR), Advanced Boiling Water Reactor (ABWR), U.S. Advanced pressurized-water Reactor, Advanced Passive 1000 Megawatt (AP1000), and Economic Simplified boiling-water Reactor, have been completed. The models run in severe accident and design-basis accident modes (containment peak pressure and source term) and provide a convenient display system for the user to define an accident sequence by introducing system malfunctions (e.g., loss-of-coolant accident) and controls (e.g., emergency core cooling system) to mitigate the consequences of the accident, as shown in Figure 3.15. In addition, the user can visually see the progression of an accident (e.g., core heatup and degradation) as the calculation is progressing. Figures 3.16 and 3.17 illustrate examples of the simulation models for the EPR and ABWR, including core degradation and available system interfaces during a short time station blackout scenario. Similar masks are currently being developed for the existing reactors (a pressurized-water reactor and a boiling-water reactor) for user training and accident analysis. To the extent possible, the development of the new MELCOR models will be done in such a manner as to provide an easy implementation within the SNAP environment for accident simulation capabilities.

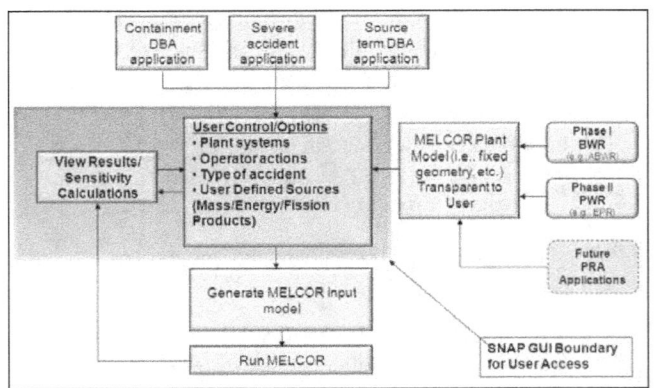

Figure 3.14 MASS design concept and applicability

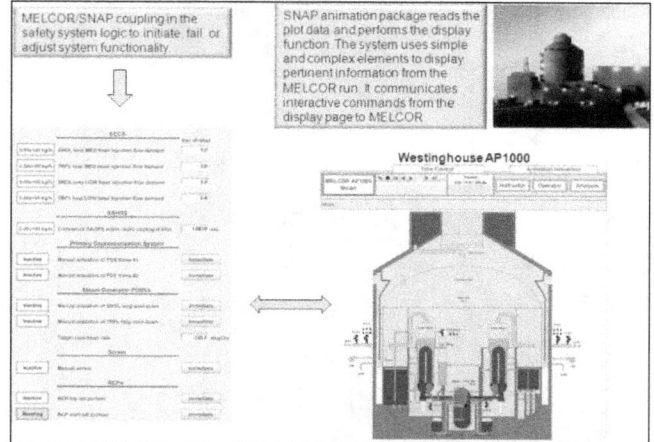

Figure 3.15 MASS user interface for AP1000

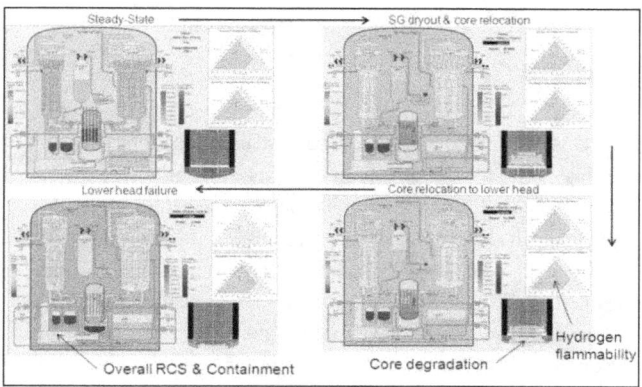

Figure 3.16 EPR accident progression

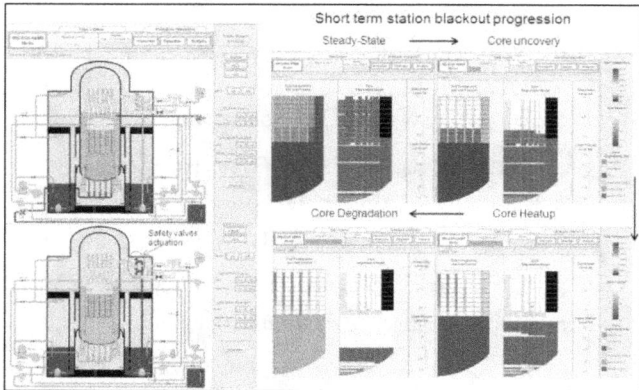

Figure 3.17 ABWR accident progression

For More Information

Contact Hossein Esmaili, RES/DSA, at Hossein.Esmaili@nrc.gov.

MELCOR Accident Consequence Code System (MACCS2)

MACCS2

The U.S. Nuclear Regulatory Commission (NRC) developed MELCOR Accident Consequence Code System Version 2 (MACCS2) to evaluate offsite consequences from a hypothetical release of radioactive material into the atmosphere. The code models atmospheric transport and deposition, emergency response actions, exposure pathways, health effects, and economic costs.

MACCS2 evolved from predecessor codes MACCS, Calculation of Reactor Accident Consequences Version 2 (CRAC2), and CRAC. MACCS was used to support NUREG-1150, "Severe Accident Risks: An Assessment for Five U.S. Nuclear Power Plants," issued December 1990. CRAC2 was used to estimate consequences in NUREG/CR-2239, "Technical Guidance for Siting Criteria Development" (also referred to as the 1982 Siting Study). CRAC was initially developed for WASH-1400, which was published in 1975. These codes were developed mainly as tools to assess the risk and consequences associated with accidental releases of radioactive material into the atmosphere in probabilistic risk assessment studies (see Figure 3.18).

MACCS Version 3.7 incorporates the following improvements:

- More cohorts for evacuation (up to 20).

- A potassium iodide ingestion model.

- More compass directions (up to 64).

- More plume segments (up to 200).

- More aerosol bins and chemical groups (up to 20).

- Multiple meteorological data intervals (15, 30, or 60 minutes).

- A diurnal mixing-height model.

- A long-range lateral plume spread model.

- An improved Briggs plume rise model.

- A plume meander based on Regulatory Guide 1.145, "Atmospheric Dispersion Models for Potential Accident Consequence Assessments at Nuclear Power Plants."

- Dynamic memory allocation.

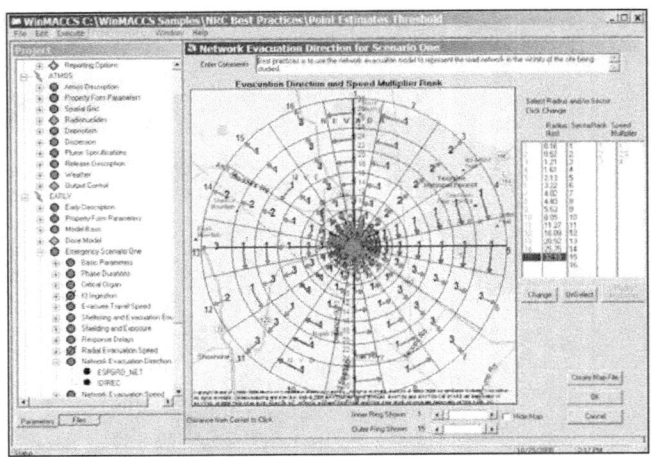

Figure 3.18 Graphical view of WinMACCS (network evacuation model is shown)

WinMACCS

The MACCS2 code has been modified and additional improvements have been added since its original release in 1997. Version 3.7 of the code has been released recently together with the graphical user interface, WinMACCS Version 3.7. WinMACCS was developed to facilitate routine use of MACCS2. The three most important modeling features implemented in WinMACCS are (1) the ability to easily evaluate the impact of parameter uncertainty, (2) the ability to manipulate input parameters for network evacuation modeling, and (3) the ability to model alternative dose-response relationships for latent cancer fatality evaluation (e.g., linear with threshold model).

WinMACCS Version 3.7 includes the following features:

- Cyclical handling of MELCOR source terms.

- Graphical manipulation of MACCS2 network evacuation parameters (e.g., direction and speed).

- Editing of grand mean and arbitrary quantile levels for uncertainty calculations.

- The option to remove the food (ingestion) pathway.

MACCS2/WinMACCS Uses

MACCS2/WinMACCS is used to evaluate the consequences of severe radiological accidents as part of the environmental reports and environmental impact statements for early site permits, to support plant-specific evaluation of severe accident mitigation alternatives required as part of the environmental assessment for license renewal, to assist in emergency planning, and to provide input to cost/benefit analyses.

New Work

Work is ongoing to update the MACCS2 code based on current technology. The new work will develop and implement an alternative economic model and an approach for treating more complex wind patterns. Other modifications will allow additional flexibility in specifying population groups (i.e., at a specific location in a defined grid area and with a finer resolution) as a function of distance from the release location. For uncertainty analyses, capabilities are being implemented to sample dose conversion factor values and to distribute numerous MACCS2 runs into a computer network cluster; this effort will include the postprocessing of the results.

MACCS2 Economic Models

MACCS2 Current Economic Model

The current economic model in MACCS2 includes the following costs:

- Evacuation and relocation costs (e.g., a per diem cost associated with displaced individuals).

- Moving expenses for people displaced (i.e., a onetime expense for moving people out of a contaminated region) and lost wages.

- Decontamination costs (e.g., labor, materials, equipment, and disposal of contaminants) if decontamination is cost effective.

- Cost from loss of land/property use (e.g., costs associated with lost return on investment and with depreciation of property that is not being maintained).

- Disposal of contaminated food grown locally (e.g., crops, vegetables, milk, dairy products, and meat).

- Cost of condemned lands (i.e., land that cannot be restored to usefulness or land for which decontamination is not cost effective).

Nearly all of the values affecting the economic cost model are user inputs and, therefore, can account for a variety of costs and can be adjusted for inflation, new technology, or changes in policy.

MACCS2 New Alternative Economic Model

The new and alternative economic model for MACCS2 is under development. The new model is based on the existing Regional Economic Accounting Tool (REAcct), which Sandia National Laboratories developed for the U.S. Department of Homeland Security (DHS). REAcct uses an economic model that is built upon the well known and extensively documented input output modeling technique initially presented by Leontief and more recently further developed by numerous contributors[1]. The model is widely accepted and used within the community of economists. In response to SECY-09-0051, "Evaluation of Radiological Consequence Models and Codes," dated March 31, 2009, the Commission approved the staff's recommendation to enhance the MACCS2 code with insights that may be learned from the DHS/National Nuclear Security Administration economic consequence model recently developed for radiological dispersion devices. Subsequently, a comparison of the new economic model for MACCS2 with the DHS Radiological and Nuclear Terrorism Risk Assessment economic consequence model was conducted.

REAcct Economic Model

REAcct is used to rapidly estimate approximate economic impacts for disruptions caused by natural (e.g., hurricanes) or man-made events. The tool estimates the following:

- Direct losses—gross domestic product (GDP) losses (within the grid) using county level economic sector (e.g., manufacturing, tourism, and agriculture) data[2].

- Indirect losses, using the national level Regional Input Output Modeling System multipliers to estimate indirect GDP losses (representing the remainder of the U.S. outside grid).

- The potential for multiple year disruption to economy in terms of present value.

The metric of concern is reduction in GDP, which can be reported at the industry, regional, and U.S. levels, as illustrated in Figure 3.19.

[1] W. Leontief, "Quantitative Input and Output Relations in the Economic System of the United States," Review of Economics and Statistics, 18:105–125 (1936).

[2] GDP data from the U.S. Bureau of Economic Analysis, Washington, DC (http://www.bea.gov/ regional/index.htm).

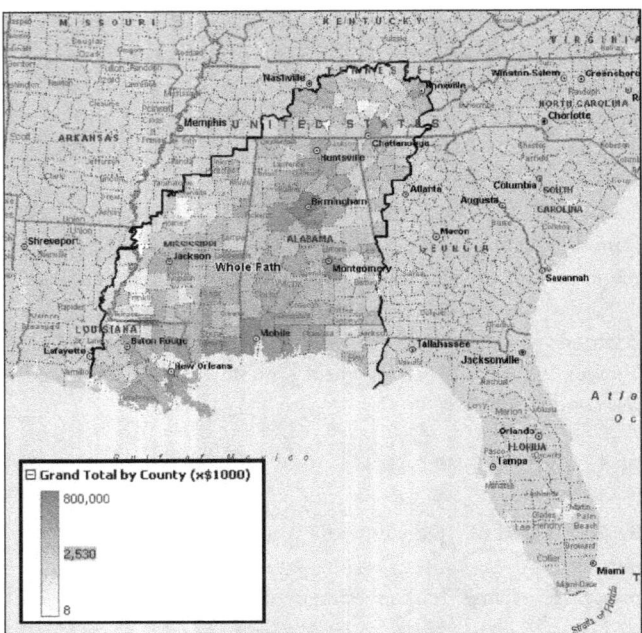

Figure 3.19 Example of total GDP loss from a natural disruption (e.g., hurricane)

The cost of decontamination in the new economic model is calculated in the same way as in the old model. It is provided as a separate output value from the loss of GDP, which accounts for the fact that property is unusable for a period of time and that people are displaced from their homes and places of work. As a result, economic activity does not take place. The GDP model does not account for decontamination/cleanup activities, which is why they are reported separately. The cost of evacuation, relocation, and condemnation of land are not provided as separate output values; however, the calculation for the loss of GDP captures the fact that the population is displaced and unable to work over a period of time.

An internal peer review for the new economic model is planned. The work is being documented in the WinMACCS draft user manual. The current schedule envisions the release of a new version of MACCS2/WinMACCS during mid-fiscal year 2014 to give the user the option to choose the new economic model.

For More Information
Contact Jonathan Barr, RES/DSA, at Jonathan.Barr@nrc.gov

State-of-the-Art Reactor Consequence Analyses

Background

The U.S. Nuclear Regulatory Commission (NRC) initiated the State-of-the-Art Reactor Consequence Analyses (SOARCA) project to develop best-estimates of the offsite radiological health consequences for potential severe reactor accidents for two pilot plants: (1) the Peach Bottom Atomic Power Station (Peach Bottom) in Pennsylvania and (2) the Surry Power Station (Surry) in Virginia. Peach Bottom is generally representative of U.S. operating reactors using the General Electric boiling-water reactor design with a Mark I containment. Surry is generally representative of U.S. operating reactors using the Westinghouse pressurized-water reactor (PWR) design with a large, dry (subatmospheric) containment.

The SOARCA project evaluated plant improvements and changes that were not reflected in earlier NRC publications. SOARCA included system improvements; improvements in training and emergency procedures; offsite emergency response; security-related improvements; and plant changes, such as power uprates and higher core burnup. To provide a perspective between SOARCA results and more conservative offsite consequence estimates, SOARCA results were compared to NUREG/CR-2239 (December 1982), "Technical Guidance for Siting Criteria Development" (also referred to as the 1982 Siting Study). The SOARCA report helps the NRC communicate its current understanding of severe-accident-related aspects of nuclear safety to stakeholders, including Federal, State, and local authorities; licensees; and the general public.

Approach

The SOARCA project took a step-by-step approach to calculate the potential consequences of the analyzed severe accidents. The project team first decided it could learn more by rigorously and realistically analyzing a relatively small number of important accident scenarios instead of carrying out less detailed modeling of many scenarios. Therefore, the team selected a threshold to help select scenarios to analyze. SOARCA aimed to assess the benefits of the mitigation measures in Title 10 of the *Code of Federal Regulations* (10 CFR) 50.54(hh), which were put in place after the terrorist attacks of September 11, 2001, for responding to fires and explosions, in other accident scenarios. The team also wanted to provide a basis for comparison to past analyses of severe accident scenarios before these mitigation measures existed. Therefore, the project analyzed the selected scenarios twice—first assuming that the event proceeds without the mitigation measures in 10 CFR 50.54(hh), called "unmitigated," and then assuming that the 10 CFR 50.54(hh) mitigation is

successful, called "mitigated." Accident progression calculations used the MELCOR computer code. For scenarios leading to an offsite release of radioactive material, SOARCA then analyzed the material's atmospheric dispersion, the surrounding area's emergency response, and potential health consequences using the MELCOR Accident Consequence Code System Version 2 (MACCS2) computer code.

Results and Conclusions

SOARCA's key results include the following:

- When operators are successful in using available onsite equipment during the accidents analyzed in SOARCA, they can prevent the reactor from melting or can delay or reduce releases of radioactive material to the environment.

- SOARCA analyses indicate that all modeled accident scenarios, even if operators are unsuccessful in stopping the accident, progress more slowly and release much smaller amounts of radioactive material than calculated in earlier studies (Figure 3.20).

- As a result, public health consequences from severe nuclear power plant accidents modeled in SOARCA are smaller than previously calculated.

- The delayed releases calculated provide more time for emergency response actions, such as evacuating or sheltering for affected populations. For the scenarios analyzed, SOARCA shows that emergency response programs, if implemented as planned and practiced, reduce the risk of public health consequences.

- Both mitigated (operator actions are successful) and unmitigated (operator actions are unsuccessful) cases of all modeled severe accident scenarios in SOARCA cause essentially no risk of death during or shortly after the accident.

- SOARCA's calculated longer term cancer fatality risks for the accident scenarios analyzed are millions of times lower than the general U.S. cancer fatality risk from all causes.

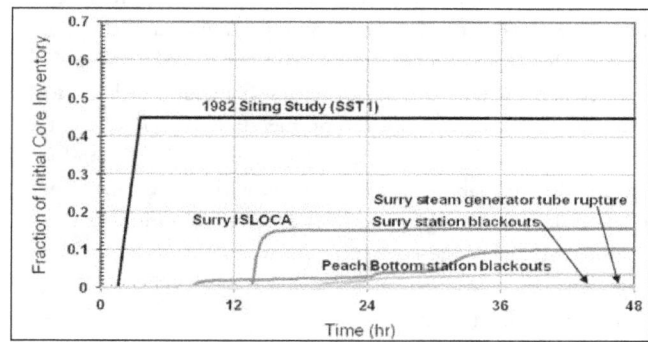

Figure 3.20 Iodine release for unmitigated cases

Status

The SOARCA project is documented in a series of NUREG reports. NUREG-1935, "State-of-the-Art Reactor Consequence Analyses (SOARCA) Report," provides a summary of the project's objectives, methods, results, and conclusions. NUREG/CR-7110, "State-of-the-Art Reactor Consequence Analyses Project," Volume 1, "Peach Bottom Integrated Analysis," and Volume 2, "Surry Integrated Analysis," provide additional technical details on the analyses conducted for each of the pilot plants. In addition, the staff developed an information brochure, NUREG/BR-0359, "Modeling Potential Reactor Accident Consequences," to facilitate communication with stakeholders, including the public.

The NRC released a draft version of NUREG-1935 for public review and comment in January 2012. The SOARCA project team then held three public meetings to discuss the project with various stakeholder groups, including members of the public. NUREG-1935 was then updated to address public comments and to include the final letters from the SOARCA independent peer review committee of subject matter experts. The NRC published the final version of NUREG-1935 in November 2012.

In addition, the NRC is conducting an uncertainty analysis (UA) for the SOARCA study. The goals of this UA are to develop insights into the overall sensitivity of SOARCA results to uncertainty in inputs; to identify the most influential input parameters for releases and consequences; and to demonstrate a UA methodology that could be used in future source term, consequence, and site Level 3 probabilistic risk assessment studies. Preliminary integrated analyses using about 40 independent MELCOR and MACCS2 parameters support the overall SOARCA results and conclusions for the selected accident scenario.

The models and methods from the SOARCA project and SOARCA UA will be used to support and inform other agency activities related to severe accidents, consequence analyses, and lessons learned from the Fukushima accident.

At the conclusion of the SOARCA project, the staff provided recommendations to the Commission for limited additional analysis that would provide additional knowledge of severe accident progression and consequences that could further inform other agency activities. Both recommendations—(1) to conduct a UA for a scenario at the Surry PWR plant and (2) to conduct a consequence analysis for a PWR plant with an ice condenser containment—were approved by the Commission subject to resources available and competing priorities.

For More Information
Contact Jonathan Barr, RES/DSA, at Jonathan.Barr@nrc.gov.

Severe Accident Analyses of Integral Pressurized-Water Reactors

Background

Currently, several designs that can be classified as integral pressurized-water reactors (iPWRs) are under pre-application review by the U.S. Nuclear Regulatory Commission (NRC). The characteristic feature of these iPWRs that differentiates them from conventional PWRs is a self-contained integral assembly comprising the reactor core, a riser, a pressurizer, and steam generators all housed within a single reactor pressure vessel.

The primary coolant flow during steady-state operation of iPWRs is driven by internal reactor coolant pumps or by natural circulation, depending on the design. The reactor core is connected to a riser that acts as a "hot leg," transporting the coolant to the steam generators, which use either a once-through or helical tube configuration with the primary coolant flowing either inside or outside the tubes. Furthermore, the proposed designs have aimed at eliminating the potential for large loss-of-coolant accidents by eliminating large pipes and other penetrations to the reactor coolant system. In addition, any pressure vessel penetrations are small and are located high in the reactor pressure vessel to eliminate the likelihood of core uncovery by coolant blowdown or drainage, or both, following any loss-of-coolant accident.

These designs often use passive innovative means for emergency core cooling systems that may include the use of depressurization systems; recirculation valves, which provide recirculation flow from the containment back to the reactor vessel; and isolation condensers. The containment of the iPWRs is also unique to each design, whereby it can be similar to the large dry containment of conventional PWRs or it can be an unconventional design in which the containment is a compact steel vessel that surrounds the reactor pressure vessel and is immersed inside a large water pool.

Because of the relatively low power output for these reactors (i.e., from 40 to 150 megawatts electric), these designs often involve multiple identical "modules" that have minimal shared systems and that are intended to be installed at a given site over a period of time on an as-needed basis. These designs are often referred to as small modular reactors, and they appear to offer improved safety while circumventing economic hurdles through modularization.

Even though the frequency of fuel/core damage in iPWRs is expected to be significantly lower than for conventional PWR plants; nonetheless, severe accidents cannot be totally eliminated from consideration because these designs may remain vulnerable to natural phenomena and other random and common-cause failures. Furthermore, the unique designs of iPWRs, as compared to conventional PWRs, may also introduce unique phenomenological challenges during severe accidents, thus requiring experimentation or the development of new or revised models for implementation into the available severe accident progression and radiological source term prediction codes (e.g., MELCOR).

Objective

The objective of this research is to identify thermal-hydraulics, melt progression, and fission product release and transport phenomena that are relevant to modeling of severe accidents in iPWRs and to provide an assessment of the applicability of the NRC-sponsored MELCOR computer code to these analyses.

Approach

MELCOR models have been developed for application to iPWRs (e.g., NuScale and mPower), and it was demonstrated that the code, without any changes to the models, can be readily applied to iPWRs. Figure 3.21 shows the results of MELCOR-calculated steady-state flow and temperature fields in the core, the region above the core, and the riser section for an example of an iPWR design. A recirculating flow pattern caused by inflow of water from the downcomer through a leakage path connecting the bottom of the riser and the downcomer is also shown in Figure 3.21.

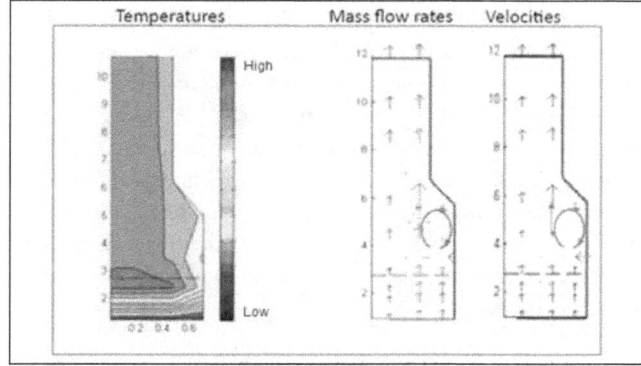

Figure 3.21 Temperature, flow rate, and velocity fields in the core and the riser section

For More Information
Contact Sergio Gonzalez, RES/DSA, at
Sergio.Gonzalez@nrc.gov.

Environmental Transport Research Program

Background

Many activities that are part of nuclear fuel cycles have the potential to expose the environment or the public to low levels of contamination from nuclear materials (Figure 3.22). Environmental assessments and protection address the vulnerability of environmental resources and public health to potential chronic exposure to radionuclides associated with nuclear facilities, including nuclear reactor, fuel cycle, waste disposal, and decommissioned facilities.

Figure 3.22 Conceptual visualization of contaminant pathways

Technical Issues

Monitoring and modeling of environmental systems at nuclear facilities are evolving in response to changing needs, increased understanding of environmental systems, and advances in technology. Issues associated with environmental monitoring include identification of potential sources and measurable indicators of system performance that can be coupled to regulatory requirements. Traditional analyses have often involved conservative assumptions that led to costly solutions. One goal of the research on environmental transport is to increase realism in current environmental assessments and to reduce unnecessary regulatory burden.

Specific Regulatory Needs

The program explicitly addresses needs imposed by risk-informed regulation. Individual research activities address needs identified by current regulatory programs, including the new reactor licensing program, the advanced reactors program, the decommissioning and uranium recovery program, and the reprocessing program.

The U.S. Nuclear Regulatory Commission (NRC) licensing staff needs improved technical bases for reviewing site characterization, monitoring, modeling, and remediation programs submitted by current and prospective licensees. Regulatory guidance is needed on environmental assessments and performance monitoring associated with new reactors and the decommissioning of nuclear facilities.

Principal Research Activities

Release of Radionuclides from Wastes or Engineered Structures and Advanced Sampling and Monitoring of Radionuclide Releases

The potential for chronic releases of radionuclides to the environment from nuclear facilities must be understood to ensure compliance with NRC regulations. Assessing long-term releases under varying chemical and physical conditions is a difficult but important aspect of ensuring that current or planned nuclear facilities conform to regulatory goals. Therefore, research activities are being conducted on how to monitor, characterize, and model the behavior of radionuclide-containing materials in the environment, including assessment of In-Situ sensors for real-time monitoring of radionuclides in the environment.

Long-Term Behavior of Engineered Materials

The expectation of the future use of engineered materials to isolate radioactive wastes or environmental contaminants results in a need for analytical tools to assess the design and performance of cement, concrete, and natural earth materials in engineered structures. Research activities include obtaining field-scale properties of soil and composite material covers that use geosynthetics to isolate radioactive waste to better understand the mechanics of contaminant release. Research is also underway to understand the long-term effectiveness of cementitious materials both for nuclear reactor and waste isolation applications. Research into the performance of reinforced-concrete or cement barriers (including the effects of degradation mechanisms, such as alkali silicate reaction) supports the assessment of reactor license renewal and the performance of engineered disposal facilities.

In the area of cementitious materials performance, the environmental transport research program has a cooperative interaction and research program with the objective of developing the next generation of simulation tools to evaluate the structural, hydraulic, and chemical performance of cementitious barriers used in nuclear applications over extended timeframes (e.g., more than 100 years for operating facilities and greater than 1,000 years for waste management applications). The program, the Cementitious Barriers Partnership, is a multidisciplinary, multi-institutional collaboration of Federal, academic, private sector, and international expertise formed to accomplish the project objectives.

Geotechnical Considerations for New and Advanced Nuclear Power Plants

Research activities address the recognition that design and construction of new or next-generation facilities can increase or inhibit the release, migration, and isolation of materials in the geosphere to improve the understanding, modeling, and monitoring of the performance of engineered features of new facilities. All of these activities involve the performance of soil-based subsurface components of the foundation system to mitigate the release of contaminants to the environment.

Advanced Modeling for Environmental Assessment

Advances in computational tools are enabling research to incorporate additional realism in the assessment of geochemical and biochemical processes that enhance or retard radionuclide transport. Additional realism significantly enhances the prospects for meaningful validation of system or subsystem models used for environmental assessment. Research on computational tools is focused on a generic framework for linking databases, models, and other analytic tools for flexible problem solving.

Decision Support for Ground Water Remediation

Technologies for the remediation of subsurface contamination have advanced significantly in recent years. Likewise, advances in understanding and manipulating subsurface biota are leading to advances in exploiting the ability of biota to remediate subsurface contamination. Research is being conducted to examine the efficacy of long-term performance of these remediation technologies and to provide tools to assist in remediation planning.

Regulatory Basis in Support of Fuel Reprocessing Facilities

A commercial nuclear fuel reprocessing plant has not been licensed in the United States for over 35 years. Consequently, the NRC's regulatory framework needs to be updated to support the licensing of such a facility. In view of recent initiatives by the U.S. Department of Energy (DOE) and commercial industry interest in developing such facilities, a multiple-office working group is developing a regulatory basis in support of rulemaking for reprocessing. The basis will address a number of issues related to reprocessing, including waste management and risk. Research activities planned to support this effort include an assessment of source term phenomena and code development.

Collaborative Efforts and Opportunities

The environmental transport research program leverages resources through cooperative interactions and special research agreements, such as the Memorandum of Understanding on Research and Development of Multimedia Environment Models, with other national and international research organizations that are pursuing related work. The technical objective is to collaborate on, or gain access to, technologies, databases, computer software, lessons learned, and methods that support the NRC's regulatory activities. Collaborators include other Federal agencies (e.g., the U.S. Department of Agriculture's Agricultural Research Service, U.S. Geological Survey, National Institute of Standards and Technology, U.S. Environmental Protection Agency, and U.S. Army Corps of Engineers), DOE national laboratories, universities, national academies, professional societies (e.g., the American Nuclear Society, American Geophysical Union, International Association of Hydrological Sciences, National Ground Water Association, and American Society for Testing and Materials), and international organizations (e.g., the Canadian Nuclear Safety Commission).

These cooperative ventures help to identify important research findings, datasets, and lessons learned for use in evaluating and testing multimedia environmental models, examining the role of engineered barrier systems in waste disposal, and evaluating the practicality of modeling chemical sorption in environmental systems. Interactions with professional societies assist in developing guidance and training programs. Knowledge management also profits from interactions with other Federal and professional organizations and from their information sources (e.g., technical journals, Web sites, and monographs).

For More Information
Contact William Ott, RES/DRA, at William.Ott@nrc.gov.

Integrated Ground-Water Monitoring and Modeling

Background

The Office of Nuclear Regulatory Research (RES) is working with U.S. Nuclear Regulatory Commission (NRC) licensing offices (the Office of Federal and State Materials and Environmental Management Programs, the Office of New Reactors, and the Office of Nuclear Reactor Regulation) and its regional offices to develop guidance for reviewing ground water monitoring programs in accordance with Title 10 of the *Code of Federal Regulations* (10 CFR) 20.1406(a), 10 CFR 20.1406(b), and 10 CFR 20.1406(c) (effective date December 17, 2012); 10 CFR 50.75(g); Appendix A, "Criteria Relating to the Operation of Uranium Mills and the Deposition of Tailings or Wastes Produced by the Extraction or Concentration of Source Material from Ores Processed Primarily for Their Source Material Content," to 10 CFR Part 40, "Domestic Licensing of Source Material"; 10 CFR 61.53, "Environmental Monitoring"; and 10 CFR 63.131, "General Requirements." In November 2007, RES issued NUREG/CR-6948, "Integrated Ground-Water Monitoring Strategy for NRC-Licensed Facilities and Sites: Logic, Strategic Approach and Discussion," which provides the technical bases for this guidance.

NUREG/CR-6948 documents the development and testing of an integrated ground water monitoring strategy. It integrates conceptual site model (CSM) confirmation with ground water monitoring through the use of performance indicators (PIs) (e.g., concentrations and water fluxes in the unsaturated and saturated zones). It outlines procedures for selecting, locating, and calibrating field instruments and methods to detect radionuclide releases in the subsurface and to determine the need and effective approaches for remediation. Recently, the Agricultural Research Service (ARS) of the U.S. Department of Agriculture developed methods for incorporating model abstraction techniques (NUREG/CR-7026, "Application of Model Abstraction Techniques To Simulate Transport in Soils," issued March 2011) into the design of subsurface monitoring and performance assessment programs. These methods are being tested through specially designed field tracer studies at ARS's Beltsville Agricultural Research Center in Maryland.

In 2007, the Nuclear Energy Institute issued its industry initiative on ground water protection that includes onsite ground water monitoring at all nuclear reactor sites. The Nuclear Energy Institute funded the Electric Power Research Institute to develop guidelines for ground water protection that were issued in January 2008 and, more recently, ground water and soil remediation guidelines for nuclear power plants. NRC Regional inspectors, working with the Office of Nuclear Reactor Regulation and RES staff, have used the RES-developed information to review these new programs in conjunction with existing offsite radiological environmental monitoring programs. This monitoring is needed to detect radionuclide releases and to evaluate the need for, and selection of, remediation approaches. In December 2010, industry guidance on ground water characterization and modeling was issued as American National Standards Institute (ANSI)/American Nuclear Society (ANS)-2.17-2010, "Evaluation of Subsurface Radionuclide Transport at Commercial Nuclear Power Plants". The NRC staff is evaluating this guidance for use by licensees and NRC staff reviewers in conjunction with radiological environmental monitoring programs and estimating offsite dose calculation models (ODCM).

Objectives

The objectives of this research are to provide the NRC with a technical basis for assessing licensees' monitoring programs and applicants' planned monitoring programs and to provide guidance on linking ground water monitoring and modeling with health physics models to estimate potential offsite doses. This research will benefit licensees and other Federal agencies in integrating ground water monitoring, modeling, and remediation programs. Monitoring is key to assessing the efficacy of remediation methods within a regulatory framework.

Approach

The strategy provides an integrated and systematic approach for monitoring subsurface flow and transport beginning at the land surface and extending through the unsaturated zone to the underlying water table aquifer (Figure 3.23). The strategy is robust and useful for determining that site- and facility-specific ground water monitoring programs do the following:

- Assess the effectiveness of contaminant isolation systems and remediation activities.

- Communicate to decisionmakers and stakeholders the monitored PIs through effective data management, analysis, and visualization techniques.

- Detect and identify the presence of contaminant plumes and preferential ground water transport pathways.

- Test alternative conceptual and numerical flow and transport models.

- Aid in the confirmation of the assumptions of the CSM and, hence, the performance of the facility through the monitoring of PIs.

- Identify uncertainties introduced by model abstraction techniques and demonstrate parameter estimation methods in conjunction with model formulation and testing, characterization techniques, and monitoring designs.

The documented strategy provides the technical bases, along with identified guidance and analytical tools, for assessing the completeness and efficacy of an integrated ground water monitoring program. It focuses on quantifying uncertainties in the hydrologic features, events, and processes using "real-time, near-continuous" monitoring data to confirm the CSM. The strategy is being updated to link the ground water monitoring program to detection requirements for early warning of releases, estimation of offsite doses, and decisions on the need and efficacy of remediation methods.

RES and ARS scientists are testing the updated strategy to demonstrate how to identify and monitor PIs of the subsurface flow and transport system behavior. Using these field studies, the updated strategy will illustrate how these methods couple subsurface characterization, monitoring, and modeling for early detection of releases to the accessible environment.

unsaturated zone) and below the water table and, as illustrated in the figure, must not introduce inadvertent pathways.

The revision to Regulatory Guide 1.21, "Measuring, Evaluating, and Reporting Radioactive Material in Liquid and Gaseous Effluents and Solid Waste," and Regulatory Guide 4.1, "Radiological Environmental Monitoring for Nuclear Power Plants," reference this research and the related guidance in the newly issued ANSI/ANS-2.17-2010.

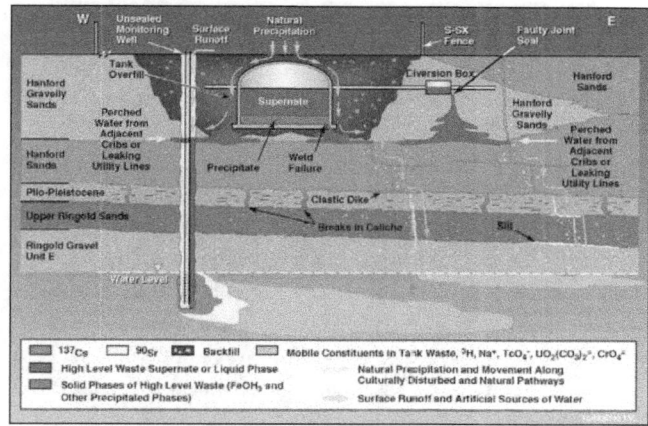

Figure 3.24 Illustration of a conceptual model of a complex field site and complex sources for which monitoring can facilitate decisionmaking for remediation (Ward et al. "A Comprehensive Analysis of Contaminant Transport in the Vadose Zone Beneath Tank SX-109," PNL-11463. Richland, Washington, February 1997.)

For More Information

Contact Thomas Nicholson, RES/DRA, at Thomas.Nicholson@nrc.gov.

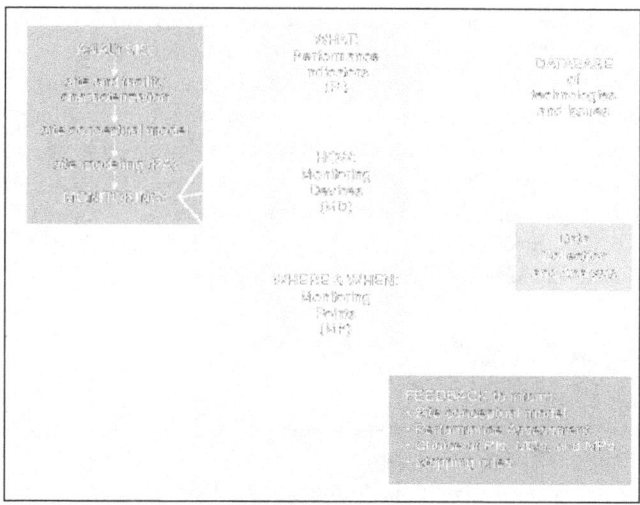

Figure 3.23 Flow chart of the integrated monitoring strategy

Research Activities

Both NUREG/CR-6948 and NUREG/CR-7026 provide the state of the practice in ground water modeling and monitoring approaches for evaluating residual subsurface radioactivity and the technical bases for determining the need for remediation. Figure 3.24 shows the natural and engineered complexities that can affect subsurface flow and radionuclide transport. Monitoring strategies need to consider these complexities in the development and testing of conceptual models. Monitoring involves detection and sampling both above (i.e., in the

In-Situ Bioremediation of Uranium in Ground Water

Background

License applications and decommissioning plans have been submitted to the U.S. Nuclear Regulatory Commission (NRC) on the use of In-Situ bioremediation of groundwater at two types of sites: (1) shallow plumes of uranium that originated with uranium processing and waste disposal operations and (2) In-Situ leach (ISL) uranium recovery sites that have been depleted and require groundwater remediation in accordance with Appendix A, "Criteria Relating to the Operation of Uranium Mills and the Deposition of Tailings or Wastes Produced by the Extraction or Concentration of Source Material from Ores Processed Primarily for Their Source Material Content," to Title 10 of the *Code of Federal Regulations* (10 CFR) Part 40, "Domestic Licensing of Source Material." In both cases, the original remediation methods have not reduced aqueous uranium concentrations to acceptable levels. As a result, a new approach (i.e., using In-Situ manipulation of native bacterial populations to alter geochemical conditions) has been proposed.

With the In-Situ bioremediation technique, electron donors (e.g., acetate and lactate) are injected through wells into the contaminated aquifer. These innocuous compounds are used by bacteria as nutrients, increasing their growth and reproduction and, in the process, using oxygen in the subsurface and generating reducing conditions. As a result, iron (Fe(III)) and uranium (U(VI)) will be chemically reduced, and U(IV) will be precipitated from solution (Figure 3.25). However, note that, although uranium is removed from solution where it was mobile as U(VI), it is precipitated as U(IV) in the solid phase or as a mineral and left in place. Eventually, many sites that have been bioremediated should be exposed to oxidizing conditions, especially at shallow sites, which may result in precipitated U(IV) being reoxidized and returned to solution.

Based on experimental and modeling results (NUREG/CR-7014, "Processes, Properties, and Conditions Controlling In-Situ Bioremediation of Uranium in Shallow, Alluvial Aquifers," issued July 2010) from the first portion of this study, the Office of Nuclear Regulatory Research staff recommended against the use of this remediation technique for a site with several uranium groundwater plumes at a 20- to 30-foot depth below ground surface.

The second portion of this study focuses on ISL sites at which conditions can be very different from those near the surface. The uranium ore zone was formed by reducing conditions, precipitating uranium from solution much as the bioremediation process does. These deposits have been stable for long periods of time, perhaps millions of years. During the extraction process, a reagent (typically oxygen) is pumped into the ore, dissolving the uranium and allowing it to be pumped up to the surface with water. In-Situ bioremediation attempts to return the site to its original reducing condition.

Approach

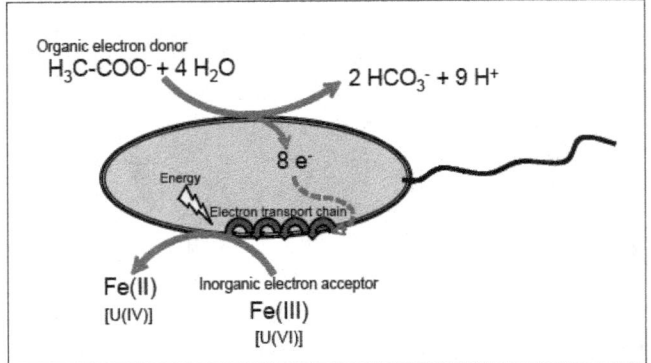

Figure 3.25 Microbial mediation of Fe(III) reduction[1]

For the ISL portion of this work, two approaches, like with the swallow site, are being used to assess the behavior of bioremediated systems. The two approaches, laboratory-scale experimental work and advanced modeling, complement each other. The U.S. Geological Survey is conducting the laboratory work in a project entitled, "Uranium Sequestration and Solid Phase Behavior during and after Bioremediation." The modeling project is being done at Pacific Northwest National Laboratory and is entitled, "Modeling the Long-Term Behavior of Uranium during and after Bioremediation."

Experimental Approach

For the experimental program, sediment was obtained from a mined ISL site from a depth of 550 to 570 feet. This unconsolidated medium-grained sand was placed in columns, and reducing conditions were established by biostimulation. **Especially important to long-term performance is the** stability of solid phase uranium and iron minerals generated by bioremediation as they are leached by various site-specific groundwaters. The behavior of uranium and other elements was followed in both the aqueous and solid phase during reduced conditions and then as oxygen containing water was introduced into the columns.

---

[1]   U(VI) is the mobile valence state of uranium, whereas reduced uranium, U(IV) has very low solubility. Addition of acetate as an electron donor stimulates dissimilatory metal-reducing microorganisms. U(VI) is reduced concurrently with Fe(III). The original concept is from Lovley et al. (1991), and the figure is from NUREG/CR-6973, "Technical Basis for Assessing Uranium Bioremediation Performance," issued August 2008.

Solid-phase analysis includes synchrotron-based methods, such as xray absorption spectroscopy, to determine the oxidation state of uranium and iron and their microscale distributions under the reduced and oxidized conditions of the columns.

The sediment from the ISL site was found to have a bacterial population that was very low and, in particular, was depleted in *Geobacteraceae*, the family of common soil bacteria that is generally active in reducing a variety of metals, including Fe(III) and U(VI). Unlike the bacterial response in materials from shallow sites (which was very rapid), a lag of several weeks with the ISL sediment occurred before the uranium began to reduce and precipitate.

A series of questions define this research. Although the fact that uranium precipitates under reducing conditions is well known, does it always precipitate as a discrete mineral or is some uranium distributed as amorphous or very small (nanoscale) particles? During bioremediation, does uranium also coprecipitate with minerals, such as mackinawite, siderite, and calcite? If so, in what oxidation state is the uranium? If aqueous uranium enters the system, how does it react with these new solids? How do major differences in microbial populations alter the processes and reagents (e.g., electron donors) needed to precipitate uranium? The experimental program will provide answers to these questions and will give details needed to estimate the long-term behavior of uranium.

Modeling Approach

The objective of the modeling work is to identify, assess, and model short and long-term chemical processes caused by bioremediation. It focuses on processes controlling uranium sequestration and changes in uranium mobility, during and after bioremediation. The approach is to use coupled models of biological, geochemical, and transport processes to determine how the chemistry in these systems changes and what the effects will be on parameters that can be monitored in the field. The modeling effort iterates through key parameters, such as flow rates, uranium concentrations, mass of iron available, carbonate concentrations, biological kinetics, alkalinity, O_2, and uranium input.

Modeling of ISL uranium recovery sites is based on the experiments conducted by the U.S. Geological Survey. Results of these experiments, the first done on material from an ISL site, showed that biological processes leading to uranium precipitation at an ISL site seem to be quite different from those at shallow sites and required significant changes in modeling. Of these, the most important were reduction in the growth rate compared to shallow sites and the dependence of uranium bioreduction rate on biomass, which increases with continuous acetate stimulation. The dual monod approach cannot account for this behavior alone.

Products of the modeling work will include a guidance document that describes approaches, criteria, and methods to predict the stability of biorestored ISL sites to help the staff in licensing reviews. It will provide modeling-based information on changes in monitorable parameters that can be expected as a result of changing conditions in the subsurface system. Ultimately, both the experimental and modeling approaches will help the NRC to do the following:

- Assess the geochemical, microbial, and groundwater conditions and processes that affect uranium transport and its potential long-term sequestration.

- Provide the technical basis to predict long-term performance for decommissioning, particularly during reoxidation after bioremediation treatments.

- Evaluate biorestoration design, performance, and stability for uranium recovery and related financial surety costs.

Final reports for the ISL work (NUREG-7176) and for the shallow aquifer work, are being reviewed and publication is expected in 2013.

References

Yabusaki, S.B., Y. Fang, P.E. Long, C.T. Resch, A.D. Peacock, J. Komlos, P.R. Jaffe, S.J. Morrison, R.D. Dayvault, D.C. White and R.T. Anderson (2007), "Uranium Removal from Groundwater via In-Situ Biostimulation: Field-Scale Modeling of Transport and Biological Processes," *Journal of Contaminant Hydrology*, 93(14):216–235.

Long, P.E., S.B. Yabusaki, P.D. Meyer, C.J. Murray, and A.L. N'Guessan, "Technical Basis for Assessing Uranium Bioremediation Performance," NUREG/CR-6973, U.S. Nuclear Regulatory Commission, Washington, DC, 2008.

Lovely, D.R., E.J. Phillips, Y.A. Gorby, and E.R. Landa, "Microbial Reduction of Uranium," *Nature*, 350(6317):413–416, 1991.

Lawrence Berkeley National Laboratory 42595, "Bioremediation of Metals and Radionuclides—What It Is and How It Works," 2nd Edition, Lawrence Berkeley National Laboratory, Berkeley, CA, 2003.

Yabusaki, S.B., Y. Fang, S.R. Waichler, and P.E. Long, "Processes, Properties, and Conditions Controlling In-Situ Bioremediation of Uranium in Shallow, Alluvial Aquifers," NUREG/CR-7014, U.S. Nuclear Regulatory Commission, Washington, DC, 2010.

For More Information
Contact Dr. Mark Fuhrmann, RES/DRA at
Mark.Fuhrmann@nrc.gov

Chapter 4: Radiation Protection and Health Effects

Radiation Protection Program

Regulatory Basis for NRC Standards for Protection Against Ionizing Radiation

Radiation Exposure Information and Reporting System (REIRS)

Analysis of Cancer Risks in Populations Near Nuclear Facilities

Report to Congress on Abnormal Occurrences

VARSKIN Skin Computer Code

Phantom with Moving Arms and Legs (PIMAL)

Radionuclide Transport, Removal, and Dose (RADTRAD)

Radiation Worker Health Studies

Radiological Toolbox

Participation in National and International Radiation Protection Activities

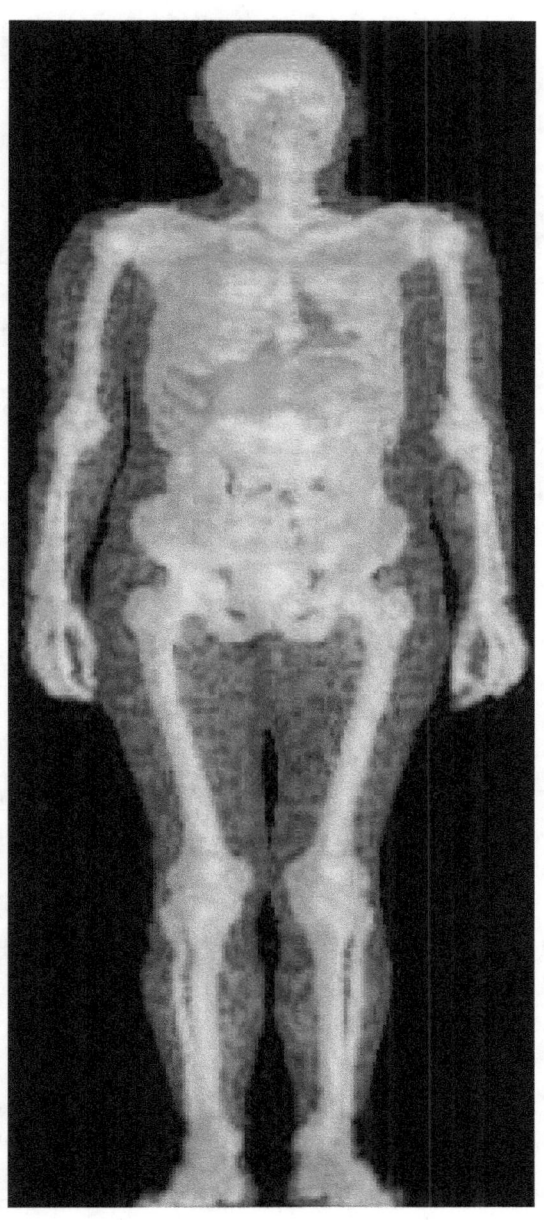

Radiation Protection Program

Background

The U.S. Nuclear Regulatory Commission's (NRC's) Radiation Protection Program is an agencywide resource that provides technical support in the areas of radiation protection, dose assessment, and assessment of human health effects for reactor and nuclear materials licensing, emergency preparedness, and nuclear security activities.

The program's scope includes the following:

- Technical basis for radiation protection regulations.
- Exposure and abnormal occurrence reports.
- Computer codes and databases development.
- Health effects and dosimetry research.

Technical and Regulatory Projects

- Regulatory Basis for NRC Standards for Protection Against Ionizing Radiation
- Radiation Exposure Information and Reporting System (REIRS)
- Analysis of Cancer Risks in Populations near Nuclear Facilities
- Report to Congress on Abnormal Occurrences
- Internal Dosimetry Research
- Radiation Protection Regulatory Guides
- VARSKIN Skin Computer Code
- Phantom with Moving Arms and Legs (PIMAL)
- Radionuclide Transport, Removal, and Dose (RADTRAD)
- Radiation Worker Health Studies
- Radiological Toolbox
- Participation in National and International Radiation Protection Activities

Internal Dosimetry Research

The Radiation Protection Branch (RPB) provides technical resources to the NRC by conducting radiation dosimetry research for regulatory applications. This research improves the agency's capability to model radiation interactions within humans, evaluate internal dosimetry codes for estimating radiation exposures, and assess worker or public exposures from licensed activities or incidents.

National and International Activities

One of the benefits of the Radiation Protection Program is the promotion of consistency in regulatory applications of radiation protection and health effects research among NRC programs, as well as those of other Federal and State regulatory agencies. The Radiation Protection Program staff collaborates with national and international experts in health physics at national laboratories, universities, and other organizations, including the following:

- Interagency Steering Committee on Radiation Standards.
- National Council on Radiation Protection and Measurements.
- National Academies.
- Nuclear Energy Agency (NEA) Information System on Occupational Exposure.
- International Commission on Radiological Protection.
- International Commission on Radiation Units and Measurements.
- International Atomic Energy Agency.
- French Institute for Radiological Protection and Nuclear Safety.

Radiation Protection Regulatory Guides

Developing and updating regulatory guides on occupational health and other radiation protection-related topics provides licensees with better methods for maintaining compliance with NRC regulations. Regulatory guides describe NRC-approved methods of meeting regulatory requirements.

VARSKIN

The NRC funded the development of the VARSKIN computer code in the 1980s to facilitate skin-dose calculations. Since then, the code has been upgraded to make it more efficient and easier to use. The NRC is currently developing a more sophisticated replacement for the code's existing photon dose algorithm, as well as further enhancements to the code's functionality.

PIMAL

The Office of Nuclear Regulatory Research (RES) and Oak Ridge National Laboratory have developed a computational "phantom" with moving arms and legs (i.e., the PIMAL) to assess the radiation dose for realistic exposure geometries. PIMAL is based on the Oak Ridge National Laboratory's mathematical phantom and can bend arms from the shoulder and elbow, and legs from the hip and knee. An accompanying graphical user interface (GUI) also has been developed to assist users with dose assessments and to reduce the analysts' time.

Radionuclide Transport, Removal, and Dose (RADTRAD)

RADTRAD is a design-basis accident (DBA) code used for commercial nuclear power plant licensing applications to determine the time-dependent dose at a specified location for a given accident scenario. The model estimates doses at the exclusion area boundary or the low population zone and in the control room and other locations. The RADTRAD code can estimate the containment release using either the NRC Technical Information Document (TID)14844, "Calculation of Distance Factors for Power and Test Reactor Sites," or NUREG-1465 ,"Accident Source Terms for Light-Water Nuclear Power Plants," source terms and assumptions, or a user-specified source term. The NRC currently is working on a JAVA-based version of RADTRAD, RADTRAD Version 4.0. In addition, the Microsoft Visual Basic GUI was replaced with a Symbolic Nuclear Analysis Package (SNAP) GUI. Placing RADTRAD in the SNAP framework allows for the use of SNAP features, including the Model Editor, to develop plant models. It also provides tools for user input checking and monitoring calculations.

Report to Congress on Abnormal Occurrences

The NRC annually publishes the Abnormal Occurrence (AO) Report to Congress. An AO is defined as an unscheduled incident or event that the NRC determines to be significant from the standpoint of public health or safety. The AO process helps identify deficiencies and ensures that corrective actions are taken to prevent recurrence. An accident or event is considered an AO if it involves a major reduction in the degree of protection of public health or safety. This type of incident or event would have a moderate or more severe on public health or safety. The AO report contains event details from both NRC and Agreement State-licensed facilities that meet the AO criteria published by the Commission.

Radiation Exposure Information and Reporting System

REIRS collects information on occupational radiation exposures to workers from certain NRC-licensed activities. Data collected in the REIRS database are used to evaluate licensee as-low-as-is-reasonably-achievable programs and is shared with national and international research counterparts. The REIRS database is also used to compile the annual report, NUREG-0713, "Occupational Radiation Exposure at Commercial Nuclear Power Reactors and Other Facilities."

Analysis of Cancer Risk in Populations Living Near Nuclear Facilities

In April 2010, the NRC staff requested the National Academy of Sciences (NAS) to perform a study on cancer mortality and incidence risks in populations living near NRC-licensed facilities. The purpose of the study is to update the 1990 National Cancer Institute report on "Cancer Risks in Populations near Nuclear Facilities." The study was divided into two-phases. In Phase 1, NAS explored the feasibility of conducting an updated study by developing modern methods to perform the analysis. The staff has reviewed the results of the Phase 1 study and the NAS recommendations for the next phase. The staff's next step will be to proceed with the NAS-recommended approach to determine the feasibility of the Phase 1 methods through pilot studies at seven sites that the NAS committee recommended: Dresden in Illinois, Millstone in Connecticut, Oyster Creek in New Jersey, Haddam Neck (decommissioned) in Connecticut, Big Rock Point (decommissioned) in Michigan, San Onofre in California, and Nuclear Fuel Services in Tennessee. Upon completion of the pilot studies, NAS will comment on whether further study is beneficial, and the NRC staff will determine whether to perform the studies at all NRC-licensed facilities (i.e., balance of operating nuclear power plants and fuel-cycle facilities).

For More Information
Contact Stephanie Bush-Goddard, RES/DSA, at
Stephanie.BushGoddard@nrc.gov.

Regulatory Basis for NRC Standards for Protection Against Ionizing Radiation

Background

The U.S. Nuclear Regulatory Commission (NRC) provides the fundamental radiological protection criteria for licensees to use in Title 10 of the *Code of Federal Regulations* (10 CFR) Part 20, "Standards for Protection against Radiation." The last major revision to 10 CFR Part 20 was completed in 1991. It was primarily based on the 1977 recommendations contained in International Commission on Radiological Protection (ICRP) Publication 26, "Recommendations of the International Commission on Radiological Protection."

Since 1991, the NRC has made minor revisions to 10 CFR Part 20, such as a reduced public dose limit that incorporates the recommendations of ICRP Publication 60, "1990 Recommendations of the International Commission on Radiological Protection," issued in 1991. However, in other NRC regulations, such as Appendix I, "Numerical Guides for Design Objectives and Limiting Conditions for Operation to Meet the Criterion 'As Low as is Reasonably Achievable' for Radioactive Material in Light-Water-Cooled Nuclear Power Reactor Effluents," to 10 CFR Part 50, "Domestic Licensing of Production and Utilization Facilities," some radiation dose criteria are based primarily on ICRP Publications 1 and 2 (the 1958 and 1959 "Recommendations of the International Commission on Radiological Protection"). In addition, NRC fuel cycle licensees have received authorization, on a case-by-case basis, to use the newer ICRP methodology (ICRP Publication 66, "Human Respiratory Tract Model for Radiological Protection," issued January 1995 and beyond) in their licensed activities. The Agreement States' requirements for their licensees are essentially identical to 10 CFR Part 20. As a result, three different sets of ICRP recommendations are in use today by various licensees.

Approach

In December 2008, the NRC staff provided the Commission with a summary of regulatory and technical options for moving—or not moving—toward a greater alignment of the NRC's radiation protection regulatory framework with ICRP Publication 103, "Recommendations of the International Commission on Radiological Protection," issued February 2008. The Commission subsequently directed the staff to begin engaging with stakeholders and interested parties to initiate development of a regulatory basis for possible revision of the NRC's radiation protection regulations, as adequate and appropriate where scientifically justified, to achieve greater alignment with the recommendations in ICRP Publication 103. In response, the NRC staff engaged a wide range of stakeholders on potential issues, conducted preliminary assessments of the impacts of implementation of ICRP's recommendations, and participated in international and national meetings. In April 2012, the NRC summarized in a paper to the Commission the staff's multiyear effort, and identified several technical and policy issues that require further study. This paper, SECY-120064, "Recommendations for Policy and Technical Direction To Revise Radiation Protection Regulations and Guidance," is available on the NRC's Web site at http://www.nrc.gov/reading-rm/doc-collections/commission/secys/2012/2012-0064scy.pdf.

Current Activities

As part of this effort, the Radiation Protection Branch (RPB) is developing technical information on the benefits and burdens associated with revising the NRC's radiation protection regulatory framework. RPB will consider (1) impacts on licensees, (2) impacts on public confidence, (3) cost-benefit issues, (4) backfit issues, (5) impacts on the NRC's materials program, and (6) other benefits and burdens of adopting ICRP Publication 103 recommendations. Currently, development of this regulatory basis comprises the four technical areas described below.

Impacts of Changing Occupational Dose Limits and Using Dose Constraints

The purpose of this task is to collect and analyze information about the actual dose distributions from industrial and medical licensees and to determine the impact of reduced dose limits from 50 to 20 millisievert (5 rem to 2 rem) per year both on an annual basis and averaged over 5 years. The staff is developing a report that provides technical information and a policy synopsis for agencywide use. RPB also contributed to the technical development of a 2011 report on dose constraints issued by the Nuclear Energy Agency (NEA) entitled, "Dose Constraints in Optimization of Radiological Protection" (NEA/CRPPH/R (2011) (see Figure 4.1). This report can be viewed on the NEA's Web site at www.oecd-nea.org.

Figure 4.1 Planned exposure situations

Occupational Dose Information and Evaluation of Potential Compliance Issues

This analysis will address potential changes to the occupational dose limit, the dose limit to an embryo or fetus of a declared pregnant woman, and the use of dose constraints. Although, there is minimal information on occupational exposures at Agreement State-licensed facilities, medical institutions, or for exposures to the embryo or fetus, the staff continues to explore additional approaches with external stakeholders to gather data needed to support this analysis. In August 2010, NRC staff issued a letter to Agreement State Radiation Control Programs requesting occupational dose information from Agreement State-licensed materials licensees. Information received from Agreement State materials licensees was analyzed for trends and impacts associated with a potential reduction in the occupational dose limit. RPB developed the July 2012 report entitled, "Occupational Radiation Exposure at Agreement State-Licensed Materials Facilities, 19972010" (NUREG-2118). This report is available on the NRC's public Web site at http://www.nrc.gov/reading-rm/doc-collections/nuregs/staff/sr2118/v1/.

Support Development of New Biokinetic and Dosimetric Models and Dose Coefficients for Occupational and Public Exposure

The purpose of this task is to support and monitor work that Oak Ridge National Laboratory (ORNL) is conducting on the development of biokinetic and dosimetric models (see Figure 4.2) and dose coefficients for occupational and public exposure to radionuclides that are based on ICRP Publication 103

recommendations. This is a multiyear effort that will continue until ICRP finalizes the numerical values associated with ICRP Publication 103.

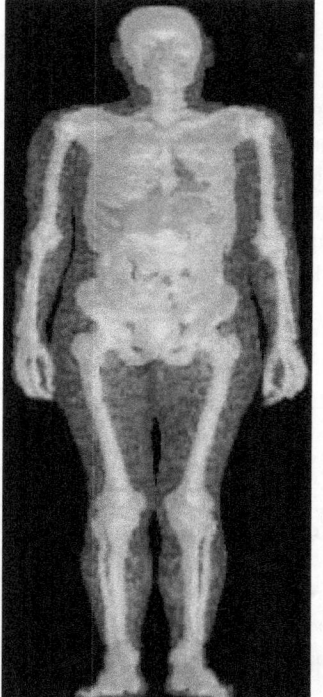

RES staff is working closely with other Federal agencies to share the cost of funding ORNL for related work, and participate in domestic and international working groups that assess potential technical and policy issues associated with the implementation of ORNL's research.

Figure 4.2 Biokinetic model

Costs and Impacts of Implementing ICRP Publication 60 in the United States

To estimate the potential costs of implementing ICRP Publication 103, the NRC is seeking information from domestic and international sources on costs for implementing ICRP Publication 60. Based on the results of initial data gathering efforts, RES staff is currently focusing on strategies that other Federal agencies and the international radiation protection community use to implement ICRP Publication 60 and more recent recommendations.

Use of Research Results

The overall goal of this effort is to obtain sufficient information to proceed with a rulemaking and to identify policy issues that require future Commission decisions. In particular, this will support the NRC staff in developing a regulatory basis, associated guidance, and proposed language for rulemaking.

For More Information
Contact Tony Huffert, RES/DSA, at Anthony.Huffert@nrc.gov.

Radiation Exposure Information and Reporting System (REIRS)

Background

The Radiation Exposure Information and Reporting System (REIRS) project collects and analyzes the occupational radiation exposure records that U.S. Nuclear Regulatory Commission (NRC) licensees submit under Title 10 of the *Code of Federal Regulations* (10 CFR) 20.2206, "Reports of Individual Monitoring."

Each year, approximately 200,000 radiation exposure reports are submitted by five categories of NRC licensees:

1. Industrial radiography.

2. Manufacturers and distributors of byproduct material.

3. Commercial nuclear power reactors.

4. Independent spent fuel storage installations.

5. Fuel processors, fabricators, and reprocessors.

The NRC does not receive radiation exposure reports from the remaining two licensee categories, low-level waste disposal facilities and geologic repository for high-level waste, because these facilities are either not under NRC jurisdiction or not currently in operation.

Approach

To maintain compliance with 10 CFR 20.2206, NRC licensees must submit their occupational radiation exposure data to the NRC by April 30 of each year. Licensees can submit this data electronically or on paper, using either NRC Form 5, "Occupational Dose Record for a Monitoring Period," or a Form 5 equivalent.

The objective of the REIRS database (see Figure 4.3) is to provide NRC staff with occupational exposure data for evaluating trends in licensee performance in radiation protection and for research and epidemiological studies. The exposure reports in this database can provide facts about routine occupational exposures to radiation and radioactive material that can occur in connection with certain NRC-licensed activities.

The analysis of REIRS data is published annually in NUREG-0713, "Occupational Radiation Exposure at Commercial Nuclear Power Reactors and Other Facilities."

Application

The radiation exposure reports that NRC licensees submit are used to meet the following NRC regulatory goals:

- Evaluate the effectiveness of licensee's as low as is reasonably achievable (ALARA) programs at commercial nuclear power plants

- Evaluate the radiological risk associated with certain categories of NRC-licensed activities

- Compare occupational radiation risks with potential public risks

- Establish priorities for the use of NRC health physics resources, such as research and development of standards and regulatory guidance

- Answer congressional and public inquiries

- Provide radiation exposure history reports to current and former occupational radiation workers who were exposed to radiation or radioactive materials at NRC-licensed or regulated facilities

- Conduct occupational epidemiological studies

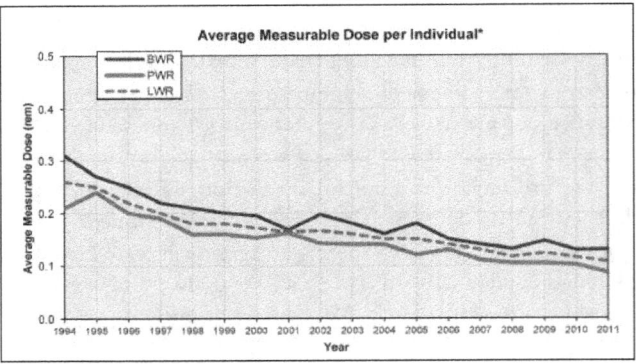

Figure 4.3 Sample data from REIRS database

Web Site

The annual NUREG-0713 reports are available on the NRC's public Web site at http://www.nrc.gov or the REIRS Web page at www.reirs.com.

REIRS Software

REMIT is a software package that allows licensees to maintain and report their exposure records to the REIRS database. REMIT allows for the electronic exchange of records from one licensee to another and the importing of records from the licensee's dosimetry processor. REIRView is another NRC-developed software package that allows licensees to validate their annual electronic submittals to the REIRS database. This saves licensees and the NRC considerable processing time because the licensee can identify and correct problems before submitting the information to the REIRS database (see Figure 4.4).

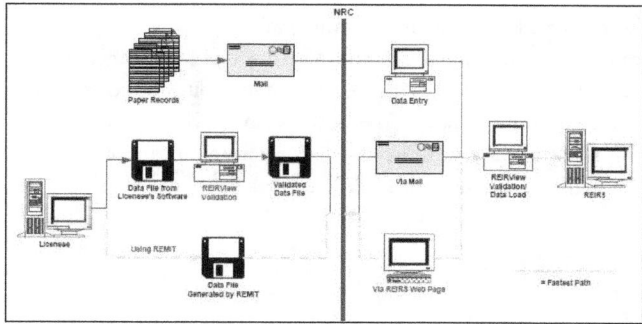

Figure 4.4 Process for submitting licensee exposure reports

For More Information
Contact Doris Lewis, RES/DSA, at Doris.Lewis@nrc.gov.

Analysis of Cancer Risks in Populations Near Nuclear Facilities

Background

Nuclear facilities that the U.S. Nuclear Regulatory Commission (NRC) licenses (Figure 4.5) sometimes release very small amounts of radioactivity during normal operations. These releases are a very small fraction of background radiation and the amount of radiation the average U.S. citizen receives in a year from all sources. NRC regulations ensure that plant operators monitor and control these releases to meet very strict radiation dose limits, and plants must publicly report these releases to the agency. Nonetheless, some communities have expressed concern about the potential impact of these releases on the health of citizens living near nuclear facilities.

To help address these concerns the NRC requested that the U.S. National Academy of Sciences (NAS) conduct a study analyzing the cancer risk of populations living near NRC-licensed facilities. This study will be used as an update to the 1990 National Cancer Institute (NCI) report, "Cancer in Populations Living Near Nuclear Facilities." The NAS is a nongovernmental organization chartered by the U.S. Congress to advise the Nation on issues of science, technology, and medicine. Through the National Research Council and Institute of Medicine, it carries out studies independent of the government using processes designed to promote transparency, objectivity, and technical rigor. More information on its methods for performing studies is available at http://www.nationalacademies.org/studycommitteprocess.pdf.

NRC staff has used the 1990 NCI study as a valuable risk communication tool for addressing stakeholder concerns about cancer mortality attributable to the operation of nuclear facilities. Stakeholders often ask the staff about perceived elevated cancer rates in populations working or residing near NRC-licensed nuclear facilities, including power reactors and fuel cycle facilities (e.g., fuel enrichment and fabrication plants). The NCI study was produced in response to concerns about elevated risk of childhood leukemia to persons near a British nuclear facility (Sellafield). NCI researchers studied more than 900,000 cancer deaths using county mortality records collected from 1950–1984. Changes in mortality rates for 16 types of cancer were evaluated. The NCI report concluded that cancer mortality rates generally are not elevated for people living in the 107 U.S. counties containing or closely adjacent to 62 nuclear facilities. However, the population data that the NCI report used is now more than 20 years old and should be updated.

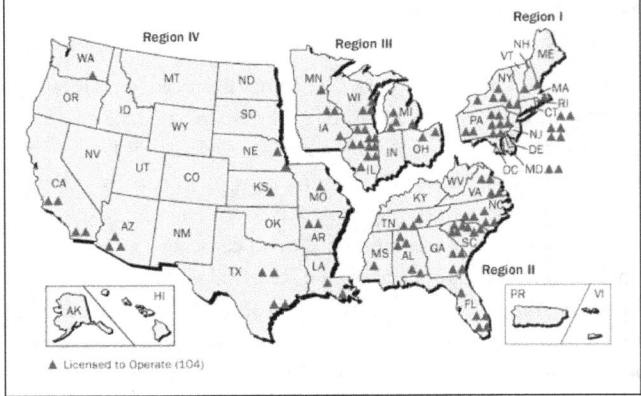

Figure 4.5 Locations of operating nuclear power facilities

Today, stakeholder interest continues about perceived elevated cancer rates in populations near reactors, including cancer *incidence* (i.e., being diagnosed with cancer, but not necessarily dying from the disease). The NRC is having NAS conduct this study to provide up-to-date information on cancer risks in populations near nuclear facilities.

Approach

The proposed study will be performed in two-phases: (1) preparation of a scoping study to determine the best methodology, the best approach, and the potential limitations for performing the cancer incidence and mortality epidemiology study and (2) conduct of the actual study. The NRC's objective is to have the latest cancer epidemiology information to communicate with its stakeholders. The study also will evaluate whether the risks are different for various age groups, including children.

Study Status—Phase 1 results and next steps

The NAS published the Phase 1 committee report on March 28, 2012, which can be accessed on the NAS Web site at: http://www.nap.edu/catalog.php?record_id=13388#toc.

The Phase 1 study committee made three recommendations to the NRC for the next phase of the study:

Recommendation 1: Two study designs were recommended subject to the feasibility assessment described in Recommendation 2.

1. An ecologic study of multiple cancer types of populations living near nuclear facilities.

2. A record-linkage based case-control study of cancers in children born near nuclear facilities.

Recommendation 2: A pilot study should be carried out to assess the feasibility of the committee-recommended dose assessment and epidemiology studies and to estimate the required time and resources.

Recommendation 3: The epidemiology studies should include processes for involving and communicating with stakeholders. A plan for stakeholder engagement should be developed before the initiation of data gathering and analysis.

The NRC has engaged with the NAS to perform the Phase 1 recommendations and expects the pilot studies to be completed in 2015.

The NCI fact sheet on the original 1990 study is available at http://www.cancer.gov/cancertopics/factsheet/Risk/nuclear-facilities.

The press release on NRC's request to NAS is available at http://www.nrc.gov/reading-rm/doc-collections/news/2010/10-060.html.

For More Information
Contact Terry Brock, RES/DSA, at Terry.Brock@nrc.gov.

Report to Congress on Abnormal Occurrences

Background

Section 208 of the Energy Reorganization Act of 1974 defines an abnormal occurrence (AO) as an unscheduled incident or event that the U.S. Nuclear Regulatory Commission (NRC) determines to be significant from the standpoint of public health or safety.

The Federal Reports Elimination and Sunset Act of 1995 (Pub. L. No. 104-66) requires the NRC to report AOs to Congress annually. The NRC initially issued the AO criteria in a policy statement published in the *Federal Register* on February 24, 1977 (42 FR 10950); several revisions followed in subsequent years.

The NRC published its most recent revision to the AO criteria in the *Federal Register* on October 12, 2006 (71 FR 60198); it took effect on October 1, 2007.

Approach

The AO process helps to identify deficiencies in the NRC's regulatory process and ensure that corrective actions are taken to prevent recurrence. An accident or event is considered an AO if it involves a major reduction in the degree of protection of public health or safety. This type of incident or event would have a moderate or more severe impact on public health or safety and could include, but need not be limited to, the following:

- Moderate exposure to, or release of, radioactive material that the Commission licenses or otherwise regulates

- Major degradation of essential safety-related equipment

- Major deficiencies in design, construction, use of, or management controls for facilities or radioactive material that the Commission licenses or otherwise regulates

Application

When an incident or event occurs, the NRC uses a generic event assessment process to assess it. This generic event assessment process includes the following actions:

- Internal coordination with NRC offices.

- Systematic review of the cause of the event.

- Followup with the reporting licensee.

- Outreach to external stakeholders, as appropriate.

- Communication of lessons learned.

Examples of AO Events

Medical Event at Presbyterian Hospital of Dallas Involving Gamma Stereotactic Radiosurgery Unit for Trigeminal Neuralgia

Presbyterian Hospital of Dallas notified the NRC of a medical event that occurred during a gamma stereotactic radiosurgery unit ("gamma knife"; see Figure 4.6) treatment for trigeminal neuralgia. The procedure prescribed the use of radiation from the cobalt-60 source to treat the patient's fifth intracranial (trigeminal) nerve. An error in entry of information into the treatment planning system caused the wrong nerve to receive treatment (the seventh instead of the fifth intracranial nerve; see Figure 4.7).

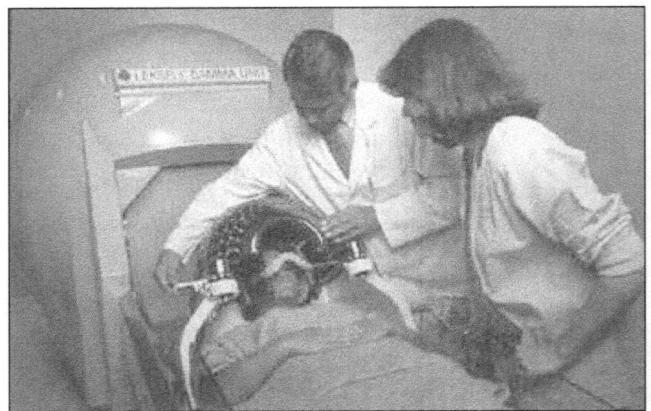

Figure 4.6 Gamma stereotactic radiosurgery unit (gamma knife) (Source US NRC TTC-TN)

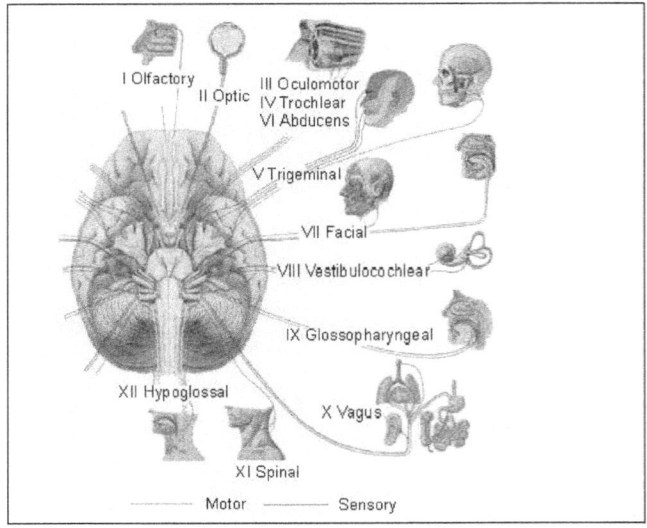

Figure 4.7 Diagram of the cranial nerves

Medical Event at Cancer Care Northwest PET Center Involving Treatment for Prostate Cancer

Cancer Care Northwest PET Center notified the NRC of a medical event that occurred with a high dose rate brachytherapy treatment (Figure 4.8) for prostate cancer containing iridium-192. During patient treatment, the aluminum connector to one of the needles (Figure 4.9) became detached from the plastic guide tube and a dose was delivered to a small area of the patient's inner thigh, which was the wrong treatment site.

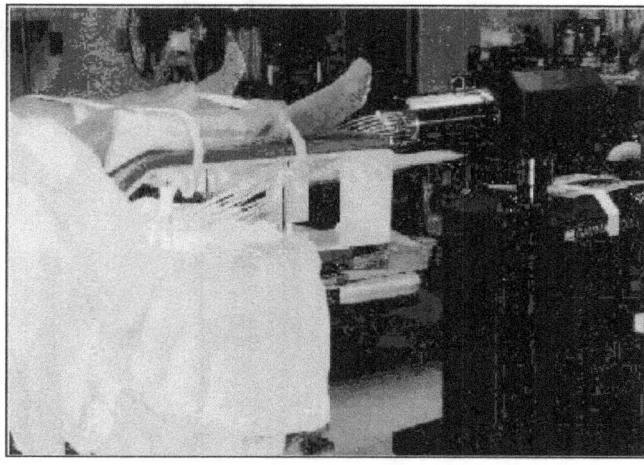

Figure 4.8 High dose rate prostate brachytherapy (Source: U.S. NRC TTC-TN)

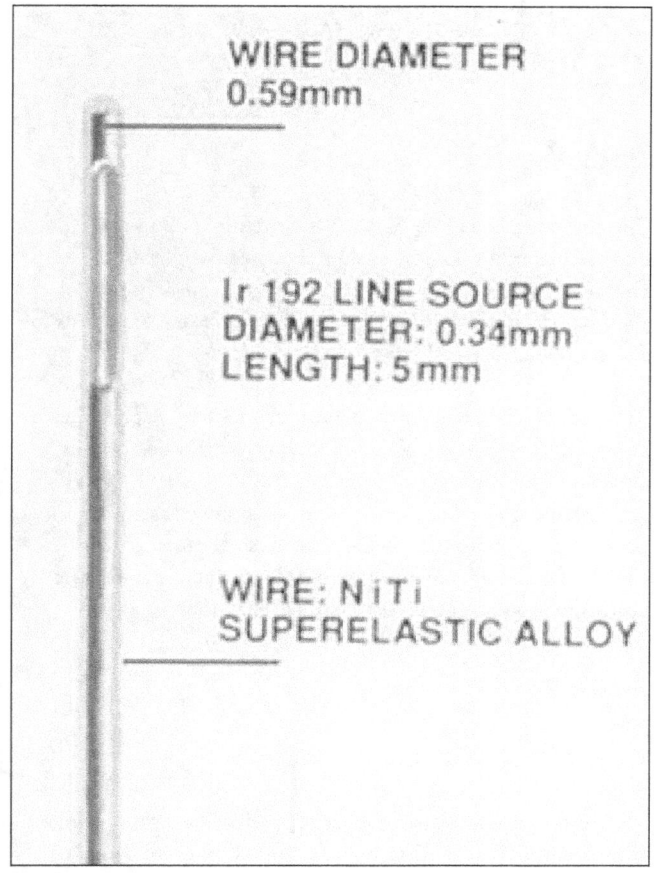

WIRE DIAMETER
0.59mm

Ir 192 LINE SOURCE
DIAMETER: 0.34mm
LENGTH: 5mm

WIRE: NiTi
SUPERELASTIC ALLOY

Figure 4.9 Diagram of iridium-192 wire

For More Information

Contact John Tomon, RES/DSA, at John.Tomon@nrc.gov.

VARSKIN Skin Computer Code

Background

The computer code VARSKIN 3 is currently used to model and calculate skin dose from skin or protective clothing contamination for regulatory requirements under Title 10 of the *Code of Federal Regulations* (10 CFR) Part 20, "Standards for Protection against Radiation."

The U.S. Nuclear Regulatory Commission (NRC) sponsored the development of the VARSKIN code to assist licensees in demonstrating compliance with 10 CFR 20.1201(c). This regulation requires licensees to have an approved radiation protection program that includes established protocols for calculating and documenting the dose attributable to radioactive contamination of the skin.

Approach

Original VARSKIN code

The initial version of the code, developed in the 1980s, fulfilled the regulatory requirement but was limited to point sources or infinitely thin disk sources directly on the skin. Soon after the initial release of VARSKIN, the industry encountered a new type of skin contaminant, which consisted of discrete microscopic radioactive particles, called "hot particles."

These hot particles differ radically from uniform skin contamination. They have a thickness and many of the exposures result from particles on the outside of protective clothing. Therefore, the code required further modifications.

VARSKIN Mod 2

VARSKIN Mod 2, developed in the early 1990s, significantly enhanced the code by adding the ability to model three-dimensional sources (cylinders, spheres, and slabs) with materials placed between the source and skin (including air gaps that attenuate the beta particles).

The code also modeled hot particle photon doses in certain cases. In addition, VARSKIN Mod 2 incorporated a user interface that greatly simplified data entry and increased efficiency in calculating skin dose.

VARSKIN 3

VARSKIN 3, released in 2004, operates in a Microsoft Windows environment and is designed to be significantly easier to learn and use than VARSKIN Mod 2.

In addition, this release enables users to calculate the skin dose (from both beta and gamma sources) attributable to radioactive contamination of skin or protective clothing.

The code also offers the ability to compute the dose at any skin depth or skin volume, with point, disk, cylindrical, spherical, or slab (rectangular) sources. It even enables users to compute doses from multiple sources. Figure 4.10 shows a typical VARSKIN 3 input screen for point source geometry.

Figure 4.10 Point source geometry screen

The input data file was also modified for VARSKIN 3 to reflect current physical data, to include the dose contribution from internal conversion and Auger electrons, and to allow a correction for low-energy electrons.

Current Status

Since the release of VARSKIN 3, the NRC staff has compared its dose calculations, for various energies and at various skin depths, with doses calculated by the Monte Carlo N-Particle Transport Code System (MCNP) developed at Los Alamos National Laboratory. The comparison shows that VARSKIN 3 overestimates the dose with increasing photon energy.

As such, the NRC is currently sponsoring further enhancement of the code to replace the existing photon dose algorithm and to develop quality assurance methods for this model.

Upgrades to VARSKIN will include the following:

• An enhanced photon dosimetry model based on Monte Carlo simulations of hot-particle contamination.

- Mathematical formulations rather than lookup tables to drive the dose estimation.

- Dose averaging to provide efficient convergence of the solution.

- Incorporation of parameters for energy, attenuation, dose-averaging area, and air gap.

- Protective clothing thickness and simple volumetric sources.

Code developers also have addressed deficiencies in the current code by creating the capability to calculate dose while accounting for attenuation and correcting the assumption that used the same effective-thickness for all materials.

Future Updates

- Correct technical issues with the beta dose model that code users report.

- Develop a quality assurance program for the beta dose model.

- Develop a training module for using the code.

VARSKIN 3 is available from the Radiation Safety Information Computational Center. For additional information, see NUREG/CR-6918, "VARSKIN 3: A Computer Code for Assessing Skin Dose from Skin Dose Contamination," issued October 2006. This document can be found in the NRC's Agencywide Documents Access and Management System at Accession No. ML063320348.

For More Information
Contact Mohammad Saba, RES/DSA, at
Mohammad.Saba@nrc.gov.

Phantom with Moving Arms and Legs (PIMAL)

Background

Modeling scenarios of radiation exposure to the human body, either internal or external, requires an extensive knowledge of fundamental particle physics and complex radionuclide biokinetics. To aid the U.S. Nuclear Regulatory Commission (NRC) staff in developing exposure models and performing the necessary dosimetry calculations for an individual, the Office of Nuclear Regulatory Research (RES) has developed humanoid phantom models ("phantoms with moving arms and legs," or PIMAL) that are now considered essential tools for radiation dose assessment.

Approach

RES partnered with Oak Ridge National Laboratory to develop two types of phantoms important for radiation dose assessment. One type, often referred to as a mathematical phantom, is an update of the Medical Internal Radiation Dose (MIRD) phantom with the substitution of movable arms and legs for the fixed ones in MIRD. Development of this updated phantom, with a graphical user interface (GUI), is near completion. A second type, now under consideration, will be a hybrid with the same arms and legs as those on the mathematical phantom but with a voxelized torso. The GUI is used to graphically set the arms and legs to the desired orientation, develop the Monte Carlo NParticle Transport Code (MCNP) input file, and display a table of organ doses and effective dose at the end of the run. The GUI allows users with only basic MCNP user skills to perform calculations using the phantom.

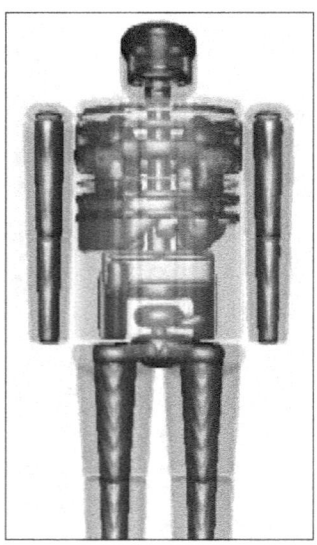

Mathematical phantoms define the external surfaces of the body and its internal organs and tissues. The main disadvantage of this approach is that the organ shapes are necessarily stylized and are therefore only approximate representations of actual organs. This is adequate for most dose calculations. Figure 4.11 graphically represents the male mathematical phantom.

Figure 4.11 Mathematically-based human phantom with articulating arms and legs

In the case of voxelized phantoms, the organs are defined by individual volumetric elements or pixels, referred to as voxels, which, depending on the resolution, may each measure a few millimeters on a side. Each part of the body is defined by an identifiable group of these pixels. Millions of pixels are required to compose a single humanoid model.

Voxel-based phantoms are excellent in applications in which extremely accurate dosimetry is needed. However, their complexity makes them computationally expensive to execute. Figure 4.12 graphically represents the male voxel phantom.

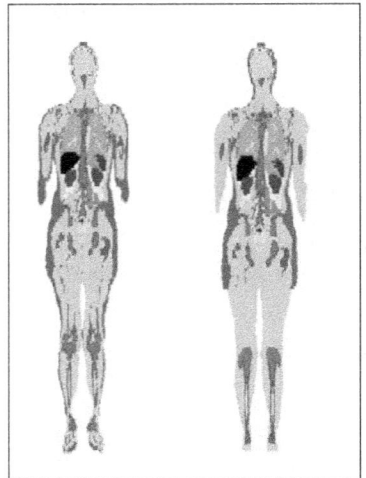

Figure 4.12 Voxel-based phantom with articulating arms and legs

The NRC staff experience using PIMAL has clearly demonstrated that state-of-the-art phantoms (Figure 4.13) and a user-friendly GUI greatly ease the burden of setting up and executing a radiation transport problem and retrieving the dosimetry results.

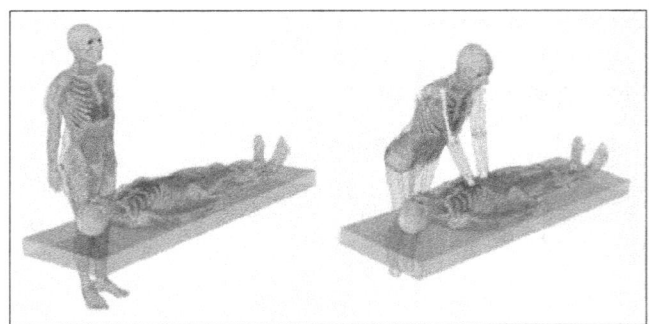

Figure 4.13 Geometrical setting for patient-physician modeling with a realistic posture (right) using PIMAL

Current Status

Work has been completed to update the 1974 MIRD5 phantom and original PIMAL GUI. Additional work to develop new mathematical and voxel-based phantoms is ongoing. An example of ongoing work is implementation of ICRP Publication103, "Recommendations of the International

Commission on Radiological Protection" (issued February 2008) tissue weighting factors, in addition to those from ICRP Publication 26, Recommendations of the International Commission on Radiological Protection" (issued in 1977) and ICRP Publication 30, "Limits for Intakes of Radionuclides by Workers"(issued in 1979) into the GUI to calculate the effective dose. The next phase of this project is to test the phantoms in a variety of exposure situations and incorporate improvements and additions to increase their utility and ease of use. Beyond this stage, the NRC will consider whether conversion to hybrid male and female voxel phantoms would be a significant addition to the MIRD set. One important advantage of such hybrids would be that they can be designed to the same specifications as the recently adopted ICRP male and female voxel phantoms, which then serve as benchmarks for the NRC's phantoms and calculations.

For More Information

Contact Mohammad Saba, RES/DSA, at Mohammad.Saba@nrc.gov.

Radionuclide Transport, Removal, and Dose (RADTRAD)

Background

The potential radiological consequences of nuclear power reactor accidents depend in part on the amount, form, and species of the radioactive material released during the postulated accident. The Radionuclide Transport, Removal, and Dose (RADTRAD) model estimates doses at the exclusion area boundary or the low population zone, and also in the control room and other locations. As radioactive material is transported through the containment, the user can account for sprays and natural deposition that may reduce the quantity of radioactive material. Material can flow between buildings, from buildings to the environment, or into control rooms through high efficiency particulate air filters, piping, or other connectors. Decay and ingrowth of daughters can be calculated over time as the material is transported. The RADTRAD code is a useful tool for supporting plant licensing reviews and other activities that require dose analysis.

Approach

Sandia National Laboratories developed the RADTRAD code for the U.S. Nuclear Regulatory Commission (NRC). The NRC uses the RADTRAD code to estimate offsite consequences from radioactive material released into the atmosphere. The RADTRAD code can be used to estimate the containment release using either NRC Technical Information Document (TID)-14844, "Calculation of Distance Factors for Power and Test Reactor Sites," or NUREG-1465, "Accident Source Terms for Light-Water Nuclear Power Plants," source terms and assumptions, or a user-specified source term. In addition, the code can account for a reduction in the quantity of radioactive material due to containment sprays, natural deposition, filters, and other natural and engineered safety features. The RADTRAD code uses a combination of tables and numerical models of source term reduction phenomena to determine the time-dependent dose at user-specified locations for a given accident scenario. The RADTRAD code can be used to assess occupational radiation exposures, typically in the control room; to estimate site boundary doses; and to estimate dose attenuation because of modification of a facility or accident sequence.

RADTRAD Version 3.03

In 1999, the code was revised to include a Microsoft Visual Basic graphical user interface (GUI) for user convenience, which is described in Supplement 1 to NUREG/CR-6604, "RADTRAD:

A Simplified Model for RADionuclide Transport and Removal And Dose Estimation." A second supplement to NUREG/CR-6604 was published in 2002, which discussed testing of RADTRAD Version 3.03.

RADTRAD Version 4.0

To improve RADTRAD's maintainability, remove platform and compiler dependencies, and add new features, RADTRAD was re-implemented in the JAVA language. This JAVA-based version of RADTRAD was named Version 4.0. In addition, the Microsoft Visual Basic GUI was replaced with the Symbolic Nuclear Analysis Package (SNAP) GUI (see Figure 4.14). SNAP uses a plugin based architecture that "wraps" all of the interfaces to an analytical code in a special file called a "SNAP plug-in". Placing RADTRAD in the SNAP framework allows for the use of SNAP features, including the Model Editor for developing plant models, and provides tools for user input checking and monitoring calculations. In 2011, the Office of Nuclear Regulatory Research (RES) performed verification testing of RADTRAD Version 4.0 and the SNAP-RADTRAD plug-in. The calculations were independently verified using MATHCAD, which is an engineering calculation software package that represents equations and data in a user readable format.

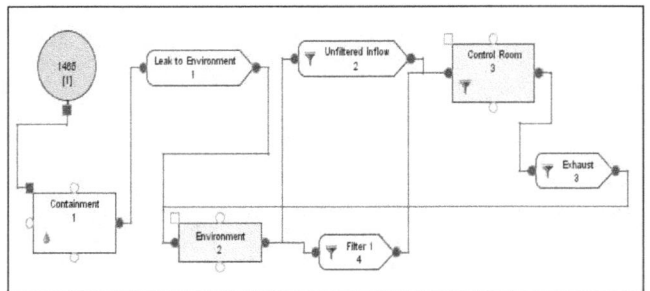

Figure 4.14 Creating RADTRAD input model using SNAP

Future Updates

Currently, RES staff is evaluating a version of RADTRAD 4.4.x (and its corresponding SNAP GUI) and efforts are directed toward finalizing the code and documentation. The RADTRAD User group has provided feedback and identified various problems in the SNAP version of RADTRAD. This has led to improvements in the code and interface. Additionally, a radionuclide editor tool has been added, which enables users to edit the source term. Finally, future work includes improving the SNAP-RADTRAD plotting package, in which data output files suitable for plotting with the APTplot plotting package are generated.

For More Information
Contact John Tomon, RES/DSA, at John.Tomon@nrc.gov.

Radiation Worker Health Studies

Background

The U.S. Nuclear Regulatory Commission (NRC) has entered into an interagency agreement with the U.S. Department of Energy (DOE) Office of Science (SC) Low Dose Radiation Research Program to study the health effects of more than 1 million radiation workers (see Figure 4.15) and atomic veterans. Supporting DOE and this multiagency effort will provide valuable new information for future radiation protection standards—setting bodies and any resultant occupational radiation dose standards.

The significance of the proposed research is considerable because it applies directly to existing concerns about standards for chronic radiation exposure. Much knowledge has been gained from the study of atomic bomb survivors, but exposure was acute and among a Japanese population living in a war-torn country. Scientific and medical committees continue to grapple with how best to estimate risks associated with the gradual exposures received from environmental, medical, and occupational radiation. Recent studies, though limited, have suggested that chronic exposures may be more hazardous than currently accepted. Governmental agencies must deal with the complex issues of compensating prior workers, veterans, and citizens who may have been potentially harmed by past exposures. Protection committees deliberate over how best to estimate and apply a "dose and dose rate effectiveness factor" to scale the risks from the A-bomb survivor data for relevant and current circumstances. Evaluation of risk among persons with intakes of radioactive substances assumes greater importance as society debates expansion of nuclear energy and deals with nuclear waste and threats of terrorist attacks with nuclear devices. The remarkable increase in population medical exposures to CT scans and other imaging technologies has raised concerns about future health consequences.

Approach

A unique opportunity exists to assemble more than 1 million radiation workers and military veterans, follow them to the present, calculate rates of mortality from cancer and other diseases, reconstruct radiation doses received (including doses from inhaled or ingested radionuclides), and provide new and essential knowledge on the level of lifetime risk from low-level ionizing radiation experienced chronically and beginning more than 60 years ago. The methodology will follow the state-of-the-art approach recently used in studying cancer and other diseases among Rocketdyne radiation workers.

Figure 4.15 Radiation worker taking measurements

Collaborating institutions: U.S. National Council on Radiation Protection and Measurements (lead), Vanderbilt University, Oak Ridge Associated Universities, Oak Ridge National Laboratory, Los Alamos National Laboratory, Landauer, Inc., Risk Assessment Corporation, Harvard Medical School, University of Southern California, and the International Epidemiology Institute.

Collaborating or cooperating agencies: DOE, U.S. Department of Defense, U.S. Department of Veteran Affairs, National Cancer Institute, and National Aeronautics and Space Administration.

Study Status

Research began on the nuclear power worker cohorts in fall 2012. The NRC expects to see the results in late 2014. The study Web page can be accessed at http://www.onemillionworkerstudy.org/.

For More Information
Contact Terry Brock, RES/DSA at Terry.Brock@nrc.gov.

Radiological Toolbox

Background

The U.S. Nuclear Regulatory Commission (NRC), in conjunction with Oak Ridge National Laboratory, developed the Radiological Toolbox (hereafter referred to as the "Rad Toolbox" or "toolbox") as a way to access information frequently needed for radiation safety and control, shielding, and dosimetry calculations.

The toolbox is essentially an electronic or digital handbook. It contains scientific and engineering data of interest to health physicists, radiologists, radiological engineers, and others working in fields involving radiation and the use of radioactive materials. Examples of data contained in the toolbox include the following:

- Nuclear transformation and decay data.

- Biokinetic and biological data.

- Internal and external dose coefficients.

- Elemental composition of many materials.

- Radiation properties and interaction coefficients.

- Kerma and stopping power coefficients.

- Transport package regulations A1/A2 table.

- Radiological risk coefficients.

The toolbox operates in Microsoft Windows 7 environments. The output data can be easily extracted or moved as ASCII files to other programs used in calculations.

Approach

The Rad Toolbox is a computer application that provides access to physical, chemical, anatomical, physiological, and mathematical data (and models) relevant to the protection of workers and the public from exposures to ionizing radiation. For the most part, the outputs of Rad Toolbox databases could be customized and obtained in International System (SI) units by users. The toolbox features (see Figure 4.16 and 4.17) many computational capabilities based on the following documentation:

- Nuclear decay data—ICRP Publication 107, "Nuclear Data for Dosimetric Calculation" (ICRP 2008) and the Japan Atomic Energy Research Institute (Endo 2001)

- Dose coefficients for photon and neutron fields—ICRP Publication 74, "Conversion Coefficients for Use in Radiological Protection against External Radiation" (ICRP 1996b).

- Organ masses values for member of the public—ICRP Publication 72, "Age-Dependent Doses to the Members of the Public from Intake of Radionuclides, Part 5, Compilation of Ingestion and Inhalation Coefficients" (ICRP 1996a).

- Reference values—ICRP Publication 89, "Basic Anatomical and Physiological Data for Use in Radiological Protection: Reference Values" (ICRP 2002).

- Radiation workers—ICRP Publications 30, "Limits for Intakes of Radionuclides by Workers, Part 3," and 68, "Dose Coefficients for Intakes of Radionuclides by Workers" (ICRP 1981, 1994).

- External irradiation—Federal Guidance Report 12, "External Exposure to Radionuclides in Air, Water, and Soil" (U.S. Environmental Protection Agency, 1993).

- Radiological cancer risk—Federal Guidance Report 13, "Cancer Risk Coefficients for Environmental Exposure to Radionuclide" (U.S. Environmental Protection Agency, 1999).

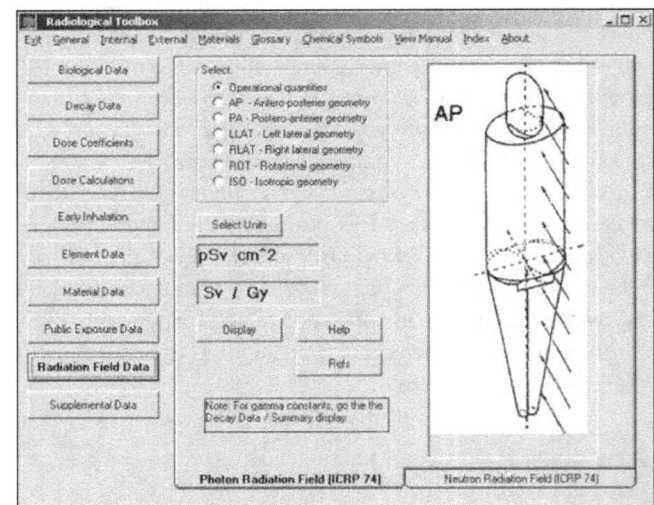

Figure 4.16 Radiological Toolbox

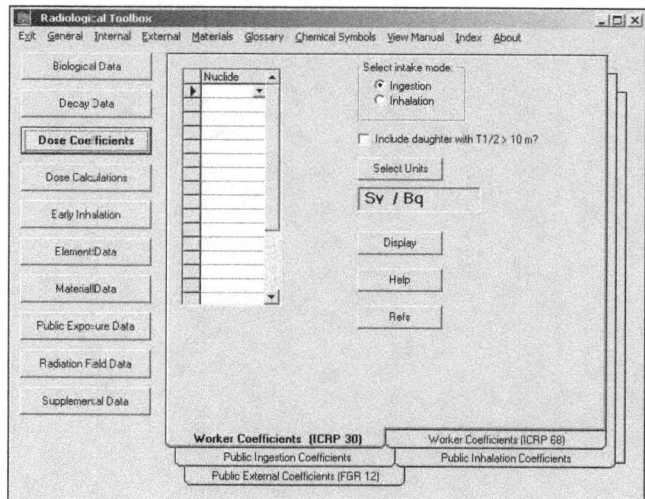

Figure 4.17 Radiological Toolbox graphical user interface

In addition, the computational modules are capable of calculating both equivalent and effective doses for dose and risk assessments and to generate radiation interaction coefficients based on user-specified materials and compositions. The software's help files provide access to textual information on topics ranging from general information to the details of mathematical models describing the translocation rates and fate of radionuclides in the body.

Toolbox Content

When the toolbox is initiated, a user screen appears. The menu bar at the top of the screen allows access to the software help files in addition to other standard functions.

The menu bar at the left of the screen allows access to all data elements included in the toolbox.

For example, the "Dose Coefficients" section of the toolbox provides access to the following sets of nuclide-specific dose coefficients:

- External dose rate coefficients for 826 radionuclides from Federal Guidance Report 12 (U.S. Environmental Protection Agency, 1993).

- Committed dose coefficients for inhalation and ingestion intakes of 738 radionuclides by workers from ICRP Publications 30 and 68 (ICRP 1978, 1994).

- Age-dependent committed dose coefficients for the inhalation and ingestion intakes of 738 radionuclides by members of the public (six ages at intake) from ICRP Publication 72 (ICRP 1996a).

For each set of coefficients, it is possible to display up to 20 nuclides at a time for a chosen route of exposure or intake.

Future Updates

Further revisions of the toolbox are planned as the NRC staff and other users identify the need for additional data.

The program and user manual can be downloaded from the NRC's public Web site at http://www.nrc.gov/about-nrc/regulatory/research/radiological-toolbox.html.

For More Information
Contact Casper Sun, RES/DSA, at Casper.Sun@nrc.gov.

References

Endo, A., and Y. Yamaguchi. *Compilation of Nuclear Decay Data Used for Dose Calculation Revised Data for Radionuclides Listed in ICRP Publication 38*, JAERI-Data/Code 2001-004. 2001.

International Commission on Radiological Protection (ICRP). *Limits for Intakes of Radionuclides by Workers*. ICRP Publication 30 *(Part 1, 2, 3, and 4)*. Annals of the ICRP. Pergamon Press, New York, NY, 1979-1988.

International Commission on Radiological Protection (ICRP). *Dose Coefficients for Intakes of Radionuclides by Workers*. ICRP Publication 68. Annals of the ICRP 24(4). International Commission on Radiological Protection, Pergamon Press, New York, NY, 1994.

International Commission on Radiological Protection (ICRP). *Age-dependent Doses to the Members of the Public from Intake of Radionuclides: Part 5 Compilation of Ingestion and Inhalation Coefficients*. ICRP Publication 72. Annals of the ICRP 26(1), Pergamon Press, New York, NY, 1996a.

International Commission on Radiological Protection (ICRP). *Conversion Coefficients for Use in Radiological Protection against External Radiation*. ICRP Publication 74. Annals of the ICRP 26(3/4), Pergamon Press, New York, 1996b.

International Commission on Radiological Protection (ICRP). *Basic Anatomical and Physiological Data for Use in Radiological Protection: Reference Values*. ICRP Publication 89. Annals of the ICRP 32(2/4), Pergamon Press, New York, 2002.

International Commission on Radiological Protection (ICRP). *Nuclear Data for Dosimetric Calculation. International Commission on Radiological Protection. ICRP Publication 107. Annals of the ICRP 38(3), Pergamon Press, New York, 2008.*

U.S. Environmental Protection Agency (EPA). External Exposure to Radionuclides in Air, Water, and Soil. U.S. Environmental Protection Agency, Federal Guidance Report 12, Washington, DC, 1993.

U.S. Environmental Protection Agency (EPA. Cancer Risk Coefficients for Environmental Exposure to Radionuclide. U.S. Environmental Protection Agency, Federal Guidance Report 13, Washington, DC, 1999.

Participation in National and International Radiation Protection Activities

Introduction

One of the benefits of the Radiation Protection Branch (RPB) program is the promotion of consistency and coherence in regulatory applications of radiation protection and health effects research among U.S. Nuclear Regulatory Commission (NRC) programs, as well as those of other Federal and State regulatory agencies. To that end, RPB staff is actively engaged in monitoring and participating in the influential organizations described below.

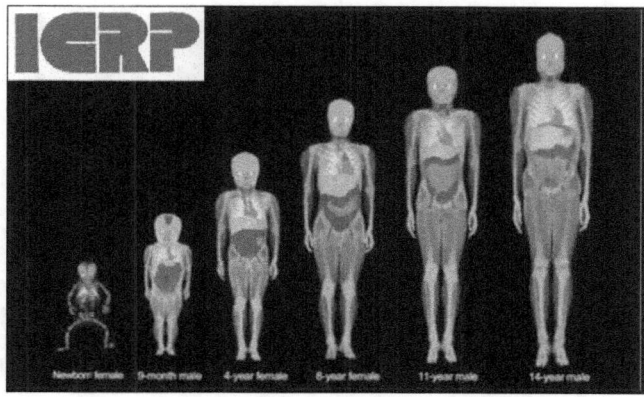

Figure 4.18 Bioklentic model

Participation

ICRP—International Commission on Radiological Protection (Terry Brock, PhD)

ICRP is an independent registered charity, established to advance the science of radiological protection for the public benefit, in particular by providing recommendations and guidance on all aspects of protection against ionizing radiation (see Figure 4.18). RPB collaborates with ICRP and stakeholders to ensure consistency in the application of radiation protection standards and dosimetry modeling.

NCRP—National Council on Radiation Protection and Measurements (Terry Brock, PhD)

The NCRP seeks to formulate and disseminate information, guidance, and recommendations on radiation protection and measurements that represent the consensus of leading scientific thinking. The Council seeks out areas in which the development

and publication of NCRP materials can make an important contribution to the public interest (see Figure 4.19).

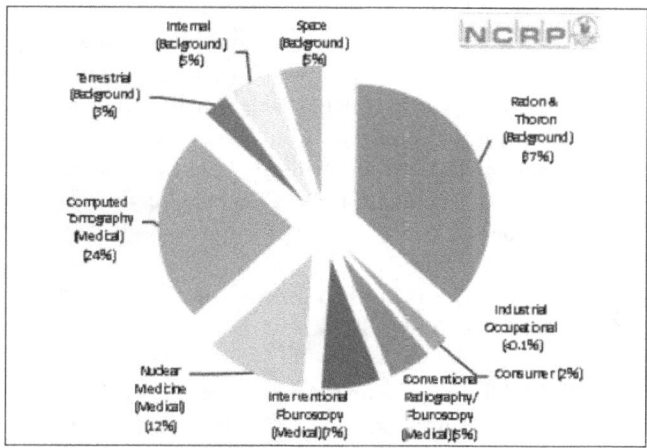

Figure 4.19 Background radiation exposure (Source: NRCP Report No. 160)

CRPPH—NEA Committee on Radiation Protection and Public Health (Stephanie Bush-Goddard, PhD)

The NEA's CRPPH is a valuable resource for its member countries. The committee is made up of regulators and radiation protection experts with the broad mission of providing timely identification of new and emerging issues, analyzing their possible implications, and recommending or taking action to address these issues to further enhance radiation protection regulation and implementation. The RPB participates in the regulatory and operational consensus developed by the CRPPH on these emerging issues, supports policy and regulation development in member countries, and disseminates good practice.

ISOE—Information System on Occupational Exposure (Doris Lewis)

ISOE was created in 1992 to provide a forum for radiation protection professionals from nuclear electricity utilities and national regulatory authorities worldwide to share dose reduction information, operational experience, and information to improve the optimization of radiological protection at nuclear power plants. The Radiation Exposure Information and Reporting System (REIRS) that RPB manages provides exposure information on domestic occupational workers for an increasingly international and global market (see Figure 4.20).

Figure 4.20 Occupational radiation exposure workers

IRSN—Institut De Radioprotection Et De Sûreté Nucléaire (Tony Huffert, CHP)

IRSN is a French public authority that conducts industrial and commercial activities. It is under the joint authority of the Ministry for Ecology, Energy, Sustainable Development and Town and Country Planning, the Ministry for the Economy, Industry and Employment, the Ministry for Higher Education and Research, the Ministry of Defense, and the Ministry for Health and Sports.

JCCRER—Joint Coordinating Committee For Radiation Effects Research (Terry Brock, PhD)

JCCRER is a bilateral government committee representing agencies from the United States and the Russian Federation tasked with coordinating scientific research on the health effects of exposure to ionizing radiation in the Russian Federation from the production of nuclear weapons. Jointly conducting radiation research with the Russian Federation provides a unique opportunity to learn more about possible risks to groups of people from long-term exposure to radiation. The RPB representative serves on the JCCRER Executive Committee, which is tasked with ensuring direct communication among the partners within the Agreement, coordinating the work of national organizations, and ensuring the effective and efficient implementation of JCCRER goals and objectives (see Figure 4.21).

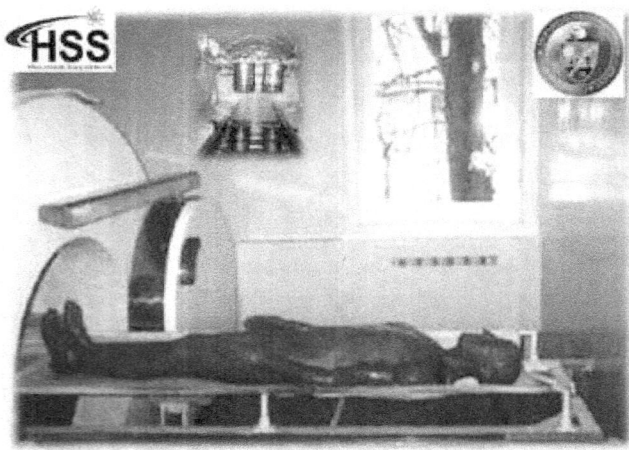

Figure 4.21 Russian Federation whole-body counting facility

For More Information
Contact Stephanie Bush-Goddard, RES/DSA, at
Stephanie.Bush-Goddard@nrc.gov.

Chapter 5: Risk Analysis

Full-Scope Site Level 3 Probabilistic Risk Assessment Project

Risk Assessment Standardization Project

Probabilistic Risk Assessment Quality and Standards

Evolutionary Methods and Models Development in Probabilistic Risk Assessment (PRA)

Treatment of PRA Uncertainties in Risk-Informed Decisionmaking

Glossary of Risk-Related Terms in Support of Risk-Informed Decisionmaking (NUREG-2122)

A Proposed Risk Management Regulatory Framework

Reactor Operating Experience Data Collection and Analysis

Accident Sequence Precursor Program

SPAR Model Development Program

SAPHIRE PRA Software Development Program

Thermal-Hydraulic Level 1 Probabilistic Risk Assessment (PRA) Success Criteria Activities

Risk-Informing Emergency Preparedness: Probabilistic Risk Analysis of Emergency Action Levels

Design-Basis Flood Determinations at Nuclear Power Plants

Assessment of Debris Accumulation on Emergency Core Cooling System Suction Strainer Performance

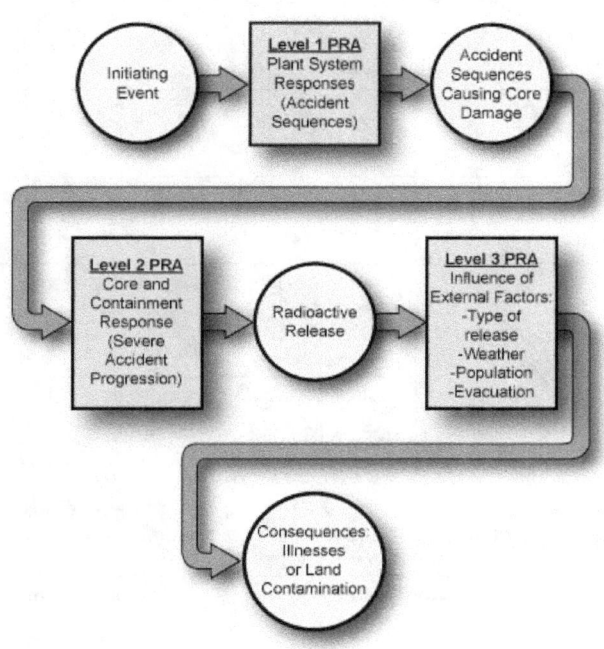

Full-Scope Site Level 3 Probabilistic Risk Assessment Project

Background

Risk and Probabilistic Risk Assessment

According to the traditional definition, risk is the product of the likelihood and consequences of an adverse event. Probabilistic risk assessment (PRA) is a systematic analysis tool consisting of specific technical elements that provide both qualitative insights and a quantitative assessment of risk by addressing the following questions, commonly referred to as the "risk triplet": (1) What can go wrong? (2) How likely is it? and (3) What are the consequences? Modern PRAs also have incorporated uncertainty analyses to address a fourth question: How confident are we in our answers to these three questions? In this way, PRAs allow the identification, prioritization, and mitigation of significant contributors to risk to improve nuclear power plant safety.

PRAs for nuclear power plants can vary in scope, depending on their intended use. The scope of a PRA is defined by the degree of coverage of the following five factors: (1) radiological hazards, (2) population exposed to hazards, (3) plant operating states, (4) initiating event hazards, and (5) level of risk characterization. Figure 5.1 summarizes the various scoping options for each factor.

Factor	Scoping Options for Operating Nuclear Power Plants
Radiological hazards	Reactor core Spent fuel Other Radioactive Sources
Population exposed to hazards	Onsite population Offsite population
Plant operating states	At-Power Low Power/Shutdown
Initiating event hazards	Traditional internal events (transients, loss-of-coolant accidents) Internal floods Internal fires
	Seismic events (earthquakes) High winds Other external hazards
Level of risk characterization	Level 1 PRA: Core damage frequency Level 2 PRA: e.g., Large early release Level 3 PRA: Early fatality risk Latent cancer fatality risk

Figure 5.1 Factors affecting the scope of PRAs for operating nuclear power plants

The Importance of Level 3 PRA

Figure 5.2 illustrates that PRAs for nuclear power plants can estimate risk measures at three different levels of characterization using sequential analyses in which the output from one level serves as a conditional input to the next. Using event trees and fault trees, a Level 1 PRA models various plant and operator responses to initiating events that challenge plant operation to identify accident sequences that result in reactor core damage. The estimated frequencies for all core damage accident sequences are summed to calculate the total core damage frequency (CDF) for the analyzed plant.

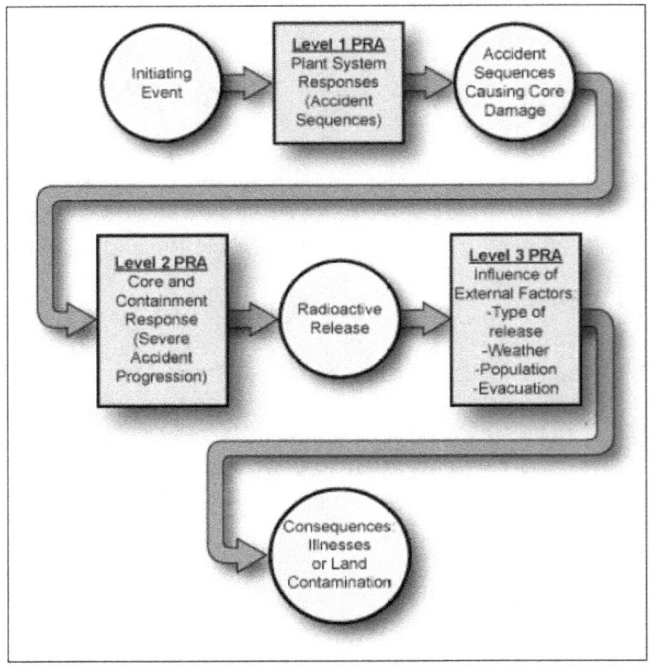

Figure 5.2 Three sequential levels of risk analysis in PRAs for nuclear power plants (Source: www.nrc.gov)

A Level 2 PRA models and analyzes the progression of "severe accidents"—those Level 1 PRA accident sequences that result in reactor core damage—by considering how the reactor coolant and other relevant systems respond, as well as how the containment responds to the accident. This analysis is based on both the initial status of structures and systems and their ability to withstand the harsh accident environment. Once the system and containment response is characterized, the frequency, type, amount, timing, and energy content of the radioactivity released to the environment—also known as source term characteristics—can be determined.

A Level 3 PRA models the release and transport of radioactive material in a severe accident and estimates the health and economic impact in terms of the following offsite consequence measures: (1) early fatalities and injuries, and latent cancer fatalities resulting from the radiation doses to the surrounding population, and (2) economic costs associated with evacuation, relocation, property loss, and decontamination. Offsite consequences are estimated based on the Level 2 PRA source term characteristics and on several other factors affecting the transport and impact of the radioactive material, including meteorology, demographics, emergency response, and land use. Combining the results of the Level 1 and Level 2 PRAs with the results of this consequence analysis, only the Level 3 PRA estimates the integrated risk (likelihood times consequences) to the public for the analyzed nuclear power plant. In fact, only a Level 3 PRA can estimate the two high-level quantitative health objectives related to early and latent cancer fatality risks that the U.S. Nuclear Regulatory Commission (NRC) identified in a 1986 safety goal policy statement on determining what level of risk is acceptable to ensure adequate protection of public health and safety.

NUREG-1150: A Landmark Study

Although Level 3 PRAs are required to directly estimate the risk to the public from nuclear power plant accidents, the NRC does not routinely use them in risk-informed regulation. In fact, NRC-sponsored Level 3 PRAs have not been conducted since the late 1980s—over 20 years ago. These Level 3 PRAs were documented in a collection of NUREG/CR reports and a single corresponding summary document, NUREG-1150, "Severe Accident Risks: An Assessment for Five U.S. Nuclear Power Plants," dated December 1990. NUREG-1150 provides a set of PRA models and a snapshot-in-time (circa 1988) assessment of the severe accident risks associated with five commercial nuclear power plants of different reactor and containment designs. The NRC has used the landmark NUREG-1150 results and perspectives in a variety of regulatory applications, including development of PRA policy statements, support of risk-informed rulemaking, prioritization of generic issues and research, and establishment of numerical risk acceptance guidelines for the use of CDF and large early-release frequency (LERF) as surrogate risk metrics for early and latent cancer fatality risks.

Since then, the NRC has ensured safety primarily by using results obtained from Level 1 and limited Level 2 PRAs—both less expensive than Level 3 PRAs—and how they relate to lower-level subsidiary safety goals based on CDF and LERF to risk-inform regulatory decisionmaking.

The Need for a New Comprehensive Site Level 3 PRA

There are several compelling reasons for conducting a new comprehensive site Level 3 PRA. First, in the two decades since the publication of NUREG-1150, there have been substantial developments that may affect the results and risk perspectives that have influenced many regulatory applications. In addition

to risk-informed regulations implemented to improve safety (e.g., the Station Blackout and Maintenance Rules), there have been plant modifications that may affect risk (e.g., the addition or improvement of plant safety systems, changes to technical specifications, power uprates, and the development of improved accident management strategies). Along with NRC and industry acquisition of over 20 years of operating experience, there have also been significant advances in PRA methods, models, tools, and data—collectively referred to as "PRA technology"—and in information technology. Finally, the state-of-the-art Reactor Consequence Analysis (SOARCA) study, which leveraged many of the same safety improvements and technological advances, integrates and analyzes two of the essential technical elements of a Level 3 PRA for some of the more likely reactor accident sequences—the severe accident progression and offsite consequence analyses. A new Level 3 PRA could, therefore, seek to leverage the methods, models, and tools used in the SOARCA analysis and capitalize on the insights gained from the application of state-of-the-art practices.

In addition to these developments, the Level 3 PRAs documented in NUREG-1150 are incomplete in scope. Figure 5.3 illustrates the scope of a complete site accident risk analysis, with the approximate scope of the NUREG-1150 PRAs shown by the gray-shaded region. These PRAs were limited to the assessment of single-unit reactor accidents initiated primarily by internal events occurring during full-power operations. The partial coverage of external events indicates that a limited set of external events (fires and earthquakes) were considered for only two of the five analyzed nuclear power plants.

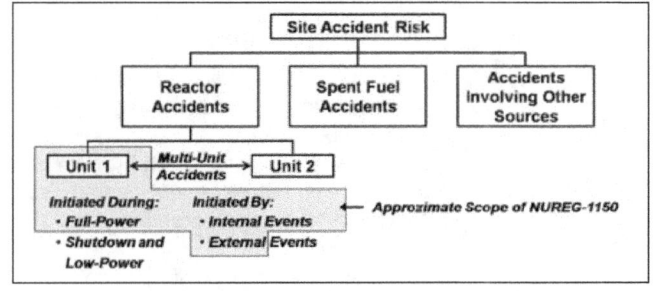

Figure 5.3 Site accident risk and approximate scope of NUREG-1150 (Source: Marty Stutzke)

During the Annual Commission Meeting on Research Programs, Performance, and Future Plans held on February 18, 2010, the staff proposed a scoping study to evaluate the feasibility of performing a new full-scope site Level 3 PRA for a nuclear power plant site. In a staff requirements memorandum (SRM)[1] on March 19, 2010, the Commission expressed conditional support for activities related to Level 3 PRA and directed the staff to provide the Commission with various options for proceeding with this work that included costs and perspectives on future

[1]   SRM M100218, "Staff Requirements—Briefing on Research Programs, Performance, and Future Plans," dated March 19, 2010 (ADAMS Accession No. ML100780578).

regulatory uses for Level 3 PRAs. On July 7, 2011, the NRC staff responded[2] to the SRM by providing three proposed options for proceeding with the Level 3 PRA development project. On September 21, 2011, the Commission directed[1] the staff to conduct a full-scope site Level 3 PRA, to be completed within 4 years. The scope of the Level 3 PRA study includes all major site radiological sources,[3] all internal and external initiating event hazards typically considered in previous internal and external event PRAs,[4] and all modes of plant operation.

Objectives

The full-scope site Level 3 PRA project includes the following objectives:

- Develop a Level 3 PRA, generally based on current state-of-practice methods, tools, and data,[5] that (1) reflects technical advances since completion of the NUREG-1150 studies, and (2) addresses scope considerations that were not previously considered (e.g., low power and shutdown,[6] multiunit risk, and spent fuel storage).

- Extract new risk insights to enhance regulatory decisionmaking and help focus limited agency resources on issues most directly related to the agency's mission to protect public health and safety.

- Enhance PRA staff capability and expertise and improve documentation practices to make PRA information more accessible, retrievable, and understandable.

- Obtain insight into the technical feasibility and cost of developing new Level 3 PRAs.

---

[2]   SECY-11-0089, "Options for Proceeding with Future Level 3 Probabilistic Risk Assessment Activities," dated July 7, 2011 (ADAMS Accession No. ML11090A039).
[3]   SRM-SECY-11-0089, "Staff Requirements—SECY-11-0089—Options for Proceeding with Future Level 3 Probabilistic Risk Assessment (PRA) Activities," dated September, 21, 2011 (ADAMS Accession No. ML112640419).
[4]   Including all reactor cores, spent fuel pools, and dry storage casks on site, but excluding fresh nuclear fuel, radiological waste, and minor radiological sources (e.g., calibration devices).
[5]   Deliberate malevolent acts (e.g., terrorism and sabotage) are specifically excluded from the scope of the study.
[6]   "State-of-practice" methods, tools, and data refer to those that are routinely used by the NRC and licensees or have acceptance in the PRA technical community.
[7]   While NUREG-1150 only addressed reactor operation at-power, the NRC subsequently sponsored two studies that addressed reactor risk for some low power and shutdown modes of operation (NUREG/CR-6143, "Evaluation of Potential Severe Accidents during Low Power and Shutdown Operations at Grand Gulf, Unit 1," July 1995, and NUREG/CR-6144, "Evaluation of Potential Severe Accidents during Low Power and Shutdown Operations at Surry, Unit 1," May 1995).

Approach

Consistent with the objectives of this project, the Level 3 PRA study will generally be based on current state-of-practice methods, tools, and data. However, there are several gaps in current PRA technology and other challenges that will require advancement in the PRA state-of-practice. The general approach to addressing these challenges for the Level 3 PRA study will be primarily to rely on existing research and the collective expertise of the NRC's senior technical advisors and contractors, and to perform limited new research only for a few specific technical areas (e.g., multiunit risk).

Based on a set of site selection criteria, and with the support of the utility, Southern Nuclear Operating Company's Vogtle Electric Generating Plant, Units 1 and 2[7,8] was selected as the volunteer site for the Level 3 PRA study. The Level 3 PRA project team will leverage the existing and available information on Vogtle and its licensee PRA, in addition to related research efforts (e.g., SOARCA), to enhance the efficiency in performing the study.

The Level 3 PRA project team plans to use the following NRC tools and models for performing the Level 3 PRA study:

- Systems Analysis Programs for Hands-on Integrated Reliability Evaluation (SAPHIRE), Version 8.

- MELCOR Severe Accident Analysis Code.

- MELCOR Accident Consequence Code System, Version 2 (MACCS2).

SAPHIRE is the NRC's standard software application for performing PRAs. This code was developed and is maintained by the NRC through contracts with Idaho National Laboratory. The latest version in use, SAPHIRE 8, has increased capability for handling large complex models and can be used to analyze both internal and external hazards and all plant operating states.

---

[8]   A. Marion, Nuclear Energy Institute, "Response to December 6, 2011 Letter Requesting Support in Identifying a Licensee Volunteer for the Full-Scope Site Level 3 Probabilistic Risk Assessment Study," February 14, 2012 (ADAMS Accession № ML12059A329).
[9]   Southern Nuclear Operating Company has received a combined construction and operating license for two additional nuclear reactors at the Vogtle site. The two new reactors are not within the scope of this study.

MELCOR is a fully integrated, engineering-level computer code whose primary purpose is to model the progression of postulated accidents in both light water reactors and in non-reactor systems such as spent fuel pools and dry storage casks. The MELCOR code routinely is used for performing thermal-hydraulic analysis to determine system success criteria and accident sequence timing and to inform severe accident progression analysis.

MACCS2 is a general-purpose tool used to evaluate the public health effects and economic costs of mitigation actions for severe accidents at diverse reactor and non-reactor facilities. The principal phenomena considered are atmospheric transport and deposition under time-variant meteorology, short- and long-term mitigation actions and exposure pathways, deterministic and stochastic health effects, and economic costs.

Both MACCS2 and MELCOR are maintained by the NRC through contracts with Sandia National Laboratory.

Besides the technical capabilities of these NRC tools and models, they offer the advantages that they are generally available, the staff is familiar with their use, and, if necessary, the staff has the ability to modify these tools. This latter advantage may be of particular importance in addressing such expanded scope items as multiunit risk, spent fuel pools, and dry storage casks.

For More Information
Contact Alan Kuritzky, RES/DRA at Alan.Kuritzky@nrc.gov

Risk Assessment Standardization Project

Background

In the U.S. Nuclear Regulatory Commission's (NRC's) Reactor Oversight Process, the NRC staff performs risk assessments of inspection findings and reactor incidents to determine their significance for appropriate regulatory response. Currently, several NRC groups are performing these risk assessments for Accident Sequence Precursor (ASP) and Significance Determination Process (SDP) Phase 3 analyses, and for Incident Investigation Program assessments under Management Directive (MD) 8.3, "NRC Incident Investigation Program," issued March 2001. Although different NRC programs have different objectives, they use the same risk tools—the Systems Analysis Programs for Hands-on Integrated Reliability Evaluation (SAPHIRE) code and Standardized Plant Analysis Risk (SPAR) models for performing risk assessments. Therefore, the NRC staff initiated the Risk Assessment Standardization Project (RASP) to establish standard procedures, improve the methods, and enhance risk models that are used in risk assessment in various risk-informed regulatory applications.

Approach

Project Objectives

The objective of RASP is to provide standard methods and tools for performing risk assessments of inspection findings or reactor incidents for the ASP program, Phase 3 analysis of the SDP, and Incident Investigation Program, while recognizing the differences in the purposes of the programs. By using these standard methods and tools, NRC analysts from various headquarters and the regional offices will achieve more consistent results when performing risk assessments of similar operational events and licensee performance issues.

RASP Activities

Major RASP activities include the following:

- Developing standard procedures and methods for the analysis of internal and external hazards, at power and shutdown events.

- Providing enhanced-quality, integrated NRC SPAR models for internal and external events, including shutdown events.

- Enhancing the SAPHIRE code for SPAR model analyses.

- Providing technical support to SDP analysts.

The NRC staff in the Office of Nuclear Regulatory Research's Division of Risk Analysis is performing these RASP activities as part of a multiyear project to enhance standard procedures, methods, and risk models that are used in risk assessment in various risk-informed regulatory applications. Staff from the Office of Nuclear Reactor Regulation, Division of Risk Assessment and Division of Inspection and Regional Support, as well as the regional senior reactor analysts, provide detailed peer review of RASP-related products, as well as feedback for future enhancements.

Specific details of the proposed work on each RASP activity are discussed below.

Development of Risk Assessment of Operational Events Handbook

The NRC staff issued the "Risk Assessment of Operational Events Handbook" ("RASP handbook") for risk assessment of internal and external events at U.S. commercial nuclear power plants. This handbook is in the form of a practical, how-to guide for the methods, best practices, examples, tips, and precautions for using SPAR models to evaluate the risk of inspection findings and reactor incidents. The handbook represents best practices based on feedback and experience from the analyses of over 600 precursors in the ASP program (since 1969) and many SDP Phase 3 analyses (since 2000).

The handbook consists of four volumes designed to address the following:

- Internal events analysis: (Volume 1).

- External events analysis (Volume 2).

- SPAR model reviews (Volume 3).

- Shutdown events analysis (Volume 4).

The scope of each of these volumes is described below. Each volume is publicly available from the Reactor Oversight Process (ROP) program documents webpage (http://www.nrc.gov/reactors/operating/oversight/program-documents.html).

Development of Standard Guidance for Internal Events Analysis. Volume 1 of the RASP handbook, "Internal Events," provides guidance on generic methods and processes to estimate the risk significance of initiating events (e.g., reactor trip, loss-of-offsite-power) and degraded conditions (e.g., a failed high-pressure injection pump, failed emergency power system) that may have occurred at a nuclear power plant. Specifically, this volume provides guidance on the following analysis methods: exposure time determination and modeling, failure determination and modeling, mission time modeling, test and maintenance outage modeling, recovery modeling of failed equipment, and multiunit considerations modeling. Volume 1 also contains an appendix that provides guidance on the process for performing risk analysis of operational events. Appendix A, "Roadmap: Risk Analysis of Operational Events," provides an overview of the risk analysis process and detailed steps on how to perform a risk analysis of an operational event.

A recent version of Volume 1 of the RASP handbook includes additional method guides, such as common-cause failure analysis in event assessment, application of SPAR-Human Reliability Analysis Method (SPAR-H) and associated human reliability analysis (HRA) technical issues in event assessment, the use of support-system initiating event models in event assessment, and the risk analysis of total loss of the offsite power initiating events.

Development of Standard Guidance for Evaluating Internal Fires and Flooding Events, External Hazards. Volume 2 of the RASP handbook, "External Events," provides methods and guidance for the risk analysis of initiating events and conditions associated with internal and external hazards. These hazards include internal fire, internal flooding, seismic events, and other external hazards, such as external flooding, external fire, high winds, tornado, hurricane, and other extreme weather-related events. This volume is intended to complement Volume 1 for internal events. The guidance for risk analysis of external hazards provides a systematic process to initiate and complete a preliminary analysis, including examples and worksheets for the required steps of the analysis method. Specifically, this volume provides guidance on the following analysis methods: internal fire modeling and fire risk quantification, internal flood modeling and risk quantification, seismic event modeling and seismic risk quantification, and other external event modeling and risk quantification.

Development of Standard Guidance for Reviews of SPAR Model Modifications. Volume 3 of the RASP Handbook, "SPAR Model Reviews," provides analysts and SPAR model developers with additional guidance to ensure that the SPAR models used in the risk analysis of operational events represent the as-built, as-operated plant to the extent needed to support the analyses. This volume provides checklists that can be used following modifications to the SPAR models for performing risk analysis of operational events. These checklists are based on NUREG/CR-3485, "PRA Review Manual," issued September 1985; Regulatory Guide 1.200, "An Approach for Determining the Technical Adequacy of Probabilistic Risk Assessment Results for Risk-Informed Activities," Revision 2, dated March 2009; and experiences and lessons learned from SDP and ASP analyses.

Development of Standard Guidance for Evaluating Shutdown Events. Volume 4 of the RASP handbook, "Shutdown Events," provides methods and practical guidance for modeling shutdown

scenarios and quantifying their core damage frequency using SPAR models and SAPHIRE software. The current scope includes the following plant operating states for boiling-water reactors and pressurized-water reactors (PWRs): hot shutdown, cold shutdown, refueling outage, and mid-loop operations (PWR only).

Enhancements to SPAR Models and the SAPHIRE Interface for SPAR Model Analyses

This activity involves enhancing SPAR models and the SAPHIRE interface to ensure that quality risk assessment tools are readily available to NRC staff performing risk assessments. The expected enhancements will include improvements in the fidelity of SPAR models for risk analysis of internal events, external events, and shutdown events. Additional description of SPAR model enhancement and development activities appears in the information sheet, "SPAR Model Development Program," in this chapter.

SAPHIRE Version 8 was made available to the staff in April 2010. This new version of the SAPHIRE software provides enhanced user interface tools, as well as improved modeling and analysis methods that support the development and use of the SPAR models. Additional description of SAPHIRE software enhancement and development activities appears in the information sheet, "SAPHIRE Development Program," in this chapter.

Technical Support for SDP Analysts

This activity involves providing technical support to SDP analysts on the efficient use of the various RASP products, such as guidance for standard risk assessment methods, enhanced SPAR models, new software tools, and the Web-based toolbox. The expected technical support will include the maintenance of RASP products and their quality, as-requested enhancements to risk assessment methods, and SPAR models.

For More Information
Contact Gary DeMoss, RES/DRA, at Gary.Demoss@nrc.gov

Probabilistic Risk Assessment Quality and Standards

Background

The NRC recognizes that probabilistic risk assessment (PRA) has evolved to the point where it can be used as a tool in regulatory decisionmaking. Consequently, confidence in the information derived from a PRA is an important issue. The accuracy of the technical content must be sufficient to justify the specific results and insights that are used to support the decision under consideration.

In 1995, the U.S. Nuclear Regulatory Commission (NRC) issued a policy statement on the use of PRA, encouraging its use in all regulatory matters. That policy statement directs that "the use of PRA technology should be increased to the extent supported by the state-of-the-art in PRA methods and data and in a manner that complements the NRC's deterministic approach." Since the NRC issued its PRA policy statement, the agency has added a number of risk-informed activities to the NRC regulatory structure (i.e., regulation and guidance, licensing and certification, oversight, and operational experience). The NRC also has developed technical documents to provide guidance on the use of PRA information to support these activities.

Objective

The PRA quality program's objective is to define PRA quality (or technical acceptability) so that there is the needed confidence in the results being used for risk-informed regulatory decisionmaking, and so that the defined technical acceptability is commensurate with the activity (or decision) under consideration.

Approach

To establish the definition of PRA technical acceptability, the NRC issued Regulatory Guide (RG) 1.200, "An Approach for Determining the Technical Adequacy of Probabilistic Risk Assessment Results for Risk-Informed Activities," Revision 2, dated March 2009, which provides the staff position on one acceptable approach for determining the technical acceptability of a PRA. RG 1.200 provides guidance on the technical acceptability of PRA in the following manner:

- Establishes the attributes and characteristics of a technically acceptable PRA.

- Endorses consensus PRA standards and the industry peer review process.

- Demonstrates technical acceptability in support of a regulatory application.

The staff position in RG 1.200 on PRA technical acceptability accomplishes the following:

- Defines the scope of a base PRA to include Level 1, 2, and 3 analyses; at-power, low-power, and shutdown operating conditions; and internal and external hazards to support operating reactors and new light water reactors (LWRs).

- Defines a set of technical elements and associated attributes that need to be addressed in a technically acceptable base PRA.

- Provides guidance to ensure that a PRA model represents the plant down to the component-level of detail, incorporates plant-specific experience, and reflects a realistic analysis of plant responses.

- Includes a process to develop, maintain, and upgrade a PRA to ensure that the model represents the as-built, as-operated (or as-designed) plant.

The staff position in RG 1.200 on consensus PRA standards and the industry peer review process does the following:

- Allows the use of consensus PRA standards and peer reviews (as endorsed by the NRC in RG 1.200) to demonstrate the technical acceptability of a base PRA.

- Provides guidance for an acceptable peer review process and peer reviewer qualifications.

- Endorses the American Society of Mechanical Engineers/ American Nuclear Society (ASME/ANS) PRA standard and the Nuclear Energy Institute (NEI) peer review guidance documents with certain objections. The endorsement of the standard and peer review guidance consists of staff objections and proposed resolutions. An application PRA needs to address the staff objections in RG 1.200, where applicable, if the PRA standard is to be considered met.

The staff position in RG 1.200 on PRA technical acceptability in support of a regulatory application does the following:

- Recognizes that the needed PRA scope (i.e., risk characterization, level of detail, plant specificity, and realism) is commensurate with the specific risk-informed application under consideration.

- Acknowledges that some applications (e.g., extension of diesel generator allowed outage time) may only use a portion of the base PRA, whereas other applications (e.g., safety significance categorization of structures, systems, and components) may require the complete model.

- Demonstrates one approach for technical acceptability of a PRA, independent of application. Inherent in this definition is the concept that a PRA need only have the scope and level of detail necessary to support the application for which it is being used, but it always needs to be technically acceptable.

RG 1.200 is also a supporting document to other NRC RGs that address risk-informed activities. Figure 5.4 shows the relationship of this RG with risk-informed activities in regulations, application-specific guidance in associated RGs, consensus PRA standards, and industry programs.

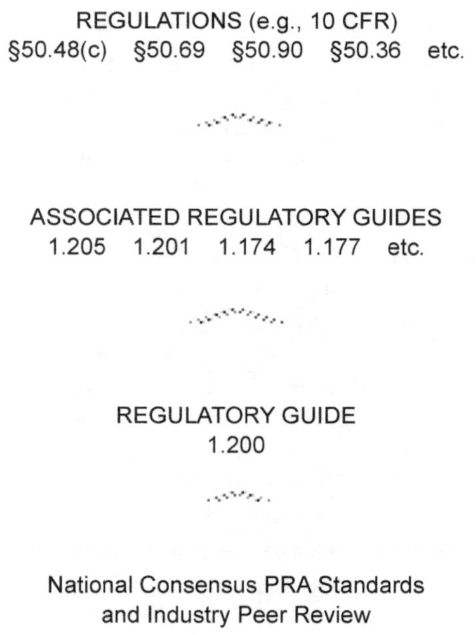

REGULATIONS (e.g., 10 CFR)
§50.48(c)   §50.69   §50.90   §50.36   etc.

ASSOCIATED REGULATORY GUIDES
1.205   1.201   1.174   1.177   etc.

REGULATORY GUIDE
1.200

National Consensus PRA Standards
and Industry Peer Review

Figure 5.4 Relationship of regulations, RGs, and standards for risk-informed activities

When used in support of an application, a major goal of RG 1.200 is to eliminate the need for an in-depth review of the base PRA by NRC reviewers, allowing them to focus their review on key assumptions and areas identified by peer reviewers as being of concern and relevant to the application. Consequently, RG 1.200 is meant to provide for a more focused and consistent review process.

Status

The status of the standards, peer review guidance, and RG 1.200 are as follows:

ASME/ANS have published ASME/ANS RA-Sa-2009 to support a PRA for operating LWRs. The scope of the standard includes a Level 1 large early-release frequency LERF PRA for at-power conditions addressing both internal and external hazards. An edition to this standard is expected to be published in 2014. This edition will address issues with internal events, internal flood, internal fires, and seismic events. Extending ASME/ANS RA-Sa-2009 to address low-power shutdown conditions and to support new LWRs is underway. Furthermore, PRA standards for Level 2 and Level 3 are under development, along with a PRA standard for non-LWRs.

NEI has published NEI-00-02, "Probabilistic Risk Assessment Peer Review Process Guidance"; NEI-05-04, "Process for Performing Follow-on PRA Peer Reviews Using the ASME PRA Standard"; and NEI-07-12, Fire Probabilistic Risk Assessment (FPRA) Peer Review Process Guidelines," which include a peer review process for Level 1 LERF PRA for internal events and internal floods, PRA updates and upgrades, and fire PRA, respectively. NEI revised NEI-07-12 in June 2010 and also published NEI-12-13, "External Hazards PRA Peer Review Process Guidelines," in August 2012.

Revision 2 to RG 1.200 provides staff endorsement of ASME/ANS RA-Sa-2009 and the NEI peer review guidance documents, except for the revised NEI-07-12 and the new NEI-12-13.

Revision 3 to RG 1.200 is expected to be published in June 2015 and endorse the next edition of ASME/ANS RA-Sa-2009. In the interim, the staff plans to endorse the revised NEI-07-12 and the new NEI-12-13 in interim staff guidance documents. The staff has no plans to endorse Addendum E to ASME/ANS RA-Sa-2009.

For More Information
Contact Mary Drouin, RES/DRA, at Mary.Drouin@nrc.gov

Evolutionary Methods and Models Development in Probabilistic Risk Assessment (PRA)

Background

Along with developing and maintaining the current generation of probabilistic risk assessment (PRA) tools, the U.S. Nuclear Regulatory Commission's (NRC's) Office of Nuclear Regulatory Research (RES) also explores evolutionary PRA methods, often in association with the Office's Long-Term Research Plan. (See the separate descriptions associated with the Systems Analysis Programs for Hands-on Integrated Reliability Evaluation [SAPHIRE] code and Standardized Plant Analysis Risk [SPAR] models.)

Objective

The objective of this exploration is to meet the agency's strategic objective of using state-of-the-art methods and tools. This enhances not only the capabilities of the agency's risk tools (such that they can be used more easily and more broadly), but also fosters the expertise needed to support the agency's program offices for regulatory reviews that use new approaches. Furthermore, activity in this general area is consistent with the agency's overall approach to risk assessment tools, as codified in the 1995 Commission policy statement on the use of PRA.

Example Area #1: PRA Quantification Methods

The most commonly used (and most efficient) PRA model quantification techniques rely on approximations and truncation to keep the calculations from becoming impractical to solve. Under most circumstances, the use of such approximations has a negligible impact on the PRA model results. However, these approximate methods are challenged by situations in which high failure rates exist. This issue most commonly manifests itself in seismic Level 1 PRAs, but it also applies to other aspects of PRA, such as shutdown Level 1 PRAs and Level 2 PRAs. The need for further development in this area has been raised by the Advisory Committee on Reactor Safeguards and others.

Recent NRC-sponsored work at Idaho National Laboratory has focused on the use of binary decision diagrams (BDDs) within SAPHIRE. The BDD method provides an exact solution for all failure combinations that are represented in the diagram. The BDD approach avoids the use of truncation and approximations that are typically used in PRA quantification methods. The BDD method has limitations that prevent its direct application to models with the size and complexity usually seen in PRAs. RES is currently supporting development of a hybrid approach that will combine traditional PRA methods with BDD quantification. RES will continue to evaluate BDDs and other PRA quantification techniques to enhance the agency's risk assessment tools.

Example Area #2: Dynamic Event Tree PRA

The agency has collaborated with the University of Maryland (UMD) for over a decade in the area of dynamic event tree Level 1 PRA, and it was previously involved in similar initiatives. The collaboration with UMD has resulted in the development and application of the Accident Dynamics Simulator using Information, Decisions, and Actions in a Crew Context (ADS-IDAC) software. [ADS-IDAC is a discrete dynamic event tree computer code that combines a nuclear plant thermal-hydraulic simulation with an operations crew cognitive decision-making model.] Ongoing work focuses on improving the usability of the code platform (parallel processing, graphical user interface), as well as application of the code for event assessment (e.g., 11th Probabilistic Safety Assessment and Management Conference paper by Coyne et al.)

In related work, the agency conducted an internal scoping study in 2009 to evaluate both methodological and implementation-oriented issues associated with the advancement of Level 2 and 3 PRA modeling techniques. The scoping study created a taxonomy of methods approach classes, which is depicted in the Figure 5.5 below. This effort included a meeting with targeted external stakeholders, and was documented in a May 2009 report entitled, "Scoping Study on Advancing Modeling Techniques for Level 2/3 PRA" (Agencywide Documents and Access Management System [ADAMS] Accession No. ML091320454). Figure 5.5 depicts the spectrum of approaches considered.

Following on the heels of the 2009 scoping study, the next phase of work began with the initiation of a methods development project at Sandia National Laboratories. This phase of the work focuses on a dynamic event tree approach (see Figure 5.6) that uses the MELCOR accident analysis program in conjunction with the previously mentioned ADS-IDAC code platform. The development of the coupled code system has been completed and the tool suite is being applied to a demonstration problem investigating a pressurized-water reactor station blackout scenario. The work is scheduled to be completed in late 2012.

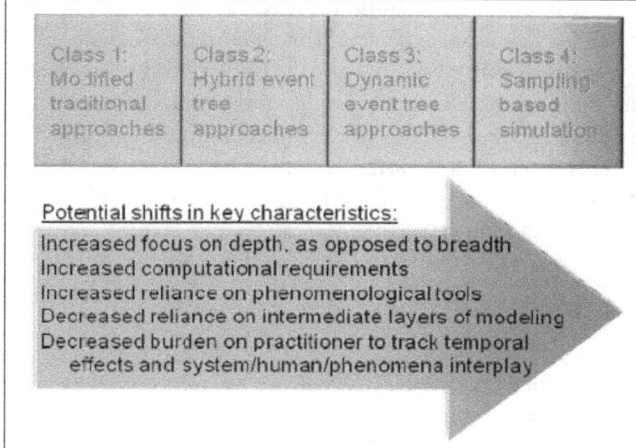

Figure 5.5 Spectrum of approach classes
This figure depicts four classes of approaches and provides thoughts on how the migration across this spectrum affects the key characteristics.

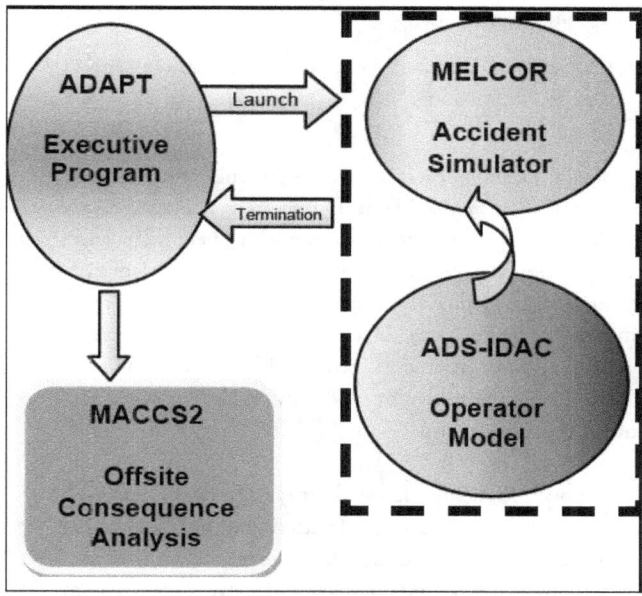

Figure 5.6 Sample high-level code coupling scheme
This figure illustrates a potential scheme for combining existing computer programs in a manner that facilitates dynamic accident simulation. Current work focuses on the area within the dashed line.

Example Area #3: PRA Uncertainty

The agency has been involved in activities related to the use of uncertainty in PRA quantification and regulatory decisionmaking for several decades. Much of the activity in this area is focused around the following:

- The American Society of Mechanical Engineers / American Nuclear Society PRA standard.

- Regulatory Guide 1.200, "An Approach for Determining the Technical Adequacy of Probabilistic Risk Assessment Results for Risk-Informed Activities," and accompanying application-specific regulatory guides.

- NUREG-1855, "Guidance on the Treatment of Uncertainties Associated with PRAs in Risk-Informed Decision Making," and the companion Electric Power Research Institute (EPRI) document, EPRI TR-1016737, "Treatment of Parameter and Modeling Uncertainty for Probabilistic Risk Assessments."

In addition to these activities, RES also supports related activities that either directly or indirectly support the above work or other aspects of PRA regulatory review. Two such activities are highlighted here.

The first activity was completed in 2010 at the Massachusetts Institute of Technology (MIT) under an cooperative agreement between the NRC and MIT. That work produced a critical review of existing methods for performing probabilistic uncertainty and sensitivity analysis for complex, computationally expensive simulation models. In the context of PRA, these models are used to (i) estimate the reliability of passive systems in the absence of operational data, (ii) inform Level 1 accident sequence development and event tree structure, (iii) establish Level 1 PRA success criteria, (iv) develop Level 2 PRA event tree structure and split fraction values, (v) perform Level 3 PRA offsite consequence analysis, and (vi) provide the simulation capacity in dynamic PRA tools. In addition to commonly-used and advanced sensitivity analysis and uncertainty propagation methods, the review also explores meta-modeling methods. The work is documented in a report available in ADAMS (Accession No. ML102350490).

The second activity is a 2012 workshop, jointly sponsored by the NRC and EPRI, on the treatment of PRA uncertainties. The purpose of the workshop was to bring together experts to gain a better understanding of the sources of uncertainty, how they manifest in the PRA, and their potential significance to the PRA model and results. More specifically, the workshop addressed uncertainties associated with risk assessments for (i) internal fires, (ii) seismic events, (iii) low-power and shutdown conditions, and (iv) the Level 2 portion of PRAs. Invited subject matter experts in each of the four topic areas were asked to give a presentation on the first day. These presentations served as a catalyst for group discussion among the workshop participants on the first and second days of the workshop. The proceedings of the workshop are available in ADAMS (Accession No. ML120680425), and an associated NUREG/CR is currently under development.

For More Information
Contact Kevin Coyne, RES/DRA, at Kevin.Coyne@nrc.gov.

Treatment of PRA Uncertainties in Risk-Informed Decisionmaking

Background

Since the issuance of the probabilistic risk assessment (PRA) policy statement, the U.S. Nuclear Regulatory Commission (NRC) has implemented or undertaken numerous uses of PRA, including modification of the agency's reactor safety inspection program and initiation of work to modify reactor safety regulations. Consequently, confidence in the information derived from a PRA is an important issue. The technical adequacy of the content has to be sufficient to justify the specific results and insights to be used to support the decision under consideration. The treatment of the uncertainties associated with the PRA is an important factor in establishing this technical acceptability. Deterministic analyses that are performed in licensing applications contain uncertainties that are addressed through defense-in-depth and safety margin. However, addressing the uncertainties associated with deterministic analyses does not address all the uncertainties associated with PRA. The NRC staff has developed guidance to address the types of uncertainties reflected in PRAs, and it has documented these in NUREG-1855," Guidance on the Treatment of Uncertainties Associated with PRAs in Risk-Informed Decisionmaking."

Objective

NUREG-1855 provides guidance on how to treat uncertainties associated with PRAs used by a licensee or applicant to support a risk-informed application to the NRC. Specifically, guidance is provided with regard to the following:

- Identifying and characterizing the uncertainties associated with PRA.

- Performing uncertainty analyses to understand the impact of the uncertainties on the results of the PRA.

- Factoring the results of the uncertainty analyses into the decisionmaking.

Furthermore, the guidance in this document is intended for both the licensee and the NRC. That is, guidance is provided with regard to (1) the approach the NRC could accept for how the licensee addresses PRA uncertainties in the context of risk-informed license application, and (2) how the impact of those uncertainties is evaluated by the NRC.

Electric Power Research Institute (EPRI), in parallel with NRC, has developed guidance documents on the treatment of uncertainties. This NUREG and the EPRI guidance have been developed to complement each other and are intended to be used as such when assessing the treatment of uncertainties in PRAs used in risk-informed decisionmaking. Where applicable, the NRC guidance refers to the EPRI work for acceptable approaches for the treatment of uncertainties.[1]

Approach

In developing the necessary guidance to meet the objectives on how to treat uncertainties associated with PRA in risk-informed decisionmaking, the guidance needs to achieve the following:

- Identify the different types of uncertainties that need to be addressed.

- Address the treatment to be performed by the licensee or applicant.

- Address how the staff accounts for the treatment in its decisionmaking.

Generally speaking, there are two main types of uncertainty; aleatory and epistemic. Aleatory uncertainty is based on the randomness of the nature of the events or phenomena and cannot be reduced by increasing the analyst's knowledge of the systems being modeled. Therefore, it is also known as random uncertainty or stochastic uncertainty. Epistemic uncertainty is the uncertainty related to the lack of knowledge about or confidence in the system or model and is also known as state-of-knowledge uncertainty.

PRA models explicitly address aleatory uncertainty that results from the randomness associated with the events of the model in the logic structure, and methods have been developed to characterize one type of epistemic uncertainty, namely parameter uncertainty. The focus of this document is epistemic uncertainty (i.e., uncertainties related to the lack of knowledge). This guidance provides acceptable methods of identifying and characterizing the different types of epistemic uncertainty and the ways that those uncertainties are treated. The different types of epistemic uncertainty are completeness, parameter, and model uncertainty

---

[1]   Electric Power Research Institute, "Treatment of Parameter and Model Uncertainty for Probabilistic Risk Assessments," EPRI 1016737, Palo Alto, CA, December 2008. Electric Power Research Institute, "Practical Guidance on the Use of PRA in Risk-Informed Applications with a Focus on the Treatment of Uncertainty," EPRI 1026511, Palo Alto, CA, 2012.

- Completeness Uncertainty—Guidance is provided on how to address one aspect of the treatment of completeness uncertainty (i.e., missing scope) in risk-informed applications. This guidance describes how to perform a conservative or bounding analysis to address items missing from a plant's PRA scope.

- Parameter Uncertainty—Guidance is provided on how to address the treatment of parameter uncertainty when using PRA results for risk-informed decisionmaking. This guidance addresses the characterization of parameter uncertainty, propagation of uncertainty, assessment of the significance of the state-of-knowledge correlation (SOKC), and comparison of results with acceptance criteria or guidelines.

- Model Uncertainty—Guidance is provided on how to address the treatment of model uncertainty. This guidance addresses the identification and characterization of model uncertainties in PRAs and involves assessing the impact of model uncertainties on PRA results and insights used to support risk-informed decisions.

The guidance for the treatment of uncertainties is organized into seven major stages.

In Stage A, guidance is provided for assessing the risk-informed activity and associated risk analysis to determine if the treatment of uncertainties should be based on the approach provided in NUREG-1855. This guidance generally involves understanding the type of application and the type of risk analysis and results needed to support the application.

In Stages B through F, guidance is provided with regard to the licensee's or applicant's treatment of uncertainties. This guidance generally involves the following:

- Stage B: Understanding risk-informed application and determining the scope of the PRA needed to support the application.

- Stage C: Evaluating the completeness uncertainties and determining if bounding analyses are acceptable for the missing scope items.

- Stage D: Evaluating the parameter uncertainties.

- Stage E: Evaluating model uncertainties to determine their impact on the applicable acceptance guidelines.

- Stage F: Developing strategies to address key uncertainties in the application.

In Stage G, an overall summary of the process used by the staff is provided with regard to the consideration of uncertainties in their decisionmaking. This process generally involves the following:

- Evaluating the PRA for technical adequacy.

- Determining if the uncertainties were adequately addressed.

- Determining if the risk element of the risk-informed decisionmaking, in light of the uncertainties, is adequately achieved in the context of the application.

- Evaluating licensee strategy for addressing the key model uncertainties result in exceeding the acceptance guideline (e.g., risk metrics).

For More Information

Contact Mary Drouin, RES/DRA, at Mary.Drouin@nrc.gov

Glossary of Risk-Related Terms in Support of Risk-Informed Decisionmaking (NUREG-2122)

Background

The final policy statement on the "Use of Probabilistic Risk Assessment Methods in Nuclear Regulatory Activities" expressed the U.S. Nuclear Regulatory Commission's (NRC's) belief that the use of probabilistic risk assessment (PRA) technology in NRC regulatory activities should be increased. Since the PRA policy statement, risk information has been used in every aspect of the NRC's work (e.g., regulation and guidance, licensing and certification, oversight, and operational experience). Some risk-related terms have been used somewhat differently. The increased development of risk-informed guidance documents, regulations, and procedures makes a common understanding fundamental to communication for the consistent and appropriate treatment of risk-informed applications by industry, as well as risk-informed regulatory actions by the NRC. It is also central to fostering clear communication between the NRC and its stakeholders. There are a variety of reasons why consistent use of terminology is not always found in the area of risk-informed activities: multiple definitions actually exist for the same term, terms are used interchangeably when they are not really synonymous, or the definition depends on scope or context.

Objective

The objective of this glossary is to identify and define terms used in risk-informed activities related to commercial nuclear power plants. This glossary provides a single source in which these terms can be found. A major goal of the glossary is also to reduce ambiguity in the definition of terms as much as possible so that a common understanding can be achieved that will facilitate communication on risk-informed activities. Among other things, this glossary will allow individuals to distinguish communication issues—erroneously perceived as technical issues—from actual technical discussions. Where terms are found to have a justifiable variety of definitions, depending on the context in which they are used, the objective of this glossary is to explain the individual definitions, along with the context, to ensure proper context-specific use of the term. Whenever possible, existing definitions are used, and redefining terms is avoided.

Approach

Two major tasks were involved in developing this glossary. The first task was identification and selection of terms. The second task was the development of the actual definitions of the selected terms.

An initial list was developed that was meant to be as broad as possible to help ensure a term was not prematurely excluded from consideration. This list was compiled in a two-step process. Terms were identified by reviewing documents related to or that support risk-informed activities. The types of documents selected for review included PRA standards, NUREGs and technical methodology documents, regulatory guides and standard review plans for risk-informed applications, risk-informed regulations, and Commission documents on risk-informed activities. The NRC staff and management also were asked to augment the initial list. Participants in this step included individuals with and without risk expertise and both junior and senior staff. Although the initial list was meant to be broad, it resulted in a list of more than 1,000 terms. With such a large list, it was necessary to prioritize the selection of the final list. A set of criteria were developed to screen terms from the glossary:

- Was the term relevant to risk-informed activities?
- Was the definition of the term easily found in the literature?
- Did the term have multiple definitions?
- Did the term have a consensually established definition?
- Was the term fundamental to risk communication?
- Did the term have policy implications?

The glossary does not recreate definitions. Consequently, where the definition of the term already exists, and there is consistency among the various sources, that definition is used as the basis for the definition provided in the glossary. However, it was determined that for some terms, an experienced risk analyst may be needed to understand the definition. In these cases, although the definition from the sources is included, a definition is developed in "plain language" (i.e., avoid the use of technical jargon). The reason the definition is written in plain language is to minimize any misunderstanding of the definitions. Furthermore, plain language helps PRA practitioners, including those who are not native English speakers, to understand the definitions with minimum language barriers.

For each term, a definition is provided, along with commentary. The commentary describes how the term is used in a PRA and provides insights into the history of the term. If the term has multiple definitions, the definitions are provided in the commentary, along with an explanation of the reasons for the differences. The sources used for each term are provided in the commentary.

To help the user, numerous terms are cross-referenced in the glossary. The authors used these cross-references when they thought that related terms were also needed to completely understand the term. The glossary also combines terms. Instead of appearing as individual terms, they are defined together in the glossary as a single term or a group of terms.

Two examples of terms are provided below.

NUREG-2122, "Glossary of Risk-Related Terms in Support of Risk-Informed Decisionmaking," has benefitted from extensive staff and external stakeholder review and comments. It will be widely transmitted to external stakeholders, especially those actively involved in consensus standards organizations (e.g., the American Society of Mechanical Engineers [ASME] and the American Nuclear Society [ANS]).

For More Information
Contact Mary Drouin, RES/DRA, at Mary.Drouin@nrc.gov

Fault Tree	
A deductive logic diagram that graphically represents the various failures that can lead to a predefined undesired event. (see Top Event, Event Tree)	In a PRA, fault trees are used to depict the various pathways that lead to a system failure. Fault trees describe how failures of top events occur because of various failure modes of components, human errors, initiator effects, and failures of support systems that combine to cause a failure of a top event in the event trees. A fault tree also has been defined as: • "A deductive logic diagram that depicts how a particular undesired event can occur as a logical combination of other undesired events." • "A fault tree identifies all of the pathways that lead to a system failure. Toward that end, the fault tree starts with the top event, as defined by the event tree, and identifies … what equipment and operator actions, if failed, would prevent successful operation of the system. All components and operator actions that are necessary for system function are considered. Thus, the fault tree is developed to a point where data are available for the failure rate of the modeled component or operator action." The following is an example of a fault tree diagram: 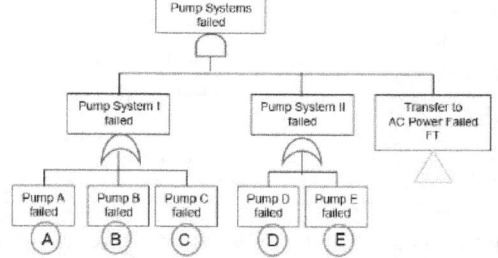

Internal Event	
Failure of equipment as a result of either an internal random cause or a human event in which either one perturbs the steady-state operation of the plant and could lead to an undesired plant condition. (see Hazard)	In a PRA, internal events result from or involve random mechanical, electrical, structural, or human failures within the plant boundary and are a specific hazard group. An example of an internal event modeled in a PRA would be the random structural failure of a reactor coolant system pipe resulting in a loss-of-coolant accident (LOCA) initiating event. Until the 2009 ASME/ANS PRA Standard revision, this term did not have a consistent definition. In some cases, a fire or flood or both occurring within the plant were considered an internal event. The ASME/ANS PRA Standard has been revised and internal flood and internal fire are not considered internal events. The ASME/ANS PRA Standard defines an internal event as "an event resulting from or involving random mechanical, electrical, structural, or human failures from causes originating within a nuclear power plant that directly or indirectly causes an initiating event and may cause safety system failures or operator errors that may lead to core damage or large early release. By historical convention, loss-of-offsite-power is considered to be an internal event, and internal fire is considered to be an external event, except when the loss is caused by an external hazard that is treated separately (e.g., seismic-induced loss-of-offsite-power). Internal floods sometimes have been included with internal events and sometimes considered as external events. For this standard, internal floods are considered to be internal hazards separate from internal events."

A Proposed Risk Management Regulatory Framework

Background

The U.S. Nuclear Regulatory Commission (NRC) has established agencywide regulations and policies to help ensure that civilian uses of radioactive materials pose no undue risk. In the 1970s, the NRC completed its first probabilistic risk assessment of two nuclear power reactors, which introduced a new way to measure nuclear safety and the effectiveness of the NRC's regulations. The Commission subsequently established a policy in 1995 on how risk assessment methods should be used to complement the NRC's established regulations in all its regulatory programs. This PRA policy, coupled with additional Commission guidance issued in 1999, has resulted in a variety of program-specific improvements.

While progress has been made, the NRC's Strategic Plan and Principles of Good Regulation make it clear that improvements in efficiency, effectiveness, and reliability continue to be agency goals. The NRC Strategic Plan notes that the expanded use of risk-informed and performance-based insights and the use of state-of-the-art technologies are the means by which the agency enhances the effectiveness and realism of NRC actions. The Principles of Good Regulation reinforce these points, noting that regulatory activities should be consistent with the degree of risk reduction they achieve. Furthermore, regulations should be based on the best knowledge available from research and operational experience.

In a memorandum dated February 11, 2011, the previous NRC Chairman Gregory Jaczko created a Risk Management Task Force (RMTF) headed by Commissioner George Apostolakis to develop a strategic vision and options for adopting a more comprehensive and holistic risk-informed, performance-based regulatory approach for reactors, materials, waste, fuel cycle, and transportation that would continue to ensure the safe and secure use of nuclear material. The RMTF was afforded the flexibility to provide options ranging from a complement to or alternative to the existing regulatory framework.

In April 2012, the RMTF issued NUREG-2150, "A Proposed Risk Management Regulatory Framework." This report describes a proposed risk management regulatory approach that could be used to improve consistency among the NRC's various programs, and it discusses implementing such a framework for specific program areas. That is, the stated objective from RMTF is: "The task force should identify the options and specific actions that the NRC could pursue to achieve a more comprehensive and holistic risk-informed, performance-based regulatory structure."

On June 14, 2012, the previous Chairman Jaczko issued a tasking memo entitled, "Evaluating Options Proposed for a More Holistic Risk-informed, Performance-based Regulatory Approach," which states:

> [T]he staff should consider the regulatory framework recommendations for power reactors provided in the RMTF report in its development of options for implementing NTTF Recommendation 1 and, in accordance with SRM-SECY-110093, should provide a notation vote paper on this matter in February 2013.

> In addition, and without impacting progress on Recommendation 1, the staff should review NUREG-2150 and provide a paper to the Commission that would identify options and make recommendations, including the potential development of a Commission policy statement. In developing its options, the staff should consider how modifications to the regulatory framework could be incorporated into important agency policy documents, such as the Strategic Plan.

> Also, the staff should seek stakeholder input on its proposed options and recommendations. This paper should be provided within six months of the staff requirements memorandum on the NTTF Recommendation 1 notation vote paper."

Objective

To develop a suggested approach for addressing the recommendations proposed by the RMTF.

Approach

The NRC has appointed a new interoffice working group to develop a proposed approach. Although the Office of Nuclear Regulatory Research (RES) has the lead for developing the requested paper (i.e., proposed approach), the recommendations will affect the regulatory structure across the agency, and therefore, could have direct impact on the processes used by each office. Therefore, the working group will be composed of individuals from each office. The individuals assigned from each office will be knowledgeable in the regulatory processes of each respective office and how risk might be integrated.

External stakeholder feedback will be solicited through public meetings and the Advisory Committee on Reactor Safeguards full and subcommittee briefings.

For More Information
Contact Mary Drouin, RES/DRA, at Mary.Drouin@nrc.gov

Reactor Operating Experience Data Collection and Analysis

Background

The collection and analysis of nuclear power plant operational data are important activities in the U.S. Nuclear Regulatory Commission's (NRC's) risk-informed regulatory programs. The results of the data collection efforts are primarily used to estimate and monitor the risk of accidents at U.S. commercial nuclear power plants. Data and information reported to the NRC are reviewed, evaluated, and coded into databases that form the basis for estimates of reliability parameters used in probabilistic risk assessment (PRA) models.

These models permit the NRC to do the following:

- Perform state-of-the-practice risk assessments of operating events and conditions.

- Assess licensee risk-related performance.

- Conduct special studies of risk-related issues, such as station blackout risk and diesel generator reliability.

- Determine trends, develop performance indicators based on operating data, and perform reliability studies for risk-significant systems and equipment.

Approach

The NRC maintains a set of PRA models for all operating U.S. commercial nuclear power plants. The staff uses these Standardized Plant Analysis Risk (SPAR) models to support risk-informed decisionmaking. For example, the Accident Sequence Precursor (ASP) program uses the SPAR models in analyses to help identify potential precursors, to support the agency's Significance Determination Process (SDP) and to confirm licensee risk analyses submitted in support of license amendment requests.

To maintain current SPAR models, the Office of Nuclear Regulatory Research (RES) collects and analyzes operating data from all nuclear power plants. The data are used to estimate the inputs required for the models. Examples of basic model inputs are initiating event frequencies, component failure probabilities, component failure rates, maintenance unavailabilities, common-cause failure parameters, and human failure probabilities.

The Reactor Operating Experience Data for Risk Applications Project collects data on the operation of nuclear power plants

as reported in licensee event reports (LERs), licensees' monthly operating reports, and the Institute of Nuclear Power Operations Equipment Performance and Information Exchange System (EPIX) (see Figure 5.7). The data collected include component and system failures, demands on safety systems, initiating events, fire events, common-cause failures, and system or train unavailabilities. The data are stored in discrete database systems, such as the Reliability and Availability Data System (RADS), Common-Cause Failure Database, and ASP Events Database.

Data input into the RADS database are used to verify and validate information used in the Mitigating Systems Performance Index (MSPI) Program. RADS data are used to review the efficiency and effectiveness of the MSPI and to suggest improvements to the index.

LERs can be individually searched by using the LERSearch program, accessible through the NRC's public Web site at: https://nrcoe.inel.gov/secure/lersearch/index.cfm.

The Computational Support for Risk Applications Project also uses the data to periodically update PRA parameters, such as initiating event frequencies, component reliabilities, maintenance unavailabilities, and common-cause failure parameters, for input into the plant-specific SPAR models. In general, the NRC uses the data to support its established regulatory programs, which help identify potential safety issues, such as the Industry Trends Program (ITF), the ASP program for evaluation of the risk associated with operating events, and the Reactor Oversight Process.

For example, RES supports the ITP by trending operating experience data and making that information available on the RES internal and public Web sites. Examples of trends that are regularly updated include thresholds for initiating events; system, component, and common-cause failures; and ASP events.

ASP analyses and the SDP use component failure probability estimates and initiating event frequencies to determine the risk significance of inspection findings. The results then are used to decide the allocation and characterization of inspection resources, the initiation of an inspection team, and the need for further analysis by other agency organizations.

The Reactor Operational Experience Results and Databases Web site (http://nrcoe.inel.gov/results/) makes current operating experience information available to the NRC staff and the public. The site also contains results from a variety of previously published studies that include initiating events, system performance, component performance, common-cause failures, fire events, and loss-of-offsite-power.

Finally, RES also supports the Baseline Risk Index for Initiating Events, a measure used to provide a risk-informed performance indicator for the initiating events "cornerstone of safety." This type of information helps the Office of Nuclear Reactor Regulation affirm that operating reactor safety is being maintained and also enhances the NRC's inspections of risk-significant safety systems.

For More Information
Contact John C. Lane, RES/DRA, John.Lane@nrc.gov.

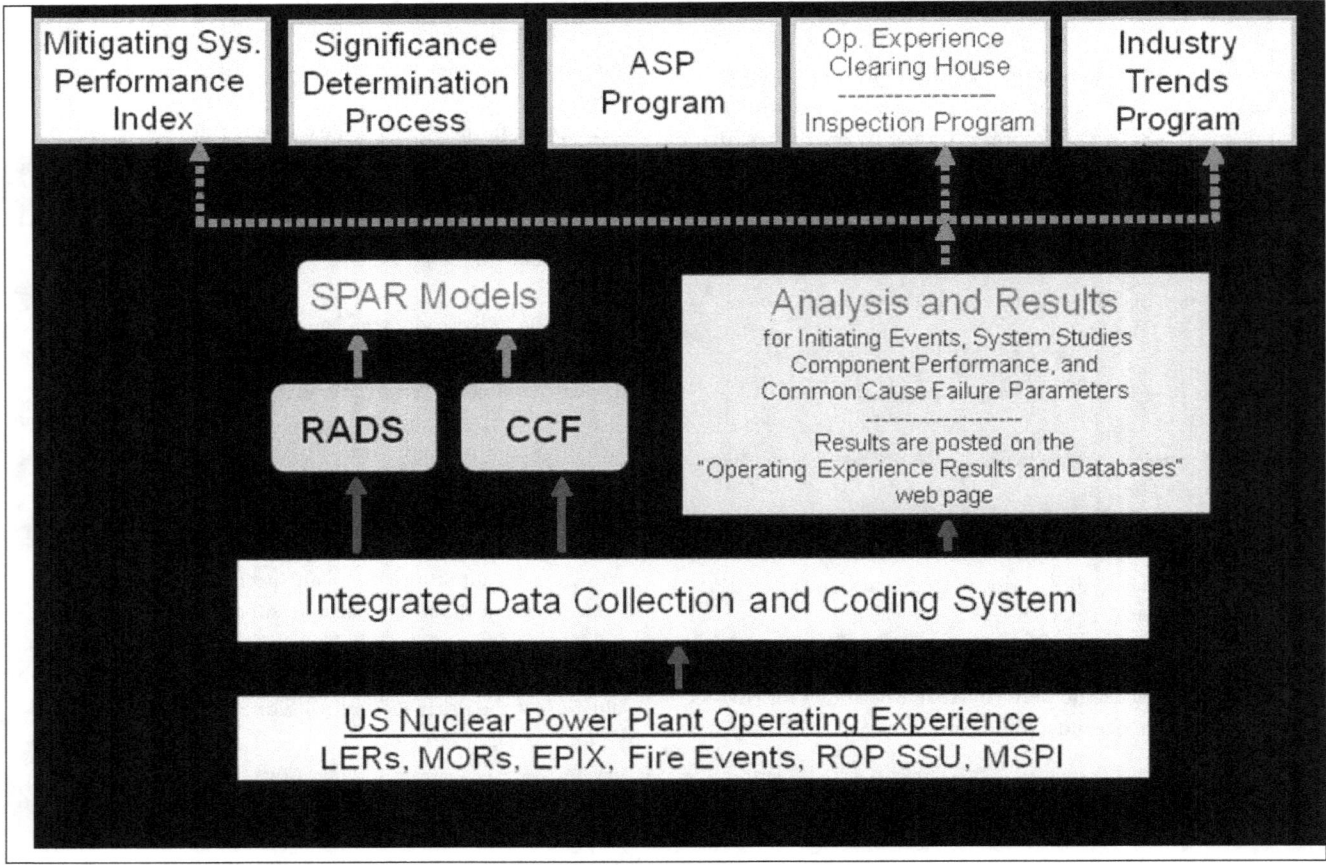

Figure 5.7 Sources and uses of operating data and analyses in NRC regulatory programs

Accident Sequence Precursor Program

Background

The Accident Sequence Precursor (ASP) Program systematically evaluates U.S. nuclear power plant operating experience to identify, document, and rank the operating events most likely to lead to inadequate core cooling and severe core damage (precursors), given the estimated probabilities of additional failures. The U.S. Nuclear Regulatory Commission (NRC) established the Accident Sequence Precursor (ASP) Program in 1979 in response to NUREG/CR-0400, "Risk Assessment Review Group Report," issued in September 1978. The Risk Assessment Review Group concluded that unidentified event sequences significant to risk might contribute a small increment to the overall risk. The review group viewed that it was important that potentially significant accident sequences and precursors be subjected to the kind of analysis contained in WASH-1400, "Reactor Safety Study – An Assessment of Accident Risks in U.S. Commercial Nuclear Power Plants," issued October 1975. Sponsored by the U.S. Atomic Energy Commission, WASH-1400 was used to estimate the public risks that could be involved in potential accidents in commercial nuclear power plants of the type in use at the time.

The ASP Program is one of three NRC programs that assess the risk significance of operational events (the other two are the Significance Determination Process (SDP) and the Incident Investigation Program defined in Management Directive 8.3, "NRC Incident Investigation Program"). Compared to the other two programs, the ASP Program assesses additional scope of operating experience at U.S. nuclear power plants. For example, the ASP Program analyzes initiating events, as well as degraded conditions where no identified deficiency occurred in the licensee's performance. The ASP Program scope also includes events with concurrent, multiple degraded conditions.

Objective

The ASP Program has the following objectives:

- Provide a comprehensive, risk-informed view of nuclear power plant operating experience and a measure for trending core damage risk.

- Provide a partial check on dominant core damage scenarios predicted by probabilistic risk assessments.

- Provide feedback to regulatory activities.

- The NRC also uses the ASP Program to monitor performance against the safety goal established in the agency's strategic plan. Specifically, the program provides input to the following performance measures:

- Zero events per year identified as a significant precursor of a nuclear reactor accident (i.e., conditional core damage probability (CCDP) or change in core damage probability (ΔCDP) greater than or equal to 1×10^{-3}).

- No more than one significant adverse trend in industry safety performance (determination principally made from the Industry Trends Program but supported by ASP results).

Approach

To identify potential precursors, the NRC staff reviews plant events from licensee event reports and inspection reports. The staff then analyzes any identified potential precursors by calculating the probability of an event leading to a core damage state. A plant event can be of one of two types: (1) an occurrence of an initiating event, such as a reactor trip or a loss-of-offsite-power event with any subsequent equipment unavailability or degradation, or (2) a degraded plant condition indicated by unavailability or degradation of equipment without the occurrence of an initiating event.

For the first type, the staff calculates a CCDP. This metric represents a conditional probability that a core damage state is reached, given an occurrence of an initiating event with any subsequent equipment failure or degradation.

For the second type, the staff calculates the ΔCDP. This metric represents the change in the probability of reaching a core damage state for the period that a piece of equipment or a combination of equipment is deemed unavailable or degraded from a nominal core damage probability for the same period for which the nominal failure or unavailability probability is assumed for the subject equipment.

The ASP Program considers an event with a CCDP or ΔCDP greater than or equal to 1×10^{-6} to be a precursor. The program defines a significant precursor as an event with a CCDP or ΔCDP greater than or equal to 1×10^{-3}.

Annual Summary of Results

Updated results from the ASP Program are published in an annual paper to the Commission. SECY-12-0133, "Status of the Accident Sequence Precursor Program and the Standardized Plant Analysis Risk Models," dated October 4, 2012 can be found at: http://www.nrc.gov/reading-rm/doc-collections/commission/secys/.

For More Information
Contact Keith Tetter, RES/DRA at Keith.Tetter@nrc.gov

SPAR Model Development Program

Background

For assessing public safety and developing regulations for nuclear reactors and materials, the U.S. Nuclear Regulatory Commission (NRC) traditionally used a deterministic approach that asked, "What can go wrong?" and "What are the consequences?" Now, the development of risk-assessment methods and tools allows the NRC to also ask, "How likely is it that something will go wrong?" These risk tools also allow the NRC to consider multiple hazards and combinations of equipment and human failures that go beyond what is traditionally considered. By making the regulatory process risk-informed (through the use of risk insights to focus on those items most important to protecting public health and safety), the NRC can focus its attention on the design and operational issues most important to safety.

In the reactor safety arena, risk-informed activities occur in five broad categories: (1) rulemaking, (2) licensing process, (3) Reactor Oversight Process, (4) regulatory guidance, and (5) development of risk analysis tools, methods, and data. Activities within these categories include revisions to technical requirements in the regulations; risk-informed technical specifications; a new framework for inspection, assessment, and enforcement actions; guidance on risk-informed in-service inspections; and improved Standardized Plant Analysis Risk (SPAR) models.

The SPAR models, Systems Analysis Programs for Hands-on Integrated Reliability Evaluation (SAPHIRE) software, and the Risk Assessment Standardization Project (RASP) handbook, developed by the Office of Nuclear Regulatory Research (RES), provide the staff with the probabilistic risk assessment (PRA) tools to support these risk-informed activities.

Objective: SPAR Model Applications

SPAR models are used to support the following activities:

Inspection Program (e.g., Significance Determination Process)

The SPAR models help determine the risk significance of inspection findings or of events to decide the allocation and characterization of inspection resources, the initiation of an inspection team, or the need for further analysis or action by other agency organizations.

Management Directive 8.3, "NRC Incident Investigation Program"

The SPAR models help estimate the risk significance of events and conditions at operating plants so that the agency can analyze and evaluate the implications of plant operating experience to compare the operating experience with the results of the licensees' risk analysis, identify risk conditions that need additional regulatory attention, identify risk insignificant conditions that need less regulatory attention, and evaluate the impact of regulatory or licensee programs on risk.

Accident Sequence Precursor Program

The SPAR models help to screen and analyze operating experience data in a systematic manner to identify those events or conditions that are precursors to severe accident sequences.

Generic Safety Issues

The SPAR models provide the capability for resolution of generic safety issues, both for screening (or prioritization) and conducting more rigorous analysis to determine if licensees should be required to make a change to their plant or to assess if the agency should modify or eliminate an existing regulatory requirement.

License Amendment Reviews

The SPAR models enable the staff to make risk-informed decisions on plant-specific changes to the licensing basis, as proposed by licensees, and provide risk perspectives in support of the agency's reviews of licensees' submittals.

Performance Indicators Verification (e.g., Mitigating System Performance Index, NUREG-1816)

The SPAR models assist in the identification of threshold values for risk-based performance indicators and in the development of an integrated performance indicator.

Special Studies (e.g., Loss-of-offsite-power and Station Blackout, NUREG/CR-6890 Volumes 1 & 2)

The SPAR models help staff perform various studies in support of regulatory decisions as requested by the Commission, Office of Nuclear Reactor Regulation (NRR), and other NRC offices.

Approach

The NRC staff uses SPAR models in support of risk-informed activities related to the inspection program, incident investigation program, license amendment reviews, performance indicator verification, accident sequence precursor program, generic safety issues, and special studies. These tools also support and provide rigorous and peer-reviewed evaluations of operating experience, thereby demonstrating the agency's ability to analyze operating experience independently of licensees' risk assessments and enhancing the technical credibility of the agency.

The SPAR models integrate systems analysis, accident scenarios, component failure likelihoods, and human reliability analysis into a coherent model that reflects the design and operation of the plant. The SPAR model gives risk analysts the capability to quantify the expected risk of a nuclear power plant in terms of core damage frequency and the change in that risk given an event, an anomalous condition, or a change in the design of the plant. More importantly, the model provides the analyst with the ability to identify and understand the attributes that significantly contribute to the risk and insights on how to manage that risk.

Currently, 79 SPAR models representing the 104 operating commercial nuclear plants in the United States are used for analysis of the core damage risk (i.e., Level 1 analysis) from internal events at operating power. The Level 1 SPAR model includes core damage risk resulting from general transients (including anticipated transients without scram), transients induced by loss of a vital alternating current or direct current bus, transients induced by a loss of cooling (service) water, loss-of-coolant accidents, and loss-of-offsite-power. The SPAR models use a standard set of event trees for each plant design class and standardized input data for initiating event frequencies, equipment performance, and human performance, although these input data may be modified to be more plant and event-specific, when needed. The system fault trees contained in the SPAR models are generally not as detailed as those contained in licensees' PRA models.

In FY 2010, the NRC revised and augmented the SPAR models to take advantage of the new features and capabilities of SAPHIRE Version 8. Model enhancements included improved modeling of common-cause failure events, handling of recovery rule linking, analysis documentation, and parameter data updates.

To more accurately model plant operation and configuration and to identify the significant differences between the licensee's PRA and SPAR logic, the staff performed detailed cut-set level reviews on all models. In addition to the internal event at-power models, the staff developed the following:

- Seventeen external event models based on the licensee responses to Generic Letter 8820, Supplement 4, "Individual Plant Examination of External Events (IPEEE) for Severe Accident Vulnerabilities," issued in 1991.

- Eight low-power/shutdown models.

- Three extended Level 1 models supporting large early release frequency (LERF) and Level 2 modeling.

These models are used to support a variety of regulatory programs, including the Significance Determination Process (SDP). In addition, the external event models were recently used to support the NRC's state-of-the-art Reactor Consequence Analysis (SOARCA) Project and to evaluate severe accident sequences for the Consequential Steam Generator Tube Rupture Project in support of the NRC's Steam Generator Action Plan.

One significant upcoming activity is the incorporation into the SPAR models of internal fire scenarios from the National Fire Protection Association (NFPA) 805, "Performance-Based Standard for Fire Protection for Light Water Reactor Electric Generating Plants," pilot applications. In addition, the staff continues to provide technical support for SPAR model users and risk-informed programs. The staff also completes about a dozen routine SPAR model updates annually.

In addition, the staff is developing design-specific internal events SPAR models for new reactor designs. The AP1000 model was completed in February 2010. The model has been optimized for SAPHIRE Version 8. Two additional models—one for the Advanced Boiling Water Reactor (ABWR) GE reactor design and one for the ABWR Toshiba design—have been completed. Staff currently is working on the addition of a Low Power Shutdown (LPSD) model for the ABWR Toshiba design. The staff also has developed a design-specific internal events SPAR model for the U.S. Advanced pressurized-water Reactor (USAPWR). A first draft of the USAPWR reactor model was provided to the Office of New Reactors for review and is currently going through validation. The staff also is developing a design-specific internal events SPAR model for the U.S. Evolutionary Power Reactor (U.S. EPR). Because design standardization is a key aspect of the new plants, it should only be necessary to develop one SPAR model for each of the new designs.

The NRC implemented a formal SPAR model quality assurance plan in September 2006. Limited scope validation and verification is accomplished by comparisons to licensee PRA models (as available) and to NRC NUREGs and analyses. Limited scope peer reviews consist of internal quality assurance review by NRC contractors, NRC PRA staff, and regional senior reactor analysts (as available). Improvements to the models on a continuing basis result from staff user feedback, peer reviews from licensees, and insights gained from special studies, such as identification of threshold values during Mitigating Systems Performance Index (MSPI) reviews and the study on loss-of-offsite-power (LOOP) and station blackout. In 2007, the NRC began a cooperative effort with the Electric Power Research Institute (EPRI) to improve PRA quality and address several key technical issues common to both the SPAR models and industry models. This cooperation resulted in the joint publication of EPRI Report 1016741, "Support System Initiating Events: Identification and Quantification Guideline," in 2008. This report documents current methods to identify and quantify support system initiating events using PRAs. Other cooperative projects include improvements to LOOP modeling (a typical LOOP event tree model is shown in Figure 5.8) and emergency core cooling system performance following boiling-water reactor (BWR) containment failure. In addition, the staff, with the cooperation of industry experts, performed a peer review of a representative BWR SPAR model and pressurized-water reactor SPAR model in accordance with American National Standards Institute/American Society of Mechanical Engineers (ANSI/ASME) RAS-2002, "Probabilistic Risk Assessment for Nuclear Power Plant Applications," and Regulatory Guide 1.200, "An Approach for Determining the Technical Adequacy of Probabilistic Risk Assessment Results for Risk-Informed Activities." The staff reviewed the peer review comments and initiated projects to address these comments, where appropriate. The staff also is reevaluating certain success criteria in the SPAR models using state-of-the-art thermal-hydraulic modeling tools.

For More Information

Contact Peter Appignani, RES/DRA at Peter.Appignani@nrc.gov

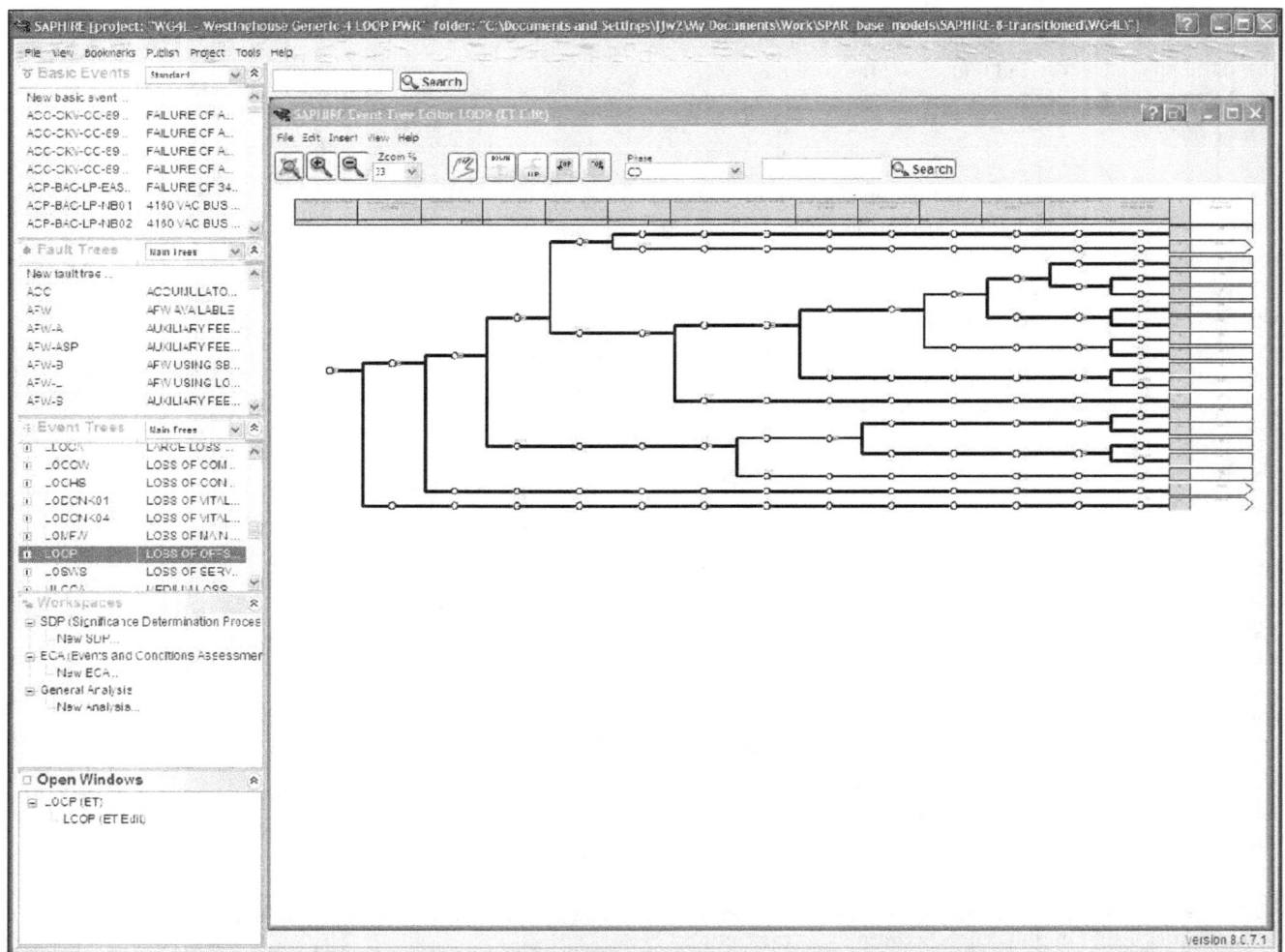

Figure 5.8 Example of loss-of-off-site-power SPAR model event tree display with SAPHIRE

SAPHIRE PRA Software Development Program

Background

Since the earliest applications of probabilistic risk assessment (PRA) to analyze nuclear reactor safety, researchers have employed computational tools to quantify measures of risk. Recognizing the important role that PRA technology plays in informing regulatory decisionmaking, the U.S. Nuclear Regulatory Commission's (NRC's) Office of Nuclear Regulatory Research (RES) has sponsored the ongoing development of PRA software applications since the 1980s. This work has led to the development of a computer software application called Systems Analysis Programs for Hands-on Integrated Reliability Evaluations (SAPHIRE). The Idaho National Laboratory developed and maintains SAPHIRE for the NRC.

SAPHIRE provides the functions required for performing a PRA. Users can supply basic event data, create and solve fault trees (see Figure 5.9) and event trees, perform uncertainty analyses, and generate reports. The NRC staff uses SAPHIRE, along with the agency's other PRA tools—the Standardized Plant Analysis Risk (SPAR) models and the Risk Assessment Standardization Project (RASP) handbook—to support the NRC's risk-informed regulatory programs, including the Accident Sequence Precursor (ASP) and the Significance Determination Process (SDP) programs.

Objective

SAPHIRE is primarily used to model a nuclear power plant's response to events that could result in core damage, quantify the associated core damage frequencies, and identify important contributors to core damage (Level 1 PRA). It can also be used to evaluate containment failure and characterize release of radioactive materials for severe accident conditions (Level 2 PRA). The objective of the SAPHIRE software development program is to provide a tool that

- Performs risk calculations accurately and efficiently.

- Reports the results in a clear and concise manner to support risk-informed decisionmaking.

RES continues to develop and improve SAPHIRE to meet these objectives and support the staff's needs.

Approach

SAPHIRE contains graphical editors for creating, viewing, and modifying fault trees and event trees. The fault tree editor includes a "drag and drop" feature that allows users to easily add basic events to fault trees. The graphical editors in SAPHIRE are used for creating the logical representations of accident scenarios that can occur at a nuclear power plant.

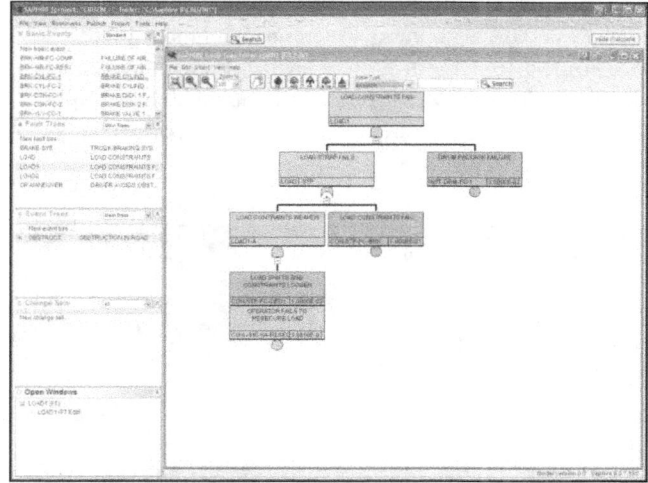

Figure 5.9 A graphical representation of a simple fault tree

SAPHIRE uses the event tree and fault tree models, along with accident sequence linking rules and postprocessing rules, to generate unique combinations of individual failures that can cause core damage (for Level 1 PRA). These unique failure combinations are called minimal cut sets. SAPHIRE quantifies the frequencies and probabilities associated with the minimal cut sets to estimate a plant's total core damage frequency. The default quantification method in SAPHIRE is the minimal cut set upper bound approximation; however, the user may also chose to use the rare event approximation or calculate the exact solution.

SAPHIRE includes many useful features to support the quantification of PRA models and identification of significant contributors to risk. SAPHIRE calculates traditional PRA importance measures such as Fussell-Vesely, risk increase ratio or interval, risk reduction ratio or interval, and Birnbaum. SAPHIRE can be used to perform uncertainty analysis. Both Monte Carlo and Latin Hypercube sampling methods are available, and uncertainty analysis can be performed on importance measures. In addition, SAPHIRE has recently been revised to use multiple computer processors to solve model objects in parallel, which helps to reduce the time needed to solve a model.

One unique aspect of SAPHIRE, in comparison to other available PRA software, is the availability of features and tools to support event and condition assessments. SAPHIRE uses analysis modules called Workspaces. These Workspaces assist the user with performing the analysis steps needed to assess the change in risk associated with the occurrence of an initiating event and/or degraded conditions. The Workspaces produce reports that document the analysis, present measures of the change in risk, and identify significant contributors to the condition being analyzed. The Workspaces were developed to assist the staff in producing accurate, consistent, and repeatable analyses to support NRC programs such as the Accident Sequence Precursor (ASP) program and the Significance Determination Process (SDP) (See Figure 5.10). The Workspace analysis features, which were specifically designed with these programs in mind, have helped SAPHIRE become an indispensable tool for supporting the NRC's risk-informed activities.

SAPHIRE is available to any NRC staff member who requires the software to support his or her work. The office provides technical support to users that need help obtaining, installing, or using SAPHIRE.

For More Information Contact:
Jeffery Wood, RES/DRA, at Jeffery.Wood@nrc.gov.

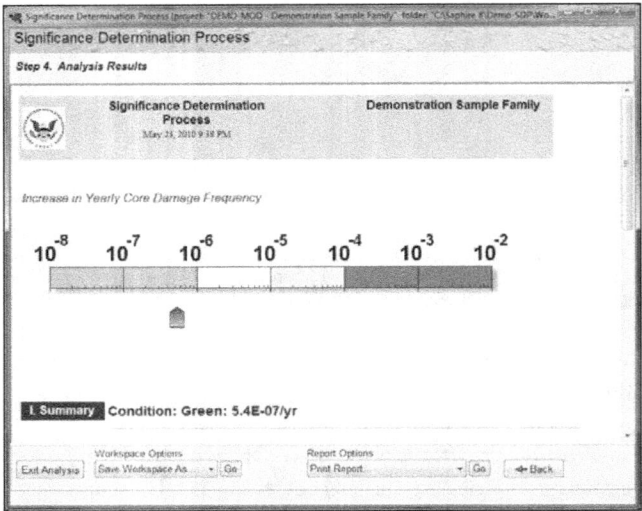

Figure 5.10 Example of Significance Determination Process (SDP) analysis results with the SAPHIRE SDP Workspace

Status and Continuing Development

RES supports the ongoing maintenance and development of the SAPHIRE software. Recent enhancements to SAPHIRE have focused on improving the quantification time needed to solve models, which is increasingly important as the size and complexity of PRA models continue to grow. Areas of continuing development include: improving the capabilities for reporting and documenting risk insights and results, exploring alternate quantification techniques for areas in which the typical approximations are challenged, and enhancing the ability to integrate different PRA model types (e.g., fire PRA, Level 2 PRA). The SAPHIRE developers have created a software quality assurance program to ensure that SAPHIRE continues to meet its requirements as new features and changes are implemented.

Thermal-Hydraulic Level 1 Probabilistic Risk Assessment (PRA) Success Criteria Activities

Background

The U.S. Nuclear Regulatory Commission's (NRC's) Standardized Plant Analysis Risk (SPAR) models are used to support a number of risk-informed initiatives. The fidelity and realism of these models is ensured through a number of processes, including cross-comparison with industry models, review and use by a wide range of technical experts, and confirmatory analysis. An ongoing activity exists to use one of the agency's mature accident simulation tools (MELCOR) to perform analyses that can be used to confirm, or to support the update of, specific aspects of the SPAR models. The aspects under consideration are the so-called, "success criteria," as well as the timing of certain key events (e.g., the depletion of a water source) that affect the estimation of the probability of success for operator actions.

What are "success criteria"?

Success criteria are criteria for establishing the minimum number of combinations of systems or components required to operate, or minimum levels of performance per component during a specific period of time, to ensure that the safety functions are satisfied.

Source: Standard for Level 1/Large Early Release Frequency Probabilistic Risk Assessment for Nuclear Power Plant Applications, American Society of Mechanical Engineers / American Nuclear Society (ASME/ANS) RA-Sa-2009

Objectives

- To perform thermal-hydraulic analyses that can update or confirm specific underlying assumptions in the agency's PRA (SPAR) models.

- To enhance inhouse expertise and knowledge transfer, for the purpose of improving the Office of Nuclear Regulatory Research's ability to consult to the program offices and regions on PRA modeling issues.

- To promote collaboration between thermal-hydraulic and PRA analysts.

Approach

Specific modeling aspects are identified, scoped and analyzed. These analyses then are used as the technical basis for making changes (as needed) to the PRA models themselves. The high-level framework for this process is depicted in the Figure 5.11 on the following page.

Examples of the type of issues that have been investigated to date include the following:

- Small-break loss-of-coolant accidents—dependency on aligning the emergency core cooling system water source to the containment sump.

- Feed and bleed decay heat removal—the minimum number of pressurizer power operated relief valves and high-head pumps needed for small loss-of-coolant accidents, loss of a direct current bus, etc.

- Spontaneous steam generator tube rupture—time available for operators to mitigate the accident before core damage.

- Station blackout—time available to recover power.

- Medium and large loss-of-coolant accidents—minimum equipment needed to prevent core damage.

Analysis for the Surry and Peach Bottom stations can be found in NUREG-1953, "Confirmatory Thermal-Hydraulic Analysis to Support Specific Success Criteria in the Standardized Plant Analysis Risk Models – Surry and Peach Bottom," September 2011.

Ongoing Activities:

As of fall 2012, ongoing activities include:

- Analysis for the Byron station, including small and medium-break loss-of-coolant accidents, loss of a direct current bus, steam generator tube rupture, and loss of decay heat removal during shutdown operations.

- Investigation of Level 1 PRA figure-of-merit issues, such as the relative conservatism in common core damage surrogates (e.g., core uncovery versus peak clad temperature of 1204 degrees Celsius [2200 degrees Fahrenheit]).

For More Information
Contact Don Helton, RES/DRA, at Donald.Helton@nrc.gov.

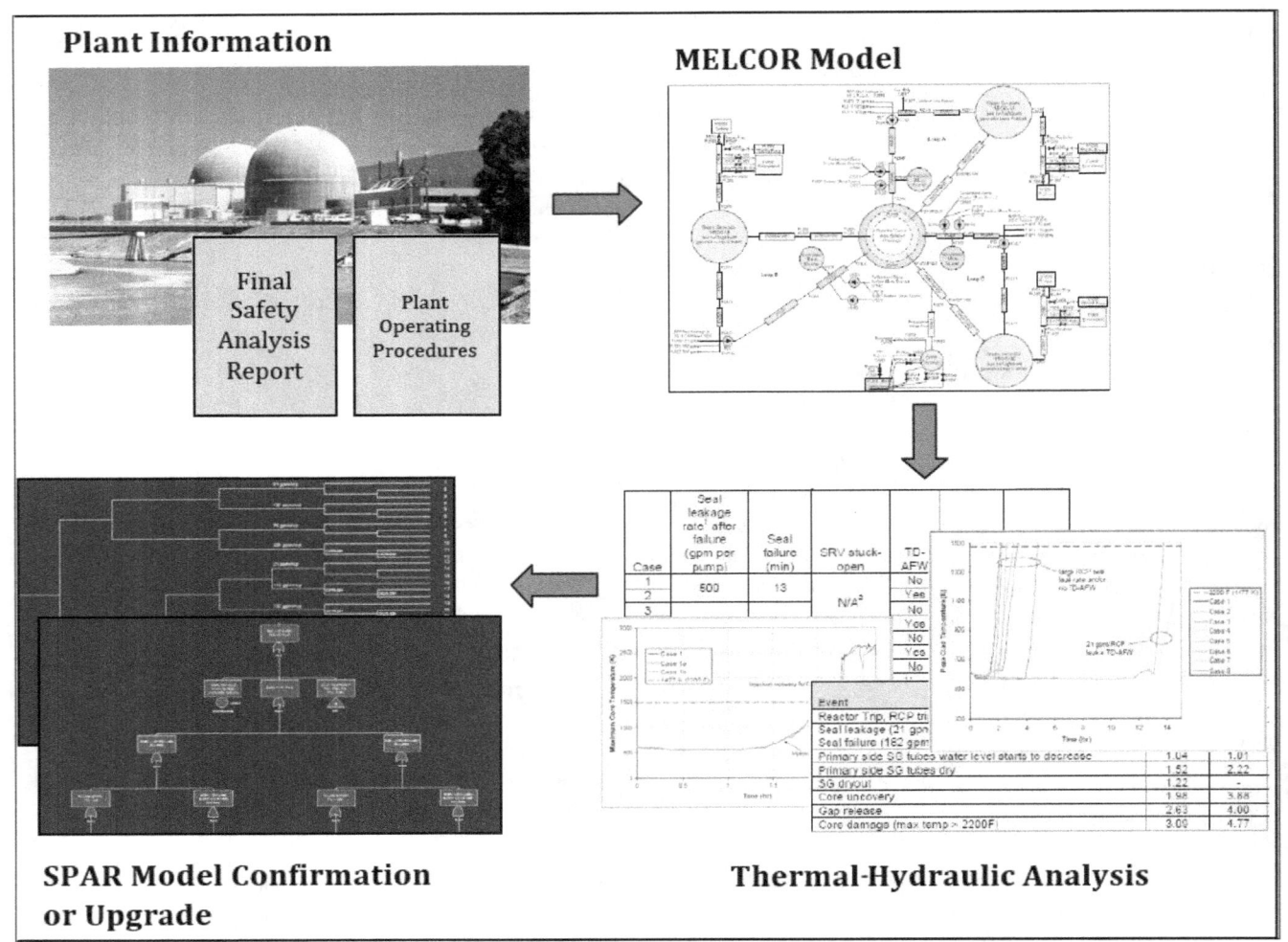

Figure 5.11 High-level overview of success criteria process

This figure shows the basic steps in the analysis, which include the translation of the actual plant design and operating features to a computer model representation, the performance of analytical studies and the generation of results, and the distillation of these results in to findings that can be used to confirm or alter the PRA model representation of the plant.

Risk-Informing Emergency Preparedness: Probabilistic Risk Analysis of Emergency Action Levels

Background

Title 10 of the *Code of Federal Regulations* (10 CFR) 50.47(a)(1) states that no initial operating license for a nuclear power reactor will be issued unless a finding is made by the U.S. Nuclear Regulatory Commission (NRC) that there is reasonable assurance that adequate protective measures can and will be taken in the event of a radiological emergency. Each operating nuclear power plant is required to include in its emergency plans a standard emergency classification (EC) and emergency action levels (EAL) scheme. An EAL is a predetermined, site-specific, observable threshold for a plant condition that places the plant in an emergency class. Both nuclear power plants and research and test reactors use the four emergency classifications listed below in order of increasing severity.

- Notification of Unusual Event—Under this category, events are in process or have occurred that indicate *potential degradation in the level of safety of the plant.* No release of radioactive material requiring offsite response or monitoring is expected unless further degradation occurs.

- Alert—If an alert is declared, events are in process or have occurred that involve an actual or potential substantial degradation in the level of safety of the plant. Any releases of radioactive material from the plant are expected to be limited to a small fraction of the U.S. Environmental Protection Agency (EPA) protective action guidelines (PAGs).

- Site Area Emergency—A site area emergency involves events in process or that have occurred and result in actual or likely major failures of plant functions needed for protection of the public. Any releases of radioactive material are not expected to exceed the EPA PAGs, except those near the site boundary.

- General Emergency—A general emergency involves actual or imminent substantial core damage or melting of reactor fuel with the potential for loss of containment integrity. Radioactive releases during a general emergency can reasonably be expected to exceed the EPA PAGs for more than the immediate site area.

The current EALs were first developed in the post–Three Mile Island era and documented in NUREG-0654, "Criteria for Preparation and Evaluation of Radiological Emergency Response Plans and Preparedness in Support of Nuclear Power Plants," dated November 1980. Although the more current approach in NEI-99-01, "Methodology for Development of Emergency Action Levels," is a significant improvement, the basic EALs and the associated emergency classes are largely unchanged from those identified in NUREG-0654. Because EALs originated from the informed judgment of staff in the early 1980s, there has never been a "first principles" analysis of damage states represented by the EALs to determine internal consistency.

In September 2008, the Commission directed the staff in staff requirements memorandum (SRM) COMDEK-08-0005, "FY 2010 NRC Performance Budget Proposal," to begin the next major enhancement in quantifying the protection that emergency preparedness plans should provide and codifying them in regulations that are transparent. This scope of work will explore the feasibility of applying risk-informed methodology to emergency response elements. If successful, this effort can result in the ability to quantify the risk associated with the different EALs, improving the NRC's ability to evaluate the licensee's emergency preparedness plans. In May 2010, a user need request originating from the Office of Nuclear Security and Incident Response (NSIR) requested that the Office of Nuclear Regulatory Research (RES) perform work to risk inform EALs.

Objective

RES started a pilot study in June 2010 that involves the use a probabilistic risk analysis (PRA) approach to evaluate the consistency of estimated risk of the initiating conditions of a given EAL for a given Emergency Classification Level to a potential reactor core damage state. The objective of this study is to explore the feasibility of using PRA to provide risk insights to improve EAL schemes. The three pilot plants selected for this study were chosen to represent a General Electric boiling-water reactor (BWR) of a BWR/4 design with a Mark I containment, a three-loop Westinghouse pressurized-water reactor (PWR) design with a large dry containment, and a four-loop Westinghouse PWR design with an ice condenser containment. The scenarios selected for the analyses are related to the following degraded conditions:

- Loss of all but one power source
- Loss of all vital direct current (dc) power.
- Simultaneous loss of all alternating current and dc.
- Loss of annunciation or indication.
- Anticipated transient without reactor scram.
- Toxic gas releases.

Approach

This study evaluates the conditional core damage probabilities (CCDPs) of the selected EAL scenarios using plant-specific Standardized Plant Analysis Risk (SPAR) models. The process is analogous to that of the Accident Sequence Precursor Program to evaluate operational events—an analyst maps specific threshold conditions that trigger an EAL of interest to the SPAR model, adjusts the failure probabilities of the affected basic events in the model, and computes the CCDP of that EAL. CCDP is then used to measure the risk significance of a specific EAL.

The following general steps are used to analyze EAL conditions:

Step 1: Gather available scenario information.

Step 2: Map the incident context into the SPAR model (scenario development).

Step 3: Use the PRA to determine scenario-specific risk measures.

The CCDP results are used to identify uniformities and inconsistencies between EALs in the same EC, and a CCDP range is established for each EC. The analysts compare the CCDP and EAL to the established CCDP range. This comparison determines if the CCDP of a specific EAL is within the established range, or falls outside the range. If the result is outside the established range, the EAL is characterized as an outlier and should be considered for future modification of EAL schemes based on the risk insights.

Results Summary

The results, in general, show a consistent relationship between the EC and the CCDP values—a higher ranking of EC generally corresponds to a higher risk, as indicated by the computed CCDP values for different EAL scenarios. However, the results also suggest that there are inconsistencies in the EC ranking of some EALs. The CCDPs of some EALs within the same EC reside outside the presumed range. The EC that these outliers fall under can be considered for reassignment—either an increase or decrease in their EC. The results and insights of this study can be used by the industry and rulemakers as part of Risk-Informing considerations to enhance EAL schemes in the future. However, the regulatory decisionmaking for EP is a complex process, and it should take into consideration information from deterministic approaches, along with the PRA insights.

The details of this study are documented in NUREG/CR-7154, Volume 1, Agencywide Documents Access and Management System (ADAMS) Accession No. ML13031A500, and Volume 2, ADAMS Accession No. ML13031A501, which was published January 2013.

For More Information
Contact Gary DeMoss, RES/DRA at <u>Gary.Demoss@nrc.gov</u>.

Design-Basis Flood Determinations at Nuclear Power Plants

Background

In 1977, the U.S. Nuclear Regulatory Commission (NRC) issued Regulatory Guide (RG) 1.59, "Design-basis Flood for Nuclear Power Plants," which detailed 1970s-era methods for determining design-basis floods at nuclear power plants. Flooding mechanisms that might need to be considered at nuclear power plants included local intense precipitation, river flooding, dam breach or failure, storm surge, seiche, tsunami, ice jams, or some of the 120 combinations of these processes. Since RG 1.59 was last updated in the late 1970s, the technical basis (data sources, analytical methods, and software tools) for flood assessment has evolved considerably, and the Office of Nuclear Regulatory Research's (RES's) Environmental Transport Branch (ETB) has recently prepared a draft revised guide (DG-1290) that is currently undergoing internal concurrence reviews by licensing offices and the Advisory Committee on Reactor Safeguards.

Objective

The research described here was undertaken to support the revision of RG 1.59.

Approach

Research activities in support of revising RG 1.59 fall into three categories: (1) overall technical basis, (2) extreme precipitation, and (3) storm surge. Staff activities included investigating potential impacts of projected climate change scenarios on flooding assessments, consulting with the U.S. Army Corps of Engineers and U.S. Bureau of Reclamation on potential dam failures, considering recently completed and ongoing research on tsunamis, and considering policy questions concerned with the revised guidance.

Technical Basis

Research in this area focused on identifying the appropriate tools (conceptual models, mathematical models, modeling software, and data sources) for conducting design-basis flood determinations. Much of this work concentrated on developing a hierarchical hazard assessment (HHA) methodology. HHA provides a roadmap for applying a hierarchy of conceptual and mathematical models for the efficient determination of design-basis flood mechanisms and levels. The appropriate blend of deterministic and probabilistic methods and the analysis of combined events also were investigated. The results of this research are described in NUREG/CR-7046 "Design-Basis Flood Estimation for Site Characterization at Nuclear Power Plants in the United States of America," published November 2011.

Extreme Precipitation

This work addresses data and methods for estimating probable maximum precipitation (PMP). Generalized PMP estimates for various areas and durations have been published in National Weather Service hydrometeorology reports. However, these estimates for much of the eastern United States have not been updated since the 1970s and do not reflect storms that have occurred since the early to mid-1970s. This is important, because the basic PMP approach begins with a catalog of observed extreme storms. Recently completed efforts have focused on a two-state pilot region comprising North Carolina and South Carolina. Although the pilot study adopted the basic approach used in the National Weather Service hydrometeorology reports, it investigated the use of radar-based precipitation estimates and included new extreme storm data sets. In particular, the impact on PMP estimates of 10 tropical cyclones that occurred in the region during 1997–2006 was investigated. Methods for addressing uncertainties in PMP estimates also were investigated. The results of this research are described in a series of three reports: NUREG/CR-7131, "Review of Probable Maximum Precipitation Procedures and Databases Used to Develop Hydrometeorological Reports," NUREG/CR-7132, "Application of Radar-Rainfall Estimates to Probable Maximum Precipitation in the Carolinas," and NUREG/CR-7133, "Synthesis of Extreme Storm Rainfall and Probable Maximum Precipitation in the Southeastern U.S. Pilot Region." These reports currently are being prepared for release as drafts for public comment.

Hurricane Storm Surge Modeling

This research investigated the application of advanced hurricane storm surge modeling methods, developed in the aftermath of Hurricane Katrina, to coastal nuclear power plant sites. The methods combine (1) high-resolution data sets for local bathymetry and topography, (2) coupled models for hurricane winds, wind-driven waves, and storm surge, and (3) a hybrid deterministic-probabilistic treatment of parameters that are input into the models. The main focus was on accurate and efficient estimation of high surge levels caused by extreme storms that have a very low probability of occurrence in any given year. A screening method useful for initial site investigations along the U.S. Gulf Coast also was developed. The results of this research are described in NUREG/CR-7134, "The Estimation of Very-Low Probability Hurricane Storm Surges for Design and Licensing of Nuclear Power Plants in Coastal Areas."

Climate Change

The ETB staff has been reviewing the current state of climate science and the scientific arguments about increased global warming and climate change over the next 90 years. The staff is assessing the possible impacts of climate change on flooding and methods for flood analysis. While a widely accepted approach for incorporating sea level rise in storm surge estimates has been identified, no generally accepted methodology currently exists to evaluate the effect of climate change on flood frequencies or amounts of extreme precipitation.

Support for Agency Actions in Response to the Fukushima Nuclear Accident in Japan

ETB staff also is participating in working groups established to implement agency actions related to flooding in response to the Fukushima accident. Activities include prioritization of the Japan Near-Term Task Force recommendations, development of the Title 10 of the *Code of Federal Regulations* (10 CFR) 50.54(f) information request letter, and development of guidance for flood protection walkdowns, Design-Basis flood reevaluations, and integrated flood protection assessments.

For More Information
Contact Joseph Kanney, RES/DRA, at Joseph.Kanney@nrc.gov.
Thomas Nicholson, RES/DRA, at Thomas.Nicholson@nrc.gov.

Assessment of Debris Accumulation on Emergency Core Cooling System Suction Strainer Performance

Background

The U.S. Nuclear Regulatory Commission (NRC) has sponsored extensive research to provide information and develop guidance for evaluating the performance of the Emergency Core Cooling System (ECCS) following a loss-of-coolant accident (LOCA) in support of resolution of Generic Safety Issue (GSI)191, "Assessment of Debris Accumulation on PWR Sump Performance." Over 30 NRC technical reports and two regulatory guides have been published documenting this effort.

Approach

To better understand the effects of debris accumulation on ECCS sump strainers, the staff initiated research in four primary areas: (1) post-LOCA chemistry, (2) sump screen headloss, (3) downstream effects, and (4) coating debris transport. The chemical research programs focused on characterizing and quantifying chemical reaction products that could form in a representative post-LOCA pressurized-water reactor (PWR) containment environment. The headloss research evaluated the pressure drop across a sump strainer attributable to the accumulation of fine particulates, insulation fibers, latent debris, and chemical byproducts observed in the chemical reaction tests. The downstream effects experiments examined the quantities of various sizes and types of insulation debris that could pass through the strainer under a variety of flow conditions, and it studied the effect of the debris on surrogate throttle valve performance and potential to clog. The coating debris transport test examined the settling and transport characteristics of coating debris in both stagnant water and water flowing at various velocities.

With the completion of the main research programs, the staff is now in the process of consolidating this knowledge in its regulatory guides and technical basis documents.

NRC regulatory guides (RGs) that provide guidance related to ECCS suction strainer performance have been revised to incorporate the lessons learned. A comprehensive state-of-the-art knowledge base report also is being prepared.

RG 1.54, Revision 2, "Service Level I, II and III Protective Coatings Applied to Nuclear Power Plants," was revised in October 2010, and RG 1.82, Revision 4, "Water Sources for Long-Term Recirculation Cooling following a Loss-of-coolant-Accident," was issued in March 2012. The revisions to these guidance documents incorporate the lessons learned during resolution of GSI191.

The NRC is preparing a comprehensive state-of-the-art report to document the ECCS strainer performance knowledge base. The intent of this report is to summarize all the NRC research activities and technical reports completed to resolve GSI191. This report also will summarize the NRC staff positions on research activities and topical reports performed by industry and licensees. There are two-phases to this project. The first phase will be to prepare a NUREG series report for the domestic fleet of plants. This report is on schedule to be completed in fiscal year (FY) 2013.

The second phase is to participate on an international team to develop a Nuclear Energy Agency/Committee on the Safety of Nuclear Installations series report for the international community. This task began in FY 2011 and is scheduled to be completed in FY 2013.

NUREG/CR-7011, "Evaluation of Treatment of Effects of Debris in Coolant on ECCS and CSS Performance in pressurized-water Reactors and Boiling Water Reactors," dated May 2010, discusses the differences in regulatory guidance for treatment of ECCS suction strainers between PWRs and boiling-water reactors (BWRs). The NRC staff is tracking the actions of the BWR owner's group to address the recommendations in this NUREG report. Confirmatory research concerning BWR chemical effects is anticipated to begin in FY 2014.

For More Information
Contact John Burke, RES/DE, at John.Burke@nrc.gov.

Chapter 6: Human Factors and Human Reliability

Human Reliability Analysis Data Repository

Human Reliability Analysis Model Differences

Using a Simulator to Improve Nuclear Power Plant Control Room Human Reliability Analysis

Human Reliability Analysis-Informed Materials for Understanding and Addressing Potential Human Errors for Medical Applications of Byproduct Materials

Human Performance for Advanced Control Room Designs

Human Performance Test Facility Research

One conceptualization of an advanced control room design

Human Reliability Analysis Data Repository

Background

Consistent with the U.S. Nuclear Regulatory Commission's (NRC's) policy statements on the use of probabilistic risk assessment (PRA) and for achieving an appropriate PRA quality for NRC risk-informed regulatory decisionmaking, the NRC has established a phased approach to PRA quality. (See SECY-04-0118, "Plan for the Implementation of the Commission's Phased Approach to Probabilistic Risk Assessment Quality," dated July 2004, and SECY-07-0042, "Status of the Plan for the Implementation of the Commission's Phased Approach to Probabilistic Risk Assessment Quality," dated March 2007.) The phased approach to PRA quality includes an action plan for stabilizing the PRA quality expectation and requirements to address PRA technical issues. Human reliability analysis (HRA) is an important PRA element. Data are key to HRA quality. The Commission identified the need for HRA data in Staff Requirements Memorandum (SRM)-M061020, "HRA Model Differences," dated November 8, 2006, and SRM-M090204B, dated February 18, 2009.

Currently, The Office of Nuclear Regulatory Research (RES) has developed the human performance data collection method (i.e., Scenario Authoring, Categorization and Debriefing Application [SACADA]) and tool, with emphasis on collecting the licensed operator simulator training data to inform the human error probability (HEP) estimations in HRA/PRA.

Objective

This project is to continue the operation of the SACADA database to collect licensed operator simulator exercise for HRA and enhance the SACADA as necessary.

Approach

The NRC staff's approach is to use the similarity-matching concept to identify the empirical data that can be used to inform the HEPs of the human failure events (HFEs) of interest. The similarity matching is based on the situational profile in challenging nuclear power plant operators in detecting the cues of plant malfunctions, understanding the situations, making correct decisions, and executing correct actions with the additional consideration of team communication and supervision. This human-centered approach differs from the traditional task-centered or component–centered approaches (e.g., turn a switch) and allows combining data of different tasks but having a similar profile in challenging human performance to inform HEP estimates. This change is expected to significantly increase the data usability.

A successful data collection program should include high data reliability and long-term data collection, resulting in the collection of a large amount of data. To achieve the objective of high data reliability, the SACADA data are entered by the plant staff (senior operator trainers and operators), and the data are entered when the information is still fresh in the individuals' memories. The key SACADA human performance data can be divided into two types. The first type of data is the performance challenge profile, which is entered by the scenario designers (i.e., operator trainers). The profile is represented by a set of factors whose states can be objectively identified. Therefore, the scenario designers could enter the data with high reliability. The second type of data is the operators' performance. The subset of this type of data includes the operators' performance in meeting the expectations, and if there are performance deficiencies, then the types of deficiencies, the causes of the deficiencies, and the remediation of the deficiencies. This type of data is entered by the plant operator crew soon after finishing their simulator exercises to ensure data reliability. For both types of data, the master set of factors are provided by SACADA for the operator trainers and operators to select the most appropriate factors and factor statuses to characterize the performance challenges and operator performance deficiencies. The approach, with the details in the narrative supplement, allows data to be entered with good consistency.

To achieve the objective of long-term data collection, the emphasis of mutual benefits to the data providers and NRC (for informing HEPs) is the key strategy. The data providers are the plants' training department and the operations department, whose main interest is improving human performance instead of estimating HEPs. Ensuring that the data provides information useful to understanding human performance issues and specifying effective measures to improve human performance is a key element to maintain engagement with data providers. The SACADA method and tool intends for the plants to replace their current practices in collecting operators' simulator performance information. Using SACADA to replace the plants' existing practices is not expected to increase plant staff effort. Furthermore, the SACADA tool would streamline data entry, which, in turn, would reduce data entry effort for other applications. The intent of using these features is to increase the likelihood that plants will collaborate by using SACADA for their daily simulator training activities. During routine operations, all data would be entered by plant staff as part of their daily practices. The NRC would only audit the data for data quality. This strategy reduces uncertainty in having a long-term data collection program.

Piloting different elements of the SACADA method and tool has been conducted under a memorandum of agreement with a U.S. nuclear power station. The first version of the SACADA tool is available for comment as of June, 2013. The goal for SACADA is to use the SACADA tool and method to broaden collaboration with U.S. nuclear power stations to collect licensed operator simulator training data and as a platform for international data exchange for HRA. In addition to work with the pilot plant to collect the plant's licensed operator simulator training data, NRC is outreaching to other plants to pilot the use of SACADA for the operator training program to collect licensed operator simulator exercise data for improving human performance and human reliability.

For More Information
Contact Y. James Chang, RES/DRA, at James.Chang@nrc.gov.

Human Reliability Analysis Model Differences

Background

The U.S. Nuclear Regulatory Commission's (NRC's) Office of Nuclear Regulatory Research (RES) is supporting the Advisory Committee on Reactor Safeguards (ACRS) to address a staff requirements memorandum (SRM)-M061020. In the memorandum, the Commission directed the ACRS to "work with the staff and other stakeholders to evaluate different human reliability models in an effort to propose a single model for the agency to use or guidance on which model(s) should be used in specific circumstances." RES is addressing this issue through collaborative work with the Electric Power Research Institute (EPRI), initiated under the RES memorandum of understanding with EPRI on probabilistic risk assessment (PRA).

Approach

To address the issue, the project is pursuing a formalization approach and a quantification tool capable of performing human reliability analysis (HRA) of nuclear power plant operators in a consistent and efficient manner. The formalization approach aims to build a foundation for HRA that uses the current understanding of human performance and is consistent with the overall PRA framework from the perspective of both failure modeling and estimation of failure probabilities. This approach relies on formulizing safety-critical task analysis of human operations. The approach introduces a crew response tree (CRT) concept, which depicts operator tasks in a manner parallel to the PRA event tree process. CRTs provide a structure for identifying the context associated with the human failure events under analysis and use a human information processing model as a platform to identify potential failures.

This approach incorporates behavioral science knowledge by providing the decompositions of human failures, failure mechanisms, and failure factors from both a top-down and bottom-up perspective. The bottom-up approach reflects findings from scientific papers documenting theories, models, and data of interest. The formalization approach provides a roadmap for incorporating the phenomena with which crews would be dealing, the plant characteristics (e.g., design, indications, procedures, training), and the plant's human performance capabilities (understanding, decision, action). The work aims to create rules and, potentially, template-based guidance for effective analysis.

The quantification tool is used to estimate human error probabilities (HEPs). The tool consists of a set of crew failure modes, each of which is associated with a decision tree that delineates relevant performance-shaping factors contributing to the human errors represented by the failure mode. The HEP for a given failure mode is estimated using various data sources (e.g., expert estimations, anchor values, simulator data, historical data) or can be modified to interface with existing quantification approaches. The grand HEP for a human failure event is the sum of the probabilities of the relevant failure modes.

The methodology has been developed, and the draft report has been in external review and comment since May 2013. The staff will incorporate the comments and finalize the draft report by September 2013. Through these collaborative efforts, the NRC also is able to take advantage of extensive domestic and international PRA and HRA expertise from recognized academics and practitioners.

For More Information
Contact Jing Xing, RES/DRA, at Jing.Xing@nrc.gov.

Using a Simulator to Improve Nuclear Power Plant Control Room Human Reliability Analysis

Background

As part of its efforts to improve human reliability analysis (HRA) performed as part of probabilistic risk assessments (PRAs), the U.S. Nuclear Regulatory Commission's (NRC's) Office of Nuclear Regulatory Research (RES) participates in and supports the International HRA Empirical Study to benchmark HRA models. In this study, different HRA models are used by different HRA teams to analyze and predict operating reactor control room crew performance responding to certain initiating events. These results are compared to actual operating reactor control room crew performance. That is, data of crew response to the simulated initiating events are gathered, analyzed, and compared to predictions of the HRA teams, which analyzed these scenarios using their models. Although the documentation of this study is not yet complete, its findings to date indicate areas for improvement in HRA methods and practices. But because the study is based on the results of simulator experiments using European crews at the Halden Reactor Project (HRP) simulator, the issue of the applicability of the study results to U.S. nuclear power plant crews has been raised.

In its staff requirements memorandum (SRM)-M090204B, the Commission directed the staff to work with industry and international partners to test the performance of U.S. nuclear power plant operating crews and to keep the Commission informed of the status of its HRA data and benchmarking projects. RES's benchmarking work is responsive to SRM-M090204B.

The NRC established a memorandum of understanding with a U.S. nuclear power plant utility that volunteered to participate in this study and offered simulator facilities, operator crews, and expertise to support the design and execution of the experimental scenario runs. As a result, a new study was initiated that the HRP staff supports with expertise in the design and execution of simulator scenario runs, as well as the collection and interpretation of crew performance data.

Objective

The objective of this study is to evaluate a specific set of HRA methods used in regulatory applications by comparing HRA predictions to crew performance in simulator experiments performed at a U.S. nuclear power plant. The results will be used to accomplish the following:

- Determine the potential limitations of data collected in non-U.S. simulators when used to evaluate U.S. applications.

- Improve the insights developed from the international HRA empirical study.

Approach

The study consists of the following four steps:

1. Experimental Design and Performance of Simulated Scenarios

The experimental design is focused on collecting information on the predictive power and consistency of HRA methods, including the following:

- A Technique for Human Event Analysis (ATHEANA).

- Standardized Plant Analysis Risk—Human Reliability Analysis Method (SPARH).

- Technique for Human Error Rate Prediction/Accident Sequence Evaluation Program (THERP/ASEP).

- Cause-Based Decision Tree (CBDT).

This effort involves analysis of crew performance in simulated nuclear power plant initiating events modeled in PRAs. It stipulates the collection of information to be used by HRA analysts to evaluate the human failure events (HFEs) involved in the scenarios and to estimate the human error probabilities (HEPs).

The study provided the following information to HRA analysts for analyses: (1) the plant status before the initiating event, (2) the initiating event, (3) and the associated plant design capabilities and operational characteristics to deal with the event. These capabilities and characteristics include procedural guidance, the predetermination and definition of the HFEs to be analyzed for each scenario and associated success criteria, the identification of human performance metrics, the development of crew performance collection protocols and questionnaires to support documentation of observed crew performance, and the development of an information package containing PRA and HRA information to be provided to the HRA teams.

The actual experiment consists of running of the accident scenarios and collecting and documenting observations about plant behavior and crew performance by experts (typically plant trainers and PRA/HRA experts). In addition to live observations, crew performance observations are collected through videotapes and debriefings of both the crews and the plant experts who observed the performance of the crews during the experiment

The experimenters evaluate crew performance by analyzing the information collected during the experiment according to predefined protocols and performance metrics. This part of the study is supported by the staff of the HRP.

2. Information Collection and Evaluation of HEPs by HRA teams

Each HRA method is applied by two or three HRA teams. The HRA teams interview plant personnel, observe operating crews in the simulator responding to simulated initiating events other than the study simulations, and collect relevant plant information. On the basis of the information collected, the teams use their selected HRA methods to perform predictive analysis and to estimate HEPs for the HFEs involved in the simulated scenarios, document the results, and submit them for review and evaluation.

One goal of the study is to understand the types of information considered by HRA teams in performing HRA analysis using a given method. Documenting this information provides insights about differences and commonalities among HRA methods; in particular, it helps staff to develop an understanding of how methods (or analysts) are using the collected information and of how the different ways of using information affect consistency among methods or analysts. Documenting information use also allows comparisons with operator crew simulator performance to examine if the appropriate factors are being considered by the teams using the different HRA methods.

3. Evaluation of the HRA Submittals

An independent group of experts reviews the submitted analyses and compares them to the observed simulator data. These experts perform method-to-method and HRA team-to-team comparisons to determine if and how method differences and analyst differences influence the HRA results. Their analysis includes both qualitative and quantitative comparisons.

Qualitative comparisons examine the extent to which HRA analysts, using their methods, were able to identify key drivers (such as misdiagnosis of equipment failures or lack of adequate procedural guidance for performing the required actions) that could influence the crew's capability to accomplish the required actions. Through such comparisons, the experts identify:

1. Method limitations with regard to guiding analysts to identify important drivers of human performance.

2. Method limitations with regard to ensuring a consistent use of the method by different analysts (intra-analyst consistency).

Quantitative comparisons involve (1) the ranking of the estimated HEPs, (2) the ranking of the human actions in terms of the level of difficulty that crews appear to have experienced during the simulation, and (3) comparison of the resulting ranking in (1) and (2). These comparisons allow the experts to examine whether or not inconsistencies in ranking stem from the following causes:

1. The extent to which the quantification tool can incorporate the important drivers of human performance identified through the qualitative analysis (e.g., the tool allows the use of only a few performance-shaping factors in the estimation of HEPs).

2. The extent to which the quantification tool can provide a consistent and traceable process to estimate HEPs.

3. The analysts' capability to correctly apply the tool.

4. Documentation of the Results

A NUREG report will (1) document the results for each method tested, including the performance characteristics of each method and potential implications for regulatory applications, and (2) assess the consistency of the methods and identify how practitioners can achieve better consistency in HRA.

RES expects this report to be published by December 2013.

For More Information
Contact Sean Peters, RES/DRA, at Sean.Peters@nrc.gov

Human Reliability Analysis-Informed Materials for Understanding and Addressing Potential Human Errors for Medical Applications of Byproduct Materials

Background

In 2011, the Office of Federal and State Materials and Environmental Management Programs (FSME) provided the Office of Nuclear Regulatory Research (RES) with a user need to 1) develop a report on understanding human error in radiation therapy, 2) publish human reliability analysis (HRA)-informed training materials, and 3) demonstrate how to use the HRA-informed job aid through illustrative examples.

This work builds on an earlier user need, provided by the Office of Nuclear Material Safety and Safeguards (NMSS) to RES, to develop HRA capability specific to materials and waste applications. This earlier work was conducted in two-phases:

- Phase 1 work consisted of feasibility studies for developing NMSS capability in HRA. The feasibility study for materials applications addressed both medical and industrial applications.

- Phase 2 work focused on the recommendations from the feasibility study, namely, the development of job aids (e.g., HRA-informed decisionmaking aids) and associated training for NRC staff on HRA-informed issues in human performance in medical applications.

In this earlier work, the final products of the Phase 2 work, a prototype HRA-informed job aid (i.e., a database of risk-relevant human performance issues and historical errors, related to treatment steps) and associated training materials for medical applications (gamma-knife based), were presented to FSME staff and delivered to the U.S. Nuclear Regulatory Commission (NRC) in December 2008.

Approach

The overall objective of the 2011 User Need is to develop three main products:

1. A NUREG on understanding human error in radiation therapy.

2. A publication of the HRA-informed training material.

3. Documentation of illustrative examples on how to use the HRA-informed job aid.

In all three cases, the products delivered to NMSS in December 2008 are the starting point for new development. However, new information and background material will be added, as needed and appropriate, for the first two products. The illustrative examples of how to use the job aid will be developed with FSME staff input and guidance.

RES is currently working on a draft of the NUREG on understanding human error in radiation therapy. RES plans to have this NUREG ready for publication in 2014.

For More Information
Contact Susan Cooper, RES/DRA, at Susan.Cooper@nrc.gov.

Human Performance for Advanced Control Room Designs

Background

The nuclear power community is currently at a stage where existing nuclear power plant (NPP) control rooms are undergoing various forms of modernization, new plants are being built with automated computer-based control rooms, and advanced reactors are being designed through international cooperation to support power generation for decades to come. The new generation of plants will differ from the existing fleet in several important ways, including the reactor technology, the design of the instrumentation and control (I&C) systems, and the types of human-system interfaces (HSI). Figure 6.1 illustrates one conceptualization of an advanced control room (CR) design.

The introduction of new NPPs will bring about a host of changes, including new technology and tools to support plant personnel and adjustments to plant staffing configurations. Moreover, the old analog control panels will be replaced by computer-based human-system interfaces that will be used for process and component control. These new digital workstations change the analog spatially dedicated and continuously visible I&C design to one that no longer has all the information control elements necessary to support operator interaction immediately available and visible at all times. This change from parallel to serial information display and component control increases the opacity of the interface, further restricting the HSI with regard to the efficiency of navigation and timely access to the required information and to the means of control. If the new technology is being used to replace tasks that were previously done by the operators, as is often the case with automation, the operators now are presented with a different job that includes supervising the automation. However, if implemented well, HSI can be enhanced by digital I&C through organizing the information presented to operators in more useful ways with better context.

Taken together, these technological advances will lead to concepts of operation and maintenance that are different from those found in currently operating NPPs. The potential benefits, as explained above, of the new technologies should result in more efficient operations and maintenance. However, if the technologies are poorly designed and implemented, there is the potential they will reduce human reliability, increase errors, and negatively impact human performance—resulting in a detrimental effect on safety. For these reasons, it is important that the potential impact of these developments is evaluated and understood by prospective operators and regulators responsible for determining the acceptability of new designs to support human performance and maintain plant safety.

Approach

To address these concerns, the U.S. Nuclear Regulatory Commission (NRC) sponsored a study to identify and prioritize human performance research that will be needed to support technical basis development and the corollary review of licensees' implementation of new technology in new and advanced NPPs. Current industry trends and developments were evaluated in the areas of reactor technology, I&C technology, HSI integration technology, and human factors engineering (HFE) methods and tools. These four broad research areas were then organized into seven HFE topic areas:

1. Role of personnel and automation.

2. Staffing and training.

3. Normal operations management.

4. Disturbance and emergency management.

5. Maintenance and change management.

6. Plant design and construction.

7. HFE methods and tools.

Next, a panel of independent subject-matter experts representing various disciplines (e.g., HFE, I&C) and backgrounds (e.g., vendors, utilities, research organizations) prioritized the areas, which resulted in 64 issues distributed among four categories, with 20 research issues placed into the top priority category. NUREG/CR-6947, "Human Factors Considerations with Respect to Emerging Technology in Nuclear Power Plants," dated October 2008, documents the results of the study. The report contains a summary of the high-level topic areas, the research issues in each topic area, the priorities for each issue, and a human performance rationale that describes the reason why each research issue is relevant. The findings from this study are being used to develop a long-term research plan addressing human performance within these technology areas for the purpose of establishing a technical basis from which regulatory review guidance can be generated. Of the 20 research projects identified as having a Priority 1 research need, six have been completed (the two most recent being the development of human factors guidance for the assessment of computerized procedures and the human factors aspects associated with the monitoring and control of multimodular plants, also referred to as small modular reactors), and three are currently underway. Descriptions of the three projects that are underway are provided below.

Advances in Human Factors Engineering Methods and Tools

The methods and tools used to design, analyze, and evaluate the HFE aspects of nuclear power plants are changing rapidly. A previous study identified the current trends in the use of HFE methodologies and tools, identified their applicability to NPP design and evaluation, and determined their role in safety reviews conducted by the NRC. The study identified seven categories of methods and tools for which additional review guidance may be needed, including (1) application of human performance models, (2) use of virtual environments and visualizations, (3) analysis of cognitive tasks, (4) rapid system development engineering models (e.g., rapid prototyping), (5) integration of HFE methods and tools, (6) computer-aided design, and (7) computer applications for performing traditional analyses. One outcome of this project to date has been the development of detailed review guidance for applying human performance models to the evaluation of NPP designs, and guidance associated with the effects of degraded digital I&C systems on HSI and operator performance. The present phase of the study is developing human factors guidance for the conduct of integrated system validation, as well as analytical methods to identify HFE-significant I&C degradations with regard to their impact on operator performance.

Roles of Automation and Complexity in Control Rooms

The overall level of automation in advanced NPPs is expected to be much higher than in plants currently operating in the U.S. It is important that the staff be cognizant of current practices and trends in the use of automation in NPP CRs and understand the influences of automation on CR design, human performance, and conduct of operations. A previous study, "Human-System Interfaces (HSIs) to Automatic Systems," developed a general framework for characterizing automation systems and developed HFE criteria for evaluating automation designs. The present study will further the state-of-the-art by examining the impact of automation on CR design, specifically the impact of automation on (1) operator performance during normal, abnormal, and emergency operations, (2) the reliability of operator's use of automation systems, including existing methods for assessing impacts, and (3) operator performance when the automation fails or is in a degraded state.

Update Existing Human Factors Engineering Regulatory Guidance

The NRC staff reviews the HFE aspects of NPPs in accordance with the guidance presented in NUREG-0800, "Standard Review Plan for the Review of Safety Analysis Reports for Nuclear Power

Plants." Detailed design review procedures for the HFE programs of applicants for construction permits, operating licenses, standard design certifications, combined operating licenses, and license amendments are provided in NUREG-0711, Revision 2, "Human Factors Engineering Program Review Model." As part of the review process, the interfaces between plant personnel and plant systems and components are evaluated using the review criteria contained in NUREG-0700, Revision 2, "Human-System Interface Design Review Guidelines." These criteria represent best practices, and when an applicant's design deviates from these criteria, a "human engineering discrepancy" is identified and evaluated for its importance to safety using the process described in NUREG-0711. NUREG-0700 also is used to conduct reviews of the applicant's design documentation, such as an HSI style guide or product specification. Thus NUREG-0700 is used to review products of the applicant's design process, as well as the detailed characteristics of the final design. NUREG-0711 and NUREG-0700 were last updated in 2004 and 2002, respectively. The guidance can benefit from further updates to keep pace with the modern I&C systems and advanced levels of automation that will be found in the next generation of NPP control rooms. The lack of up-to-date guidance leads to uncertainty, both for vendors and plant owners who worry about the acceptability of such systems to the regulators, as well as for the regulators who lack the technical basis on which to judge the acceptability of the new highly-integrated control rooms employing state-of-the-art digital system designs. This study will update NUREG-0711 and NUREG-0700 with HFE criteria developed from the most recent and best available technical bases. The availability of up-to-date HFE review guidance will help to ensure that the NRC staff has the latest knowledge, information, and tools to safely and efficiently perform its regulatory tasks.

Figure 6.1 One conceptualization of an advanced control room design

For More Information

Contact Stephen Fleger, RES/DRA, at Stephen.Fleger@nrc.gov.

Human Performance Test Facility Research

Background

The nuclear industry is entering an era in which presently operating plants are undergoing modernizations, and applications for new plants are being submitted to and reviewed by the U.S. Nuclear Regulatory Commission (NRC). This resurgence has introduced new designs, new computer systems to support work of plant personnel, and new concepts of operations. It also brings with it modern technologies, such as digital (rather than analog) instruments and control systems, computerized human-system interfaces, and increased automation.

The NRC staff is responsible for reviewing and determining if the new designs adequately support safe plant operations. Given that humans are a vital part of plant safety, the NRC staff needs to understand the potential impact of new designs on human performance to make sound regulatory decisions.

There are several ways to determine the impact of new designs, technologies, and concepts of operations, including reviewing literature from nuclear and other domains, reviewing and assessing operational experience from the nuclear industry, and conducting human performance and reliability research in a nuclear environment.

Objective

The objective of this work is to conduct research assessing the impact of new designs on human performance, with a larger and lower cost research subject pool as a supplement to the research being performed at the Halden Reactor Project.

To meet this objective, the NRC Office of Nuclear Regulatory Research recently procured two copies of a desktop computer based nuclear control room simulator to conduct this research; one copy is housed at NRC headquarters and the other is at the University of Central Florida (UCF) (Figure 6.2), under contract with the NRC.

The simulators have the following characteristics:

- Westinghouse, 3-Loop.
- "Sanitized" model of existing U.S. plant reference simulator.
- RETACT thermal-hydraulics code.
- Reprogrammable analog panel, soft controls, digital interfaces.
- Supporting documents (e.g., procedures, tech specs).

Figure 6.2 NRC simulation facility at the University of Central Florida

The NRC and UCF are working together to design and conduct human-in-the-loop experiments. This research is expected to produce nuclear-specific human performance data that aids in the evaluation of prioritized issues identified in NUREG/CR-6947, "Human Factors Considerations with Respect to Emerging Technologies in Nuclear Power Plants." These issues include the impact that new designs, technologies, and concepts of operations have on human performance. The information gained will be incorporated in updates to the NRC staff's human factors review guidance, NUREG-0700, "Human-System Interface Design Review Guidelines," NUREG-0711, "Human Factors Engineering Program Review Model," and in updates to the NRC's Human Reliability Analysis method development initiatives.

For More Information
Contact Amy D'Agostino, RES/DRA, at
Amy.DAgostino@nrc.gov.

Chapter 7: Fire Safety Research

Fire Probabilistic Risk Assessment Methodology for Nuclear Power Facilities

Fire Human Reliability Analysis Methods Development

Fire Modeling Activities

Cable Heat Release, Ignition, and Spread in Tray Installations During Fire

Direct Current Electrical Shorting in Response to Exposure Fire (DESIREEFIRE).

Fire Effects on Electrical Cables and Impact on Nuclear Power Plant System Performance: Phenomena Identification and Ranking Table and Expert Elicitation Programs

Advancements in Understanding Fire-Induced Effects on Electrical Circuits

Beyond-Design-Basis Fires for Spent Fuel Transportation: Shipping Cask Seal Performance Testing

Training Programs for Fire Probabilistic Risk Assessment, Human Reliability Analysis, and Fire Modeling

Fire Research and Regulation Knowledge Management

Evaluation of Very Early Warning Fire Detection System Performance for Fire Probabilistic Risk Assessment (PRA) Applications

Joint Analysis of Arc Faults (Joan of Arc) OECD International Testing Program for High Energy Arc Faults (HEAF)

Fire Probabilistic Risk Assessment Methodology for Nuclear Power Facilities

Background

The results of the Individual Plant Examination of External Events (IPEEE) program conducted in the 1990s and actual fire events indicate that fire can be a significant contributor to nuclear power plant (NPP) risk, depending on design and operational conditions. In particular, these studies show that failures of fire protection defense-in-depth features (i.e., failure to prevent fires, failure to rapidly suppress fires, or failure to protect plant systems to provide stable, safe shutdown) can lead to risk-significant conditions. Fire probabilistic risk assessment (PRA) provides a structured, integrated approach to evaluate the impact of failures in the fire protection defense-in-depth strategy on safety. Figure 7.1 illustrates a simplified fire PRA event tree representing different sets of fire damage and plant response. The fire PRA directly addresses technical issues, such as fire ignition frequency, detection and suppression, fire damage to diverse and redundant trains of core cooling equipment, circuits (i.e., spurious actuations), and plant response.

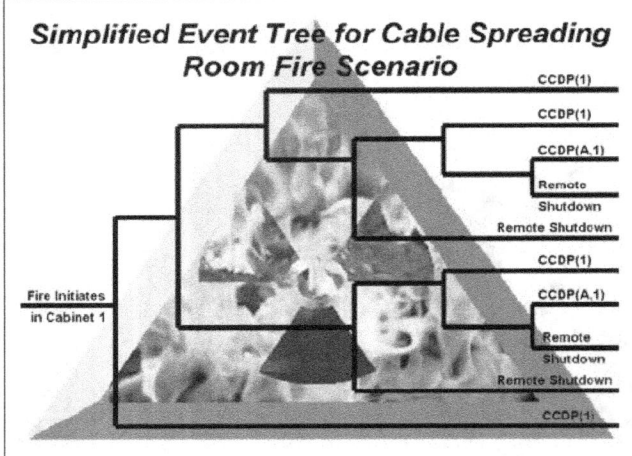

Figure 7.1 Simplified fire PRA event tree representing different sets of fire damage and plant response

The conditional core damage probability (CCDP) shown in Figure 7.1 is a combination of the following: (1) fire-induced failure only of the cabinet PLUS random failures of trains A and B, (2) fire-induced failures of the cabinet AND train A PLUS random failure of train B, (3) fire-induced failures of all three—all of the above, along with failures of any remaining mitigative measures that may still be available, that would thereby lead to core damage.

In 1995, the U.S. Nuclear Regulatory Commission (NRC) adopted a policy statement on PRA with the intent to increase the use of this technology in all regulatory matters to the extent supported by the state-of-the-art in PRA methods and data. PRA enhances safety by allowing the licensee to gain insights that supplement the NRC's traditional approach of maintaining defense in depth and safety margin and its overall engineering judgment. In 2004, the NRC amended its fire protection requirements to allow existing reactor licensees to voluntarily adopt the risk-informed, Performance-Based requirements in Title 10 of the *Code of Federal Regulations* (10 CFR) 50.48(c). This rule endorses National Fire Protection Association (NFPA) Standard 805[1], "Performance-Based Standard for Fire Protection for Light Water Reactor Electric Generating Plants," as an alternative to the existing prescriptive fire protection requirements. Licensees will need a fire PRA to realize the full benefits of making the transition to the risk-informed, Performance-Based standard.

Objective

The primary objective of this research is to advance the state-of-the-art in fire PRA methods, tools, and data for use in regulatory decision making.

Approach

In 2001, the Electric Power Research Institute (EPRI) and the NRC's Office of Nuclear Regulatory Research (RES) embarked on a cooperative project to improve the state-of-the-art in fire risk assessment to support this new risk-informed environment in fire protection. This project produced a consensus fire PRA document (NUREG/CR-6850 (EPRI TR-1011989), "EPRI/NRC-RES Fire PRA Methodology for Nuclear Power Facilities," issued September 2005) that addresses NPP fire risk for at-power operations.

Pilot plants making the transition to the rule, 10 CFR 50.48(c), rely on NUREG/CR-6850 (EPRI TR-1011989) to develop their fire PRAs, whereas the NRC uses it to support reviews. The NRC, with participation by EPRI, has produced interim solutions to all 15 fire PRA issues raised by the pilot plants and EPRI related to NUREG/CR-6850 (EPRI TR-1011989) in the NFPA Standard 805 frequently asked questions (FAQ) program and issued it as Supplement 1 to NUREG/CR-6850 in September 2010.

---

[1]   *Approval of incorporation by reference.* National Fire Protection Association (NFPA) Standard 805, "Performance-Based Standard for Fire Protection for Light Water Reactor Electric Generating Plants, 2001 Edition" (NFPA 805), which is referenced in this section, was approved for incorporation by reference by the Director of the Federal Register pursuant to 5 U.S.C. 552(a) and 1 CFR part 51.

Additionally, RES and EPRI are working jointly to update and improve the fire events database used for NUREG/CR-6850 (EPRI TR-1011989). Initially, RES and EPRI will update fire ignition frequencies; however, they envision other applications. RES has also developed fire PRA methods for low power and shutdown with EPRI serving as peer reviewers and supporting two table-top plant exercises. (See NUREG/CR-7114, "A Framework for Low Power/Shutdown Fire PRA.) Overall, this joint work is producing a significant convergence of technical approaches.

Future Work

A revision to the joint report is in the planning stages as the methodology continues to mature and other fire research programs advance the state-of-the-art knowledge.

For More Information
Contact Nicholas Melly, RES/DRA, at Nicholas.Melly@nrc.gov.

Fire Human Reliability Analysis Methods Development

Background

The Individual Plant Examination of External Events program and the experience from actual fire events found that, depending on design and operational conditions, fire can be a significant or dominant contributor to nuclear power plant risk. Human errors have been shown to be a significant contributor to overall plant risk (including the risk from fires) because of the significant role that operators play in the fire protection strategy for reactor safety. Figure 7.2 illustrates operators in a nuclear power plant (NPP) control room. Human reliability analysis (HRA) is the tool used to assess the implications of various aspects of human performance on risk. Currently, the U.S. Nuclear Regulatory Commission (NRC) is expanding existing HRA methods to evaluate the impact of human failures in the fire protection defense-in-depth safety strategy.

Figure 7.2 Operators in a NPP control room

In 2004, the NRC amended its fire protection requirements to allow existing reactor licensees to voluntarily adopt the risk-informed, Performance-Based rule in Title 10 of the *Code of Federal Regulations* (10 CFR) 50.48(c). This rule endorses National Fire Protection Association Standard 805, "Performance-Based Standard for Fire Protection for Light Water Reactor Electric Generating Plants," as an alternative to the existing prescriptive fire protection requirements. To realize the full benefits of making the transition to the risk-informed, Performance-Based standard, plants will need to have a fire probabilistic risk assessment (PRA) that includes quantitative

HRA for post-fire mitigative human actions modeled in a fire PRA.

The Electric Power Research Institute (EPRI) and the NRC's Office of Nuclear Regulatory Research (RES) embarked on a cooperative project to improve the state-of-the-art in fire risk assessment to support this new risk-informed environment in fire protection. This project produced a consensus document (NUREG/CR-6850 (EPRI TR-1011989), "EPRI/NRC-RES Fire PRA Methodology for Nuclear Power Facilities," issued September 2005) that addresses fire risk for at-power operations. This report provides high-level qualitative guidance and quantitative screening guidance for conducting a fire HRA. However, this document does not provide a detailed quantitative methodology to develop best-estimate human error probabilities (HEPs) for human failure events under fire-generated conditions.

Objective

The overall objective of the effort is to develop fire HRA methods beyond those currently in NUREG/CR-6850 (EPRI TR-1011989) and to develop an HRA methodology and approach suitable for use in a fire PRA.

The intent of the fire HRA guidance developed through this effort is to support plants making the transition to 10 CFR 50.48(c) and NRC reviewers evaluating the adequacy of submittals from licensees making that transition. It may also be used as a general fire PRA tool for HRA.

Approach

RES has worked collaboratively with EPRI to develop a methodology and associated guidance for performing quantitative HRAs for post-fire mitigative human actions modeled in a fire PRA. The NRC issued NUREG-1921 (EPRI 1023001), "EPRI/NRC-RES Fire Human Reliability Analysis Guidelines—Final Report," in July 2012 (Figure 7.3). It provides the following three approaches to quantification: (1) screening, (2) scoping, and (3) detailed HRA. Screening is based on the guidance in NUREG/CR-6850 (EPRI TR-1011989) with some additional guidance for scenarios with long time windows. Scoping is a new approach to quantification developed specifically to support the iterative nature of fire PRA quantification. The intent of scoping is to provide less conservative HEPs than screening; however, scoping requires fewer resources than a detailed HRA. For detailed HRA quantification, the NRC has developed guidance on how to apply existing methods to assess post-fire HEPs.

Future Work

The NRC has added an HRA module to the NRC-RES/EPRI Fire PRA Workshop to provide in-depth training on the use of this methodology. The joint fire HRA methodology development team delivered the fire HRA training at the 2010, 2011, and 2012 workshops and plans to deliver this training at future fire PRA workshops. In addition to delivering the fire HRA training the fire HRA methodology development team has been tasked with providing NRR with expert fire HRA consulting as needed on NUREG-1921 methodologies as NRR performs reviews of licensee submittals. The team will also assist NRR with the development of responses to NFPA 805 Frequently Asked Questions (FAQs) regarding HRA and will provide support for other future activities that require fire HRA expertise.

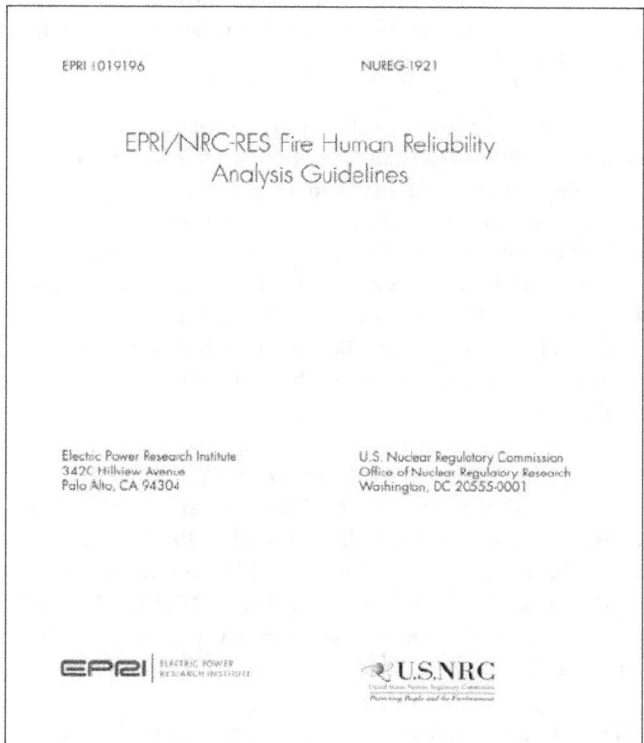

Figure 7.3 NUREG-1921 cover page

For More Information
Contact Kendra Hill, RES/DRA, at Kendra.Hill@nrc.gov or Susan E. Cooper, RES/DRA, at Susan.Cooper@nrc.gov.

Fire Modeling Activities

Background

The results of the Individual Plant Examination of External Events program and actual fire events indicate that fire can be a significant contributor to nuclear power plant (NPP) risk, depending on design and operational conditions. Fire models can be used to evaluate fire scenarios in risk assessments, determine damage to cables and other systems and components important to safety, and characterize the progression of fire beyond initial targets. Used in these ways, fire models are important tools in determining the contribution of fire to the overall risk in NPPs.

Objective

The objective of this program is to provide methodologies, tools, and support for the use of fire modeling in NPP applications.

Approach

In 2004, the U.S. Nuclear Regulatory Commission (NRC) amended its fire protection requirements to allow existing reactor licensees to voluntarily adopt the fire protection requirements in National Fire Protection Association (NFPA) Standard 805, "Performance-Based Standard for Fire Protection for Light Water Reactor Electric Generating Plants," which allows licensees to use fire models as part of their fire protection programs. However, the fire models are subject to verification and validation (V&V), and the NRC must find them acceptable to ensure the quality and integrity of the modeling. To this end, the NRC Office of Nuclear Regulatory Research (RES), the Electric Power Research Institute (EPRI), and the National Institute of Standards and Technology (NIST) conducted an extensive V&V study of fire models used to analyze NPP fire scenarios. This study resulted in the seven-volume report NUREG-1824, "Verification and Validation of Selected Fire Models for Nuclear Power Plant Applications," issued May 2007.

The NRC and its licensees use the results in NUREG-1824 to provide confidence in the predictive capabilities of the various models evaluated. These insights are valuable to fire model users who are developing analyses to support a transition to NFPA Standard 805 to justify alternatives to existing prescriptive regulatory requirements and to conduct significance determination process reviews under the Reactor Oversight Process.

The NRC completed a phenomena identification and ranking table study of fire modeling (NUREG/CR-6978, "A Phenomena Identification and Ranking Table (PIRT) Exercise for Nuclear Power Plant Fire Modeling Applications," issued November 2008) that identified important fire-modeling capabilities needed to improve the agency's confidence in the results. This study helps define future research priorities in fire modeling.

Fire risk assessments often need to determine when cables will fail during a fire in NPPs. As part of the Cable Response to Live Fire (CAROLFIRE) program, the NRC and NIST have developed a simple cable damage model named Thermally Induced Electrical Failure (THIEF). NUREG/CR-6931, "Cable Response to Live Fire (CAROLFIRE)," issued April 2008, documents the test results and model. Volume 3 of CAROLFIRE describes how the THIEF model uses empirical information about cable failure temperatures and calculations of the thermal response of a cable to predict the time to cable damage. The NRC benchmarked and validated the THIEF model against real cable failure and thermal data acquired during the CAROLFIRE program.

NIST used the THIEF model in both two-zone and computational fluid dynamics models. In addition, the NRC incorporated the THIEF model in its fire dynamics tools spreadsheets. (See NUREG-1805, "Fire Dynamics Tools (FDTs) Quantitative Fire Hazard Analysis Methods for the U.S. Nuclear Regulatory Commission Fire Protection Inspection Program," issued December 2004.) The THIEF spreadsheet is a useful tool for inspectors and licensees to quickly determine the likelihood of cable damage from a fire or to indicate the need for further analysis.

Recently, the NRC completed another joint project with EPRI and NIST to develop technical guidance to assist in the conduct of fire-modeling analyses of NPPs. NUREG-1934, "Nuclear Power Plant Fire Modeling Analysis Guidelines (NPP FIRE MAG)," issued November 2012, expands on NUREG-1824 by providing users with best practices from experts in fire modeling and NPP fire safety.

This application guide contains five commonly available fire modeling tools (FDTs, Fire-Induced Vulnerability Evaluation (Revision 1), Consolidated Fire Growth and Smoke Transport Model, MAGIC, and Fire Dynamics Simulator (FDS)) that were developed by nuclear power stakeholders or that were applied to NPP fire scenarios. Previously, RES, EPRI, and NIST used these same models in the V&V study documented in NUREG-1824. Figure 7.4 illustrates an isometric view of a room in an NPP and shows the temperature profile above an electrical cabinet fire in a fire dynamics simulation.

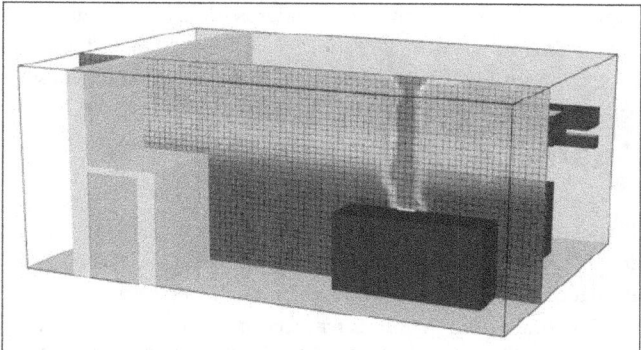

Figure 7.4 Graphical output from FDS/Smokeview fire model

NUREG-1934 will assist both the user performing the calculation and the reviewers. The report includes guidance on selecting appropriate models for a given fire scenario and on understanding the levels of confidence that can be attributed to the model results. The report will also form the foundation for future fire model training under development by RES and EPRI.

Future Work

The NRC is continuing to update the fire modeling tools, expand the V&V effort, and develop additional model input data. An updated, expanded edition of NUREG-1824 is in the developmental stages with EPRI. The NRC is finalizing Supplement 1 to NUREG-1805 to be issued in 2013. Supplement 1 will include the THIEF model and updated versions of the spreadsheets that are currently documented in NUREG-1805.

For More Information
Contact David Stroup, RES/DRA, at David.Stroup@nrc.gov.

Cable Heat Release, Ignition, and Spread in Tray Installations During Fire

Background

Fire can be a significant contributor to nuclear power plant risk. In 1975, a serious fire involving electrical cables occurred at the Browns Ferry Nuclear Power Plant, operated by the Tennessee Valley Authority. Nuclear power plants typically contain hundreds of miles of electrical cables. The burning behavior of cables in a fire depends on a number of factors, including their constituent materials and construction and their location and installation geometry. Burning cables can propagate flames from one area to another, or they can add to the amount of fuel available for combustion. Burning cables also produce smoke containing toxic and corrosive gases. The lower the heat exposure required to ignite the electrical cables, the greater the fire hazard in terms of ignition and flame spread. Electrical cables exposed to fire can lose physical integrity (i.e., melting of the insulation) and insulation resistance, thus leading to an electrical breakdown, a short circuit, or the spread of fire to other cables or combustibles.

The amount of experimental evidence and analytical tools available to calculate the effects of cable tray fires is relatively small when compared to the vast number of possible fire scenarios. Many of the large-scale fire tests conducted with cables are qualification tests in which the materials are tested in a relatively realistic configuration and qualitatively ranked on a comparative basis. This type of test typically does not address the details of fire growth and spread and does not provide useful data for realistic fire-risk and -model calculations.

Objective

The Cable Heat Release, Ignition, and Spread in Tray Installations during Fire (CHRISTIFIRE) experimental program is an effort to quantify the mass and energy released from burning electrical cables. The program includes fire tests on grouped electrical cables to enable better understanding of the fire hazard characteristics, including the ignition, heat release rate, and flame spread. The U.S. Nuclear Regulatory Commission (NRC) will use this type of quantitative information to develop more realistic models of cable fires for use in fire probabilistic risk assessment (PRA) analyses, such as those performed using the methods in NUREG/CR-6850 (Electric Power Research Institute (EPRI) TR-1011989), "EPRI/NRC-RES Fire PRA Methodology for Nuclear Power Facilities," issued September 2005, in applications under National Fire Protection Association Standard 805, "Performance-Based Standard for Fire Protection for Light Water Reactor Electric Generating Plants."

Approach

Phase 1 of CHRISTIFIRE included experiments ranging from micro-scale to full scale. Small samples of cable jackets and insulation were burned within a calorimeter to measure the heat of combustion, pyrolysis temperature, heat release capacity, and residue yield. Meter-long cable segments were slowly fed through a small tube furnace, and a variety of spectrometric techniques measured the composition of the effluent. The standard cone calorimeter test measured the heat release rate per unit area for a variety of cable types at several external heat fluxes.

A large radiant panel apparatus (Figure 7.5), specially designed for this test program, measured the burning rate of cables when installed in ladderback trays. Finally, a series of 26 multiple-tray full-scale experiments assessed the effect of changing the vertical tray spacing, tray width, and tray fill (Figure 7.6).

Figure 7.5 Radiant panel cable tray fire test (side view of burning cables in a tray exposed to a radiant heat source)

Figure 7.6 Burning cables during cable tray fire test (side view of burning cables in trays during a multiple-tray test after ignition using a small gas burner)

During Phase 1, the NRC developed a simple model of flame spread in horizontal tray configurations (called Flame Spread over Horizontal Cable Trays (FLASHCAT]) that makes use of semi-empirical estimates of lateral and vertical flame spread and measured values of combustible mass, heat of combustion, heat release rate per unit area, and char yield. The NRC completed Phase 1 in 2011 and documented the results in NUREG/CR-7010, "Cable Heat Release, Ignition, and Spread in Tray Installations during Fire (CHRISTIFIRE)—Phase 1: Horizontal Trays," Volume 1, issued July 2012.

Figure 7.7 Burning cables in vertical trays (side view of burning cables in two vertical trays after ignition using a small gas burner)

Phase 2 of the CHRISTIFIRE project examined flame spread on cables in trays oriented in the vertical direction and the impact of an enclosure on cable flame spread in multiple horizontal trays. A series of 17 experiments were conducted using two vertical cable trays that were installed adjacent to each other (Figure 7.7). A series of 10 experiments were conducted using multiple horizontal trays located in a simulated hallway relatively close to the wall and ceiling (Figure 7.8). The results of these experiments, along with additional cone calorimeter measurements, will be used to extend application of the FLASHCAT model. The NRC is in the process of documenting the Phase 2 test results and will publish them later in 2013 in Volume 2 of NUREG/CR-7010, "Cable Heat Release, Ignition, and Spread in Tray Installations during Fire (CHRISTIFIRE) – Phase 2: Vertical Shafts and Corridors."

Figure 7.8 Burning cables in hallway enclosure (end view of burning cables in two horizontal trays in hallway enclosure)

Future Work

CHRISTIFIRE was the first attempt at developing a more realistic understanding of the burning behavior of grouped cables. Based on its success, future phases of the project will examine the effectiveness of various methods of protection for electrical cables. The FLASHCAT model will be validated and extended to other configurations. The first two-phases are complete. Additional phases of the project are currently under development.

For More Information
Contact David Stroup, RES/DRA, at David.Stroup@nrc.gov.

Direct Current Electrical Shorting in Response to Exposure Fire (DESIREEFIRE)

Background

The Individual Plant Examination of External Events program results and actual fire events indicate that fire can be a significant contributor to nuclear power plant (NPP) risk. The question on how to determine risk resulting from fire damage to electrical cables in NPPs has been of concern since the Browns Ferry Nuclear Power Plant (BFN) fire in 1975. In earlier years, it was generally believed that any system that depended on electric cables passing through a compartment damaged by fire would be unavailable for its intended safety function. The BFN fire and recent testing have prompted wider understanding that short circuits involving an energized conductor can pose considerably greater risk. The resultant "hot shorts" (Figure 7.9) can cause systems to malfunction and inadvertently reposition motor-operated valves and start or stop plant equipment. Plant safety analyses need to account for this risk.

A consensus on the likelihood of hot shorts given fire-damaged cables did not exist in the late 1990s. The Nuclear Energy Institute and the Electric Power Research Institute (EPRI) conducted a testing program in 2001, and the U.S. Nuclear Regulatory Commission (NRC) conducted follow-on testing in its Cable Response to Live Fire (CAROLFIRE) program in 2006. Volumes 1–3 of NUREG/CR-6931, "Cable Response to Live Fire (CAROLFIRE)," issued April 2008, document the CAROLFIRE results. These programs produced a vast amount of data and knowledge related to fire-induced circuit failures of alternating current (ac) circuits. However, none of the previous testing explicitly explored the fire-induced circuit failure phenomena for direct current (dc). Both current operating plants and the proposed new reactor designs use dc circuits to operate numerous safety-related systems.

Some recent tests performed by industry indicate that the results for ac circuits may not be fully representative of what might occur from fire-induced damage to dc circuits. Because of the differences in the operating voltages and circuit design between ac and dc, the previous data gathered for ac circuits may not be applicable to dc circuits.

Objective

The Direct Current Electrical Shorting in Response to Exposure Fire (DESIREEFIRE) testing (Figures 7.10 and 7.11) of risk-

significant dc circuits will allow the fire protection community to better understand dc circuit failure characteristics.

Approach

The NRC staff elected to perform fire testing of dc circuits using configurations that are representative of safety-significant circuits and components used in NPPs to better understand the probability of spurious actuations and the duration of those actuations in dc circuits.

The DESIREEFIRE testing program used small- and intermediate-scale tests to evaluate the response of dc circuits to fire conditions. The tests included several different circuits, as follows:

- Direct current motor starters.
- Pilot solenoid-operated valve coils.
- Medium-voltage circuit breaker control.

The DESIREEFIRE project is another Office of Nuclear Regulatory Research fire research project established under a memorandum of understanding to perform collaborative research with EPRI. This agreement has provided various components and cabling to the DESIREEFIRE testing program at little or no cost to the NRC.

Figure 7.9 Example of a dc electrical cable hot short

It also provided industry expert advice on the various aspects of the dc power system and circuit design. Testing is complete, and NUREG/CR-7100, "Direct Current Electrical Shorting in Response to Exposure Fire (DESIREEFIRE)," issued April 2012, documents the results.

Figure 7.10 Intermediate-scale dc fire tests

Figure 7.11 Battery bank for dc fire tests

Future Work

The determination for future cable testing programs will be based on the outcome of the fire effects on electrical cables and impact on NPP system performance phenomena identification and ranking table (PIRT) and expert elicitations that are currently underway. Preliminary areas for future research identified by the PIRT panel include evaluating the fire-induced effects on instrumentation circuits, electrical panel/cabinet wiring, surrogate ground path failure mode, current transformers, and high conductor count trunk cables.

For More Information
Contact Gabriel Taylor, RES/DRA, at
Gabriel.Taylor@nrc.gov.

Fire Effects on Electrical Cables and Impact on Nuclear Power Plant System Performance: Phenomena Identification and Ranking Table and Expert Elicitation Programs

Background

Beginning in 1997, the U.S. Nuclear Regulatory Commission (NRC) staff noticed a series of licensee event reports related to potential plant-specific problems involving fire-induced electrical cable circuit failures. The staff issued Information Notice 99-17, "Problems Associated with Post-Fire Safe-Shutdown Circuit Analyses," dated June 3, 1999, to alert the industry. Under the leadership of the Nuclear Energy Institute (NEI), the industry performed a joint series of fire tests with the Electric Power and Research Institute (EPRI) to better understand the issue. The industry used an expert elicitation to review the results and to provide recommendations on their use in probabilistic risk assessments (PRAs). EPRI 1003326, "Characterization of Fire-Induced Circuit Faults—Results of Cable Fire Testing," issued December 2002, documents the testing and expert panel results.

On February 19, 2003, the NRC sponsored a facilitated public workshop to discuss the results of the NEI/EPRI tests. Following the workshop, the NRC issued Regulatory Issue Summary 2004-03, "Risk-informed Approach for Post-Fire Safe-Shutdown Circuit Inspections," dated December 29, 2004. This report identifies a number of areas that require additional testing. The NRC Office of Nuclear Regulatory Research (RES) initiated the Cable Response to Live Fire (CAROLFIRE) test program to address these concerns and documented the results in Volumes 1–3 of NUREG/CR-6931 "Cable Response to Live Fire (CAROLFIRE)," issued April 2008. In 2006, a licensee performed independent testing of ungrounded direct current (dc) circuits and obtained unexpected results.

In 2009–2010, the NRC and EPRI initiated the Direct Current Electrical Shorting in Response to Exposure Fire (DESIREEFIRE) testing program to better understand the performance of dc circuits. This testing program used small- and intermediate-scale tests to evaluate the response of dc electric cables and circuits to fire conditions. Several different circuits were tested, including dc motor starters, pilot solenoid-operated valve coils, and medium-voltage circuit breaker control circuits. NUREG/CR-7100, "Direct Current Electrical Shorting in Response to Exposure Fire (DESIREEFIRE): Test Results," issued April 2012, documents the results of this testing. Additionally, RES issued, for public comment, NUREG-2128, "Electrical Cable Test Results and Analysis during Fire Exposure (ELECTRAFIRE)," issued June 2012. This report consolidates the three major fire-induced circuit and cable failure experiments performed between 2001 and 2011.

Objective

Following the development of circuit failure probabilities in NUREG/CR-6850 (EPRI TR-1011989), "EPRI/NRC-RES Fire PRA Methodology for Nuclear Power Facilities," issued September 2005, the NRC added the following two major fire testing programs on cable hot shorting: (1) CAROLFIRE in 2008 and (2) DESIREEFIRE in 2011. The objective of these phenomena identification and ranking table (PIRT) and expert elicitation programs is to advance the state-of-the-art in regard to understanding and predicting hot shorting when cables are exposed to fire conditions.

Approach

The NRC convened two separate expert panels. The first was comprised of electrical engineering experts who reviewed all currently available testing data. This panel followed the NRC's PIRT process to determine the state-of-the-art in predicting hot shorting when cables are exposed to fire conditions. The results of this work are documented in NUREG/CR-7150, Vol. 1, "Joint Assessment of Cable Damage and Quantification of Effects from Fire (JACQUE-FIRE)," issued October 2012.

The second expert panel will comprise fire PRA experts to explore and advance the state-of-the-art in determining realistic probabilities of hot shorting when cables are exposed to fire conditions. This panel will follow the NRC's Senior Seismic Hazard Analysis Committee (SSHAC) process for conducting expert assessments (NUREG-2117, "Practical Implementation Guidelines for SSHAC Level 3 and 4 Hazard Studies," issued April 2012). The results from this work will be documented in Volume 2 of JACQUE-FIRE.

Figure 7.12 below illustrates a typical PIRT panel discussion in progress.

Figure 7.12 A typical expert panel discussion

Future Work

The determination for future cable electrical functionality testing will be based on these PIRT and expert elicitations.

For More Information

Contact Gabriel Taylor, RES/DRA at Gabriel.Taylor@nrc.gov, or Nicholas Melly, RES/DRA at Nicholas.Melly@nrc.gov.

Advancements in Understanding Fire-Induced Effects on Electrical Circuits

Background

The results of the Individual Plant Examination of External Events program and actual fire events indicate that fire can be a significant contributor to nuclear power plant risk. Fire probabilistic risk assessment (PRA) provides a structured, integrated approach to evaluate the impact of fire to safety systems and safe operation of nuclear power plants. In 1995, the U.S. Nuclear Regulatory Commission (NRC) adopted a policy statement on PRA with the intent to increase the use of this technology in all regulatory matters to the extent supported by the state-of-the-art in PRA methods and data. In 2004, the NRC amended its fire protection requirements to allow existing reactor licensees to voluntarily adopt the risk-informed, Performance-Based rule in Title 10 of the *Code of Federal Regulations* (10 CFR) 50.48(c). This rule endorses National Fire Protection Association (NFPA) Standard 805, "Performance-Based Standard for Fire Protection for Light Water Reactor Electric Generating Plants," as an alternative to the existing prescriptive fire protection requirements. Licensees will need a fire PRA to realize the full benefits of making the transition to the risk-informed, Performance-Based standard. In September 2005, the NRC Office of Nuclear Regulatory Research (RES), in joint collaboration with the Electric Power Research Institute (EPRI), published NUREG/CR-6850 (EPRI TR-1011989), "EPRI/NRC-RES Fire PRA Methodology for Nuclear Power Facilities." This report and its associated supplements provide detailed methodology for developing a fire PRA. This methodology provides a comprehensive guide for developing a fire PRA using the state-of-the-art knowledge and science at the time of its development.

Although the EPRI/NRC-RES method is state-of-the-art, additional test data and improved tools could further advance the methodology in several areas. In 2011–2012, the NRC worked on a number of areas to help advance the state-of-the-art fire PRA methods. These efforts included testing fire retardant cable coatings, testing a unique cable type known as Kerite-FR™, and analyzing fire-induced cable failure results from three major testing programs. The results of this work assisted in advancing the state of knowledge and improving the realism of fire PRA to better understand plant risk from the effects of fire.

Objective

The following three specific objectives were accomplished with support from Sandia National Laboratories (SNL) and EPRI:

1. The Kerite Analysis in Thermal Environment of Fire (KATEFire) test results document the performance of cable types manufactured by Kerite-FR.

2. A study of cable fire retardant coating performance was conducted to evaluate the coatings effects on delay time to functional failure and delay cable ignition.

3. A comprehensive review of the three major fire-induced cable damage programs was conducted to provide a reference to the parameters that influence cable failure.

Approach

The Kerite-FR testing program used small-scale radiant and larger scale open flame thermal exposures to damage several types of Kerite-FR electrical cables and monitored the cables electrical response during the severe thermal exposure. The results demonstrated that Kerite-FR should not use the generic thermoset-insulated cable threshold of 330 degrees Celsius, but could use a damage threshold slightly above the generic thermoplastic threshold of 205 degrees Celsius. Figure 7.13 shows the Kerite-FR material cracking that occurred during testing and that was part of the failure mechanisms that led to cable failure. The other types of Kerite tested (FRII, FRIII, and HTK) performed in excess of the generic thermoset threshold. The NRC documented the results of this work in NUREG/CR-7102, "Kerite Analysis in Thermal Environment of Fire (KATEFire): Test Results," issued December 2011.

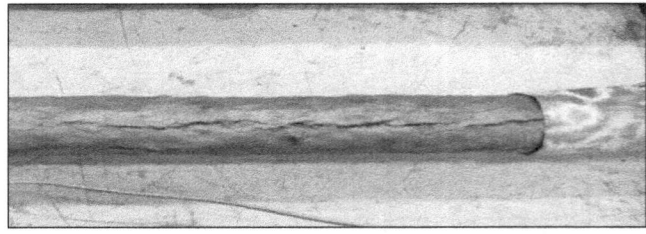

Figure 7.13 Photo of Kerite-FR cable after testing, showing the cracking of insulation material

The cable fire retardant coating testing followed a similar approach; however, only the small-scale radiant testing apparatus was used. Both coated and uncoated cable assemblies were tested, and the delay in time to damage and time to ignition were evaluated. Figure 7.14 provides an illustration of a coated electrical cable. The preliminary results showed that coatings provide little to no benefit in delaying cable functional damage.

Additional testing is planned for 2013 with a NUREG/CR report documenting the results which is expected to be published in 2014.

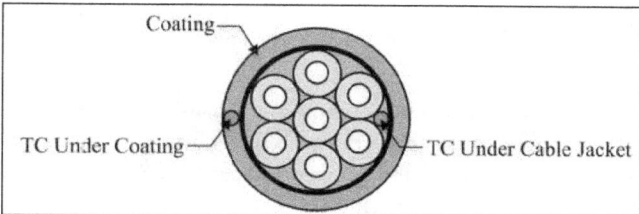

Figure 7.14 Illustration of cables with coating applied and thermocouples installed

In support of the phenomena identification and ranking table (PIRT) exercise on fire-induced damage to electrical cables, the NRC, in collaboration with EPRI and SNL, performed a comprehensive review of the three major fire-induced cable damage testing programs. The work used a graphical analysis approach to display the data in a manner that would identify trends on spurious operation likelihood and spurious operation duration. The analysis also shows that multiple cable shorts to ground can cause spurious operations resulting from an ungrounded and compatible power supply. NUREG-2128, "Electrical Cable Test Results and Analysis during Fire Exposure (ELECTRAFIRE)," issued for public comment in June 2012, documents the results of this work. The NRC expects to publish a final version of NUREG-2128 in 2013.

Future Work

Future work is currently being planned to explore the effects of cable tray attributes, such as tray cover effects on time to cable damage. The prioritization for future cable testing programs will be based on the outcome of the fire-induced cable PIRT and expert elicitations. In addition, the results of the PIRT and expert elicitation projects will also be used to update the state-of-the-art fire PRA methods and data in NUREG/CR-6850 (EPRI TR-1011989). The results of the PIRT report were published in October 2012 in NUREG/CR-7150, "Joint Assessment of Cable Damage and Quantification of Effects from Fire (JACQUEFIRE)—Final Report," Volume 1, "Phenomena Identification and Ranking Table (PIRT) Exercise for Nuclear Power Plant Fire-Induced Electrical Circuit Failure," issued October 2012. The expert elicitation results will be published in 2013.

For More Information
Contact Gabriel Taylor, RES/DRA, at Gabriel.Taylor@nrc.gov.

Beyond-Design-Basis Fires for Spent Fuel Transportation: Shipping Cask Seal Performance Testing

Background

The U.S. Nuclear Regulatory Commission (NRC) is obtaining data to determine the performance of seals in spent fuel transportation packages during beyond-design-basis fires, similar to the Baltimore Tunnel Fire in 2001. The performance of the package seals is important for determining the potential release of radioactive material from a package during a beyond-design-basis accident. The seals have lower temperature limits than other package components and are a vital part of the containment barrier between the environment and the cask contents.

NUREG/CR-6886, "Spent Fuel Transportation Package Response to the Baltimore Tunnel Fire Scenario," issued November 2006, describes, in detail, an evaluation of the potential release of radioactive materials from three different spent fuel transportation packages. This evaluation used estimates of temperatures resulting from the 2001 Baltimore Tunnel Fire as boundary conditions for finite-element models to determine the temperature of various components of the packages, including the seals. For two of the packages, the model-estimated temperatures of the seals exceeded their continuous-use rated service temperature, which means that the release of radioactive material could not be ruled out with available information. However, for both of those packages, the analysis determined through a bounding calculation that the maximum expected release would be well below the regulatory safety requirements in Title 10 of the *Code of Federal Regulations* (10 CFR) Part 71, "Packaging and Transportation of Radioactive Material," for a release from a spent fuel package during these beyond-design-basis accident conditions. The study concluded that neither spent nuclear fuel particles nor fission products would be released from a spent fuel transportation package carrying intact spent fuel because the peak fuel cladding temperature is conservatively predicted to remain below the short-term limit of 570 degrees Celsius (C) (1,058 degrees Fahrenheit).

The Office of Nuclear Material Safety and Safeguards evaluates the integrity of spent fuel packages using finite-element computer models. NUREG/CR-6886 states that, during beyond-design-basis accident conditions, the failure of the seals could not be ruled out and that the decision was made to perform small-scale testing to quantify the performance envelope of O-ring seals. These data can be later used in the evaluation and analysis of finite-element computer models of spent fuel packages.

Objective

The objective of this test program was to quantify the performance envelope of O-ring seals under beyond-design-basis accident conditions and to estimate package leakage rates under these conditions.

Approach

The experimental apparatus comprised a small-scale vessel fabricated to standard specifications under American Society of Mechanical Engineers B16.5-2009, "Pipe Flanges and Flanged Fittings NPS ½ through NPS 24 Metric/Inch Standard," issued 2009, with an internal cavity of 100-milliliter nominal internal volume filled with helium to a pressure of 5 bars (72.5 pounds per square inch) for the 12 metallic seal tests and 2 bars (29 pounds per square inch) for the 2 polymeric seal tests. An electric furnace with an internal dimension of 25.4 centimeters (cm) x 25.4 cm x 40.64 cm (10 inches x 10 inches x 16 inches) was used to uniformly heat the vessel. Pressure and temperature were monitored for several days before the test to ensure that the vessel had no leaks, during the test to monitor for leakage, and for several days after the test to achieve cool down and pressure stability.

No catastrophic vessel leakage (e.g., loss of all vessel pressure) was observed in any of the tests. Small leaks did occur in several of the tests. The metallic seals tested at 800 degrees C began to experience a small leak several hours into the test; however, the seals did not lose all pressure even after several days of cool down and pressure monitoring (see Figure 7.15).

The polymeric seal, which was tested at 450 degrees C for 25 hours, also continued to hold pressure after cool down. This result was surprising because the integrity of the ethylene propylene seal was compromised (i.e., the seal had transformed into a powderlike material). The pressure boundary was most likely maintained because of tight clearances between the test vessel body and head.

Results

During 2010 and 2011, the Office of Nuclear Regulatory Research Division of Risk Analysis/Fire Research Branch and the National Institute of Standards and Technology (NIST) performed small-scale thermal tests to gather data on the performance of O-ring seals used in spent nuclear fuel transportation packages in beyond-design-basis thermal excursions. Under severe transportation accident conditions, O-ring seals are generally the first components of spent fuel transportation packages to reach their operational temperature limits.

The final test report on the first phase of testing was published in April 2012 as NUREG/CR-7115, "Performance of Metal and Polymeric O-ring Seals in Beyond-design-basis Temperature Excursions" (Agencywide Documents Access and Management System Accession No. ML12110A066.)

Future Work

The next phase of small scale testing is currently in process with NIST and will include further characterization of different material of polymeric seals and testing of double O-ring seal configurations in order to investigate the effect of multiple seals in failure times and temperature exposures.

For More Information
Contact Felix E. Gonzalez, RES/DRA, at Felix.Gonzalez@nrc.gov.

Figure 7.15 Pictures of the small-scale test vessel after 800 degrees C exposure for 9 hours (small-scale test vessel [top left], vessel head after disassembly [top right], and vessel body and metallic seal after disassembly (bottom left and bottom right])

Training Programs for Fire Probabilistic Risk Assessment, Human Reliability Analysis, and Fire Modeling

Background

In 1995, the U.S. Nuclear Regulatory Commission (NRC) adopted a policy statement on probabilistic risk assessment (PRA) that was intended to increase the use of PRA technology in all regulatory matters to the extent supported by the technical merit of the PRA methods and data. In 2004, the NRC amended its fire protection requirements to allow existing reactor licensees to voluntarily adopt the risk-informed, Performance-Based requirements in Title 10 of the *Code of Federal Regulations* (10 CFR) 50.48(c). This rule endorses National Fire Protection Association (NFPA) Standard 805, "Performance-Based Standard for Fire Protection for Light Water Reactor Electric Generating Plants," as an alternative to current prescriptive fire protection requirements. Approximately one-half of the current licensed nuclear power plants (NPPs) plan to make the transition to this new rule. To realize the full benefits of making the transition to the risk-informed, Performance-Based standard, plants will need to perform a fire PRA. The fire protection inspection program also uses fire PRAs to perform other regulatory activities, such as the significance determination process for inspection findings. Many NPPs use the joint Electric Power Research Institute (EPRI) and NRC document NUREG/CR-6850 (EPRI TR-1011989), "EPRI/NRC-RES Fire PRA Methodology for Nuclear Power Facilities," issued September 2005, to create fire PRAs for at-power operations. The NRC staff uses NUREG/CR-6850 (EPRI TR-1011989) to support reviews of license amendment requests that a licensee submits when transitioning its fire protection program to NFPA Standard 805. As part of a pilot plant's transition to 10 CFR 50.48(c), the NRC and EPRI have jointly produced interim solutions to fire PRA issues concerning the implementation of NUREG/CR-6850 (EPRI TR-1011989) in the NFPA Standard 805 frequently asked questions (FAQ) program.

The staff also published NUREG-1921 (EPRI 1023001), "EPRI/NRC-RES Fire Human Reliability Analysis Guidelines—Final Report," issued July 2012, for use in developing human reliability analysis (HRA) components of fire PRAs. The Office of Nuclear Regulatory Research (RES) (again in partnership with EPRI) also finalized NUREG-1934 (EPRI 1019195), "Nuclear Power Plant Fire Modeling Analysis Guidelines (NPP FIRE MAG)—Final Report," issued November 2012. The NRC uses this report as the basis for fire model training in the semiannual EPRI/NRC-RES Fire PRA training workshops.

Objective

This program supports the NRC's policy to increase the use of PRA technology by providing training for 10 CFR 50.48(c) and other fire protection programs in fire PRA, circuit analysis, HRA, and fire modeling.

Approach

Since 2005, the NRC and EPRI have jointly conducted training sessions in fire PRA. These sessions, hosted in alternate years by RES and EPRI, are available at no charge to all interested stakeholders. In 2005 and 2006, 3 days of general training covered fire PRA topical areas, including PRA, fire models, and fire circuit analysis. In 2007, training was expanded to 2 weeks per year. The courses offered detailed discussions and hands-on examples for each topical area in parallel for 4 days per week. The 2008 training sessions (Figure 7.16) were video recorded and documented along with their training materials in Volumes 1–3 of NUREG/CP-0194, "Methods for Applying Risk Analysis to Fire Scenarios (MARIAFIRES)," issued July 2010 (Figure 7.17), thus enabling self-study for persons unable to attend the course. This detailed instruction continued through 2009 when it was expanded in 2010 to provide an introduction to fire HRA in NUREG-1921. The fifth class entitled, "Advanced Fire Modeling," was added in 2011. This class uses NUREG-1934 as the text for instruction.

Figure 7.16 Photo from the 2008 NRC-RES/EPRI fire PRA workshop

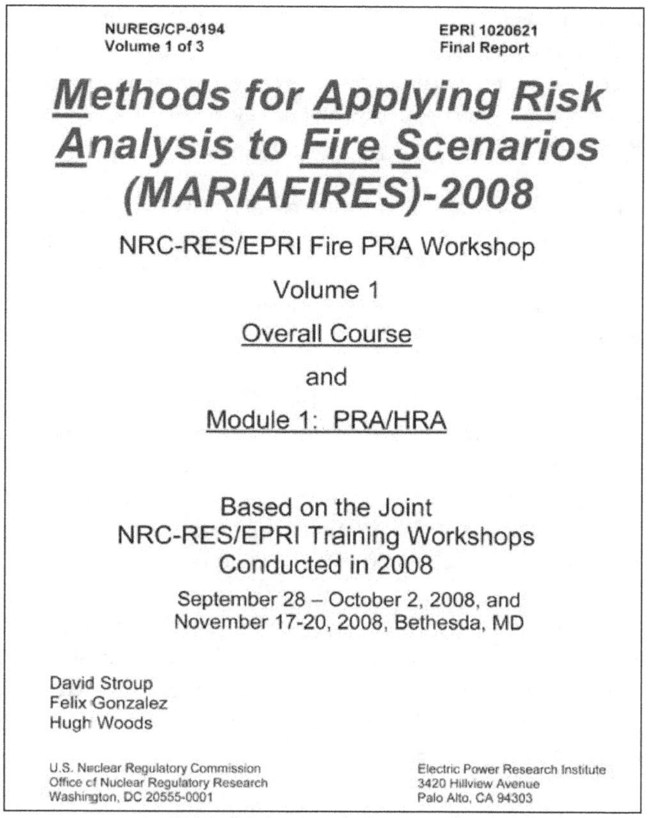

Figure 7.17 NUREG/CP-0194, Volume 1 of 3, cover page (Video recording of the training sessions covered in each volume are included on a DVD in that volume)

In 2009, the NRC endorsed the American Society of Mechanical Engineers/American Nuclear Society PRA standard in Regulatory Guide 1.200, "An Approach for Determining the Technical Adequacy of Probabilistic Risk Assessment Results for Risk-informed Activities." Therefore, the 2010 training has also been updated to include the relationship between NUREG/CR-6850 (EPRI TR-1011989) and the fire PRA standard. Overall, this joint work is producing a higher level of understanding of fire PRA methods that will likely enhance the efficiency of NRC and industry efforts in fire PRA.

Future Work

The fire PRA, HRA, and fire-modeling programs are scheduled to continue into the future. A MARIAFIRES-2010 is in the finalization stages. It will include two volumes, with Volume 1 documenting the four basic concepts modules that were presented on the first day of the 2010 NRC-RES/EPRI Fire PRA Workshop (Basic Concepts of Fire Analysis, Basic Concepts of Fire Human Reliability Analysis, and Basics of Nuclear Power Plant PRA), and Volume 2 documenting the HRA module that was added to the course in 2010. An updated training video was also implemented for the 2012 training class. The training continues to be well attended by all stakeholders. The

2012 course had over 200 participants in the two training sessions. Participants came from a diverse range of backgrounds, including NRC headquarters and regional staff; NPP industry employees and consultants; international regulators and power plant operators; national research laboratories; universities and other Federal agencies, such as the Bureau of Alcohol, Tobacco, Firearms and Explosives; National Institute of Standards and Technology; National Aeronautics and Space Administration; Naval Surface Warfare Center; and Defense Nuclear Facilities Safety Board.

For More Information
Contact
Nicholas Melly, RES/DRA, at Nicholas.melly@nrc.gov for fire PRA;
Kendra Hill, RES/DRA, at Kendra.Hill@nrc.gov for fire HRA;
David Stroup, RES/DRA, at David.Stroup@nrc.gov for fire analysis and fire modeling; and
Gabriel Taylor, RES/DRA, at Gabriel.Taylor@nrc.gov for electrical analysis.

Fire Research and Regulation Knowledge Management

Background

The results of the Individual Plant Examination of External Events program and actual fire events indicate that fire can be a significant contributor to nuclear power plant (NPP) risk, depending on design and operational conditions. During the last 30 years, the U.S. Nuclear Regulatory Commission (NRC) has undertaken many studies to better understand fire hazards, fire events, and fire risk in NPPs. The Fire Research Branch (FRB) in the Office of Nuclear Regulatory Research (RES) initiated the Fire Research and Regulation Knowledge Base Project to assemble the collection of NRC fire-related publications issued over the past 30 years. FRB has also undertaken a similar project to document and preserve the history of the influential Browns Ferry Nuclear Power Plant (BFN) fire of 1975 and has published NUREG/BR-0364, "A Short History of Fire Safety Research Sponsored by the U.S. Nuclear Regulatory Commission, 1975–2008," issued June 2009, to document the agency's research activities.

Objective

The objective of this research is to support the NRC's knowledge management initiative in the fire protection area by identifying relevant information to be documented.

Approach

NUREG/BR-0465, "Fire Protection and Fire Research Knowledge Management Digest," issued February 2010

The Fire Research and Regulation Knowledge Base is a user-friendly database that provides information needed during such activities as inspections and reviews. The database includes publicly available documents, such as Title 10 of the *Code of Federal Regulations* (10 CFR) Part 50, "Domestic Licensing of Production and Utilization Facilities"; guidelines for fire protection in NPPs; fire inspection manuals; fire inspection procedures; generic letters; bulletins; information notices; circulars; administrative letters; regulatory issue summaries; and regulatory guides. The technical knowledge includes NRC technical publications (i.e., NUREGs) that serve as background information to the regulatory documents. It includes reports of NRC-sponsored fire experiments, studies, and probabilistic risk assessments (PRAs). These documents often provide the

technical bases and insights for fire protection requirements and guidelines. NUREG/KM-0003 is being prepared and will supersede NUREG/BR-0465. This new report will update and expand the information currently available in NUREG/BR-0465 Rev. 1.

NUREG/KM-0002, "The Browns Ferry Nuclear Plant Fire of 1975 Knowledge Management Digest," supersedes NUREG/BR-0361, "The Browns Ferry Nuclear Plant Fire of 1975 and the History of NRC Fire Regulations," issued February 2009

In 1975, a fire occurred at BFN that challenged the operators' ability to safely shut the plant down. The fire prompted a new series of fire protection regulations and is a formative event in the history of fire protection regulations for NPPs. The NUREG/KM and DVD on the BFN plant fire of 1975 (Figure 7.18) contain all major public documents, publications, regulations, and presentations pertaining to the BFN fire in a one-stop information resource with a user-friendly format to provide a well-informed perspective about the BFN fire. Combined, these sources create a well-rounded picture of the event for varied types and levels of users; individually, they paint a detailed picture of specific aspects of the event.

Figure 7.18 Screenshot of NUREG/KM-0002 (DVD main menu)

NUREG/BR-0364 , "A Short History of Fire Safety Research Sponsored by the U.S. Nuclear Regulatory Commission, 1975-2008"

The knowledge management program NUREG/BR-0364, is divided into four separate areas:

1. <u>1975–1987</u>. The Fire Protection Research Program investigated the effectiveness of changes made to the NRC's fire protection regulations after the 1975 BFN fire.

2. 1987–1993. Early fire PRAs were conducted (e.g., the LaSalle Risk Methods Integration and Evaluation Program (RMIEP]).

3. 1993–1998. Incremental improvements were made to the RMIEP methods.

4. 1998–present. Methods were developed to better apply the Commission's PRA technology policy to fire risk technology (to be used, where practical, in all regulatory matters).

Future Work

An update to NUREG/BR-0465 to a new category of NUREG reports is in process. Similarly to NUREG/BR-0361 (now NUREG/KM-0002), the report will be revised to the NUREG/KM, which will be the classification for knowledge management (KM) reports. The NRC conducted this work in 2012, and it continues into the future.

For More Information
Contact David Gennarto, RES/DRA, at David.Gennarto@nrc.gov, or Felix Gonzalez, RES/DRA, at Felix.Gonzalez@nrc.gov.

Evaluation of Very Early Warning Fire Detection System Performance for Fire Probabilistic Risk Assessment (PRA) Applications

Background

Fire protection programs in U.S. nuclear power plants (NPPs) use the concept of defense-in-depth to achieve the required degree of fire safety by using echelons of protection from fire effects. The three echelons for fire protection are:

- Prevent fire from starting.

- Rapidly detect, control, and promptly extinguish those fires that do occur

- Protect structures, systems, and components important to safety so that a fire not promptly extinguished by the fire suppression activities will not prevent the safe shutdown of the plant.

Fire detection systems provide a fundamental means of detecting fire combustion products such that automatic or manual suppression activities can be initiated.

A common fire detector type is the smoke detector which, at its basic level, is a particle detector. Conventional smoke detectors work on one of two technology types–ionization or photoelectric–and are typically of a point-type design (spot detector). Another type of smoke detector actively samples air from a protected space and transports the samples back to a centralized detector unit where the samples are analyzed for combustion products. These types of detectors are referred to as aspirating smoke detectors (ASDs) and have become a popular detection technology in sensitive building areas and telecommunications facilities.

Currently, about one-half of the U.S. commercial NPP fleet is transitioning from a deterministic to a performance-based fire protection program under Title 10 of the Code of Federal Regulations (10 CFR) 50.48(c). The rule endorses National Fire Protection Association Standard 805, "Performance-Based Standard for Fire Protection for Light Water Reactor Electric Generating Plants." Under this approach, several licensees have proposed using very early warning fire detection (VEWFD) systems as enhanced fire detection systems to reduce the likelihood of a fire resulting in core damage.

In accordance with the NRC NFPA 805 frequently-asked-question (FAQ) process, the NRC staff issued interim guidance FAQ 08-0046 "Incipient Fire Detection Systems" on use of ASD VEWFD systems and associated fire probabilistic risk assessment (PRA) values. Concurrently, with the issuance of this guidance, the NRC began a confirmatory test program to evaluate the performance of these VEWFD systems and determine the correct values to be assigned to the system for consideration in the fire PRA.

Objective

The research effort related to the testing and evaluation of VEWFD systems will allow the fire protection community to better understand how these systems can be used to rapidly detect actual and potential fire sources in NPP applications.

Approach

The NRC staff elected to sponsor testing, conduct literature reviews, and visit both nuclear and nonnuclear sites to support its evaluation of this technology.

The testing (see Figure 7.19) included evaluating conventional spot type detectors (ionization and photoelectric) and ASD configured as VEWFD systems tested in three different scales (Laboratory bench scale, small room, and large open areas). Variables in test parameters that impact detector response such as smoke source, ventilation rate, device location and system configuration were evaluated during each scale of testing.

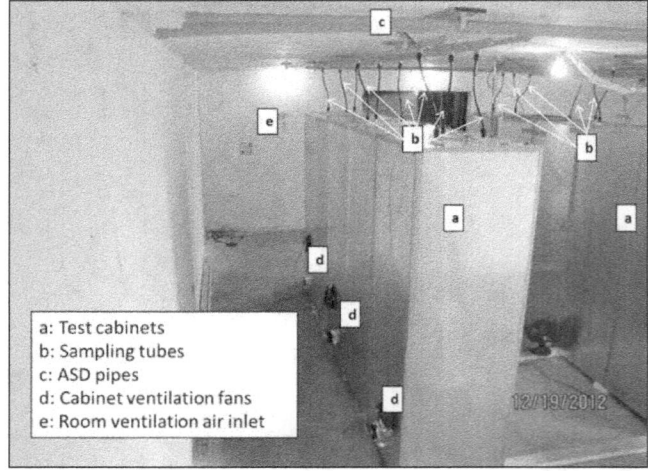

a: Test cabinets
b: Sampling tubes
c: ASD pipes
d: Cabinet ventilation fans
e: Room ventilation air inlet

Figure 7.19 Fire test room configuration

In addition to the confirmatory testing, site visits and a comprehensive literature search were conducted to support an evaluation of the factors that affect the performance of ASD VEWFD system technology and any associated values assigned to the systems in fire PRA to evaluate preventing or detecting and suppressing fires.

The specific values used in the fire PRA as presented in the interim guidance makes an assumption that these systems will detect fires in their incipient stages when smoldering and flaming combustion have not yet begun (see Figure 7.20). This allows additional time for operators to locate the potential fire source and remove power prior to a fire becoming a potential threat to reactor safety. Because of the human involvement in the fire PRA success scenario, human factors and human reliability engineering experts have been supporting this project and will provide guidance in system design and estimates on the human failure probability of preventing fire damage in the final NUREG report.

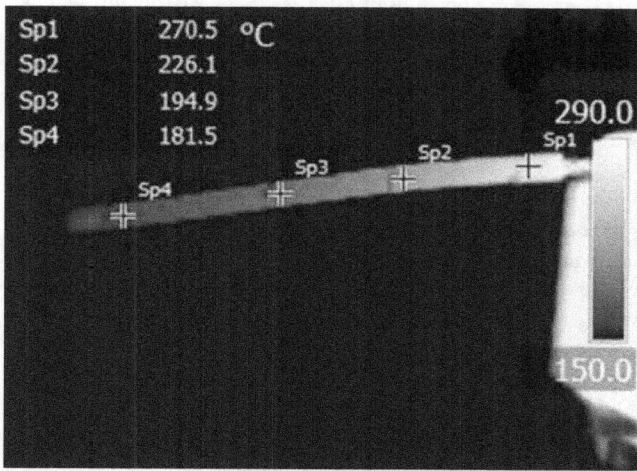

Figure 7.20 Thermal image of overheated electrical conductor

For More Information

Contact Gabriel Taylor, RES/DRA, at Gabriel.Taylor@nrc.gov.

Joint Analysis of Arc Faults (Joan of ARC) OECD International Testing Program for High Energy Arc Faults (HEAF)

Background

This project was identified as part of the Organisation for Economic Co-operation and Development (OECD) fire events database program. Catastrophic failures of energized electrical equipment referred to as high energy arcing faults (HEAF) have occurred in nuclear power plant (NPP) components throughout the world. HEAF typically occur in 480V and higher electrical equipment and cause large pressure and temperature increases in the component electrical enclosure. These increases in pressure and temperature could ultimately lead to serious equipment failure and secondary fires and could put the NPP at risk. Figure 7.21 shows an example of HEAF damage.

Most recently, the United States has experienced events at Palo Verde Nuclear Generating Station in 2013, H.B. Robinson Steam Electric Plant in 2010, and Columbia Generating Station in 2009. Discussions at the OECD Fire Incidents records exchange meetings indicate similar HEAF events have recently occurred in Canada, France, Germany, and most recently at Japan's Onagawa NPP during the earthquake and tsunami of 2011. OECD Fire Project – Topical Report No.1, "Analysis of High Energy Arcing Fault (HEAF) Fire Events," NEA/CSNI/R (2013)6 published in June 2013, documents these international events.

Figure 7.21 HEAF damage

Objective

HEAF have the potential to cause extensive damage to the failed electrical component and distribution system along with adjacent equipment and cables located in close proximity. This area is identified as the zone of influence. The significant electrical energy released during a HEAF event can act as an ignition source to other components in this zone.

The primary objective of this project is to perform experiments to obtain scientific fire data on the HEAF phenomenon known to occur in NPPs through carefully designed experiments. The goal is to use the data from these experiments and past events to develop a mechanistic model to account for the failure modes and consequence portions of HEAFs. These experiments will be designed to improve the state of knowledge and provide better characterization of HEAF in the fire probabilistic risk assessment and National Fire Protection Association 805 license amendment request applications.

Initial impact of the arc to primary equipment and the subsequent damage created by the initiation of an arc (e.g., secondary fires) will be examined.

Figure 7.22 illustrates a 480 V load center undergoing HEAF testing.

Figure 7.22 480 V load center undergoing HEAF testing: before arcing (left); after arcing (right)

Approach

To meet the goals of this test program, experiments will be conducted to explore the basic configurations, failure modes, and effects of HEAF events. Figure 7.23 shows typical electrical enclosure failure modes.

The equipment to be tested in this study primarily consists of switchgears and bussing components.

The project will be operated as part of a larger international OECD/Nuclear Energy Agency effort. The Nuclear Regulatory Commission will be leading the physical testing and instrumentation of equipment at the designated test laboratory. International member countries participating in the project will provide equipment to be tested as well as technical expertise.

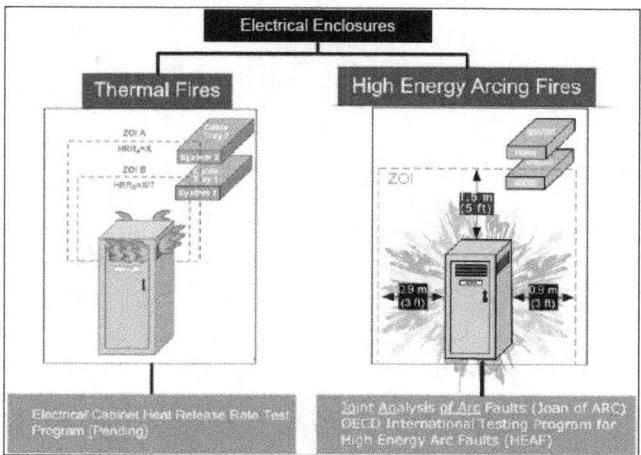

Figure 7.23 Typical electrical enclosure failure modes

For More Information

Contact:

Nicholas Melly, RES/DRA, at Nicholas.Melly@nrc.gov or
Gabriel Taylor, RES/DRA, at Gabriel.Taylor@nrc.gov

Chapter 8: Seismic and Structural Research

Advances in Seismic Hazard Assessment for the Central and Eastern United States

Tsunami Research Program

Seismic Isolation Technology Regulatory Research

Risk-Informed Assessment of Containment Degradation

Structural Analyses in Regulatory Applications

Post-Tensioned Concrete Containment: Grouted Tendons vs. Ungrouted Tendons

Concrete Degradation Issues

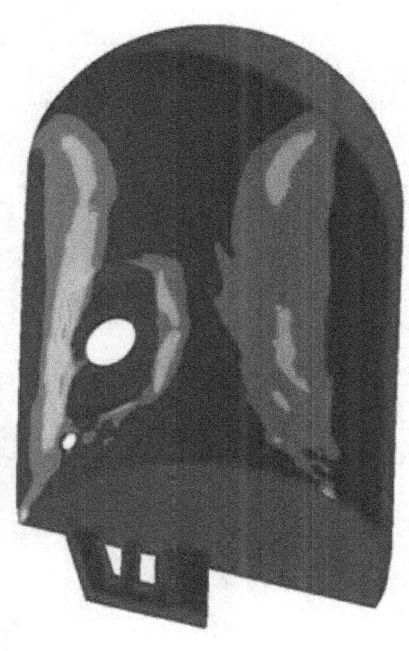

Analysis of PWR prestressed concrete containment vessel under beyond design basis pressurization for severe accident conditions using three-dimensional, nonlinear finite element analysis: cutout view of finite element model and contours of maximum tensile strains in the liner showing strain concentrations near the equipment hatch

Advances in Seismic Hazard Assessment for the Central and Eastern United States

Background

Seismic safety in the design and operation of nuclear facilities has been evolving since the development of the first rules and guidance for seismic design by the Atomic Energy Commission. In 1998, the U.S. Nuclear Regulatory Commission (NRC) issued a policy decision to move toward a risk-informed and performance-based regulatory framework. Risk-informed frameworks use probabilistic methods to assess not only what can go wrong, but also how likely it is to go wrong. Over the last few decades, significant advances have been made in the ability to assess seismic hazard. To date, the NRC has sponsored and continues to sponsor projects in support of both an updated assessment of seismic hazard in the central and eastern United States (CEUS) and an enhancement of the overall framework under which the hazard characterizations are developed. Particularly, as shown in Figure 8.1, the NRC sponsored two projects in this area that have recently been completed, and the agency continues to sponsor the Next Generation Attenuation Relationships for the Central and Eastern North America (NGA-East) Project. Together, these projects will result in an advanced regional model for probabilistic seismic hazard analysis (PSHA) for critical facilities in the CEUS.

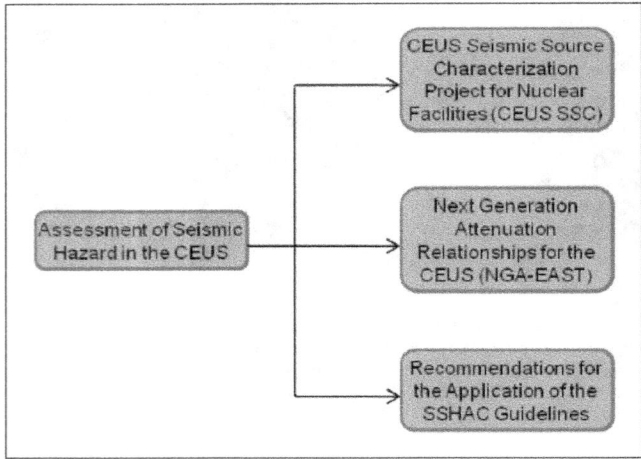

Figure 8.1 Projects supporting seismic hazard assessment

A PSHA requires two key inputs, namely a seismic source characterization (SSC) model, which characterizes the seismic sources that may impact a site, and a ground motion characterization (GMC) model, which predicts ground motion at a site for a particular scenario earthquake. This research program focuses on the GMC input and will develop new state-of-the art ground motion prediction equations (GMPEs) for the CEUS by following up on the NGA relationships project that was completed for the western United States.

NGA-East

Similar to the recently completed CEUS SSC for Nuclear Facilities Project (see report at http://www.ceus-ssc.com and Agencywide Documents Access and Management System [ADAMS] Accession No. ML12048A776), the NGA-East Project is being conducted as a Senior Seismic Hazard Analysis Committee (SSHAC) Level 3 study using the original SSHAC guidance, as well as new SSHAC implementation guidance. Namely, this guidance is found in NUREG/CR-6372, "Recommendations for Probabilistic Seismic Hazard Analysis: Guidance on Uncertainty and Use of Experts," dated April 1997, and NUREG-2117, "Practical Implementation Guidelines for SSHAC Level 3 and 4 Hazard Studies," dated April 2012, respectively. Such a study includes the development of new and complete databases; the full assessment and incorporation of variability and uncertainty; the inclusion of the center, body, and range of technically defensible interpretations of the available data, models, and methods; the development of exhaustive documentation; and a thorough, robust, and participatory peer review, thereby ensuring the necessary regulatory stability and transparency of the resulting NGA-East models.

The new GMC model resulting from this project will replace the Electric Power Research Institute (EPRI) GMC model currently used for new nuclear power plants in the CEUS. In addition to the tectonic region covered by the CEUS SSC for Nuclear Facilities Project, the tectonic region of interest for the NGA-East Project reaches across into Canada. As such, the new GMC model will be applicable to the larger central and eastern North America region. A large number of earthquake records used in this project were provided with support from the Geological Survey of Canada (GSC). Additionally, GSC staff is also participating in the project.

The NGA-East project is being sponsored cooperatively by the NRC, U.S. Department of Energy, EPRI, and the U.S. Geological Survey. The project is being led by the Pacific Earthquake Engineering Research (PEER) Center. However, the project involves a large number of participating researchers from dozens of organizations in academia, industry, and government, including the sponsoring organizations. Together, these participants fulfilled a variety of roles, shown schematically in Figure 8.2. This project is expected to end in 2015.

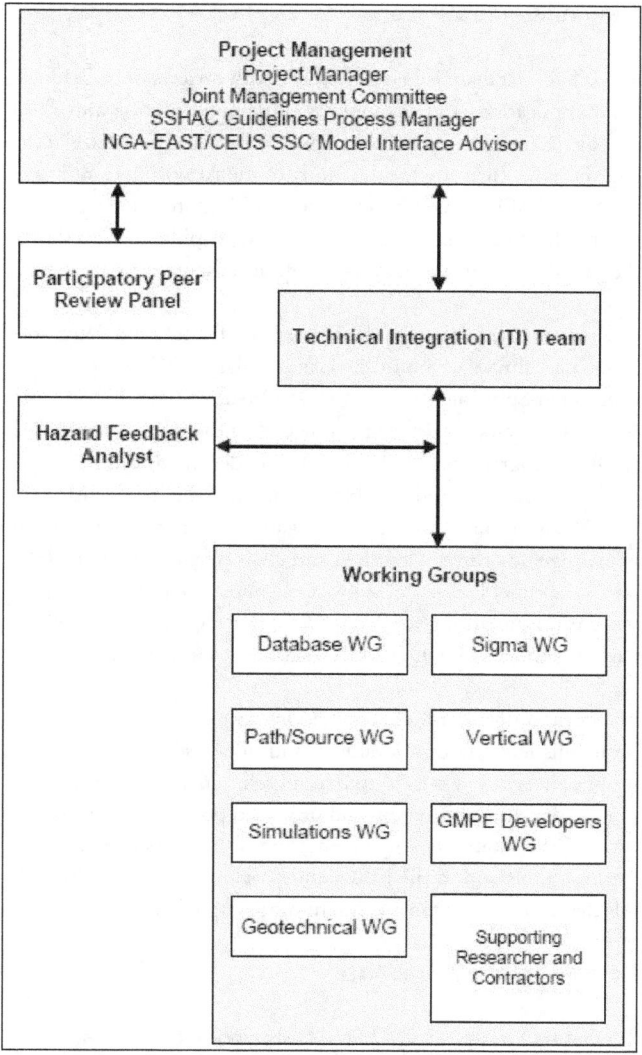

For More Information
Contact Annie Kammerer, RES/DE, at
Annie.Kammerer@nrc.gov, or Richard Rivera-Lugo, RES/DE, at
Richard.Rivera-Lugo@nrc.gov.

Figure 8.2 Participants and organization of the NGA-East project

Probabilistic Seismic Hazard Analysis (PSHA) Software

The NRC is also developing a PSHA software package working through a commercial contract that will allow the NRC to perform enhanced computation of the probabilistic seismic hazard at any arbitrary location in the CEUS. This software will incorporate the newly developed CEUS-SSC model and the latest ground motion attenuation. Eventually, this software will use the attenuation relationships from the ongoing NGA-East project. Other features of this PSHA software package will include the implementation of custom USGS National Hazard Mapping Models, and the development of software to incorporate a Monte Carlo approach to estimate the uncertainty in seismic hazard for CEUS.

Tsunami Research Program

Background

Since the 2004 Indian Ocean tsunami, significant advances have been made in the ability to assess tsunami hazards globally. The U.S. Nuclear Regulatory Commission's (NRC's) current tsunami research program was initiated in 2006, and it focuses on bringing the latest technical advances to the regulatory process and exploring topics unique to nuclear facilities. The tsunami research program focuses on several key areas: landslide-induced tsunami hazard assessments, support activities associated with the licensing of new nuclear power plants in the United States, development of probabilistic methods, and development of the technical basis for new NRC guidance.

This program, which includes cooperative work with the United States Geological Survey (USGS) and the National Oceanic and Atmospheric Administration (NOAA), has already resulted in several important publications on tsunami hazard assessments on the Atlantic Coast of the United States.

Approach

Tsunamigenic Source Characterization

The NRC tsunami research program includes assessment of both seismic- and landslide-based tsunamigenic sources in both the near and the far fields. The inclusion of tsunamigenic landslides, an important category of sources that impact tsunami hazard levels for the Atlantic and Gulf Coasts, is a key difference between this program and most previous tsunami hazard assessment programs. The USGS conducted the initial phase of work related to source characterization, which consisted of collection, interpretation, and analysis of available offshore data, with significant effort focused on characterizing offshore near-field landslides and analyzing their tsunamigenic potential and properties. A publicly available USGS report to the NRC, titled "Evaluation of Tsunami Sources with the Potential to Impact the U.S. Atlantic and Gulf Coasts," ten Brink et al., 2008, Agencywide Documents Accession and Management System (ADAMS) Accession No. ML082960196, which is currently being used by both NRC staff and industry, summarizes this work. In addition, eight papers have been published in a special edition of Marine Geology dedicated to the results of the NRC research program ("Tsunami Hazard along the U.S. Atlantic Coast," Marine Geology, Volume 264, Issues 1–2, 2009).

Tsunami Generation and Propagation Modeling

The USGS database is being used for both reviews of individual plant applications and as input for tsunami generation and propagation modeling being conducted by the experts at USGS and the Joint Institute for the Study of the Atmosphere and Ocean (JISAO) at the University of Washington. The goal of this modeling is to better understand the possible impacts that the identified sources could have on the coasts.

To undertake modeling of the impact of a flank failure landslide of the La Palma volcano in the Canary Islands, NOAA's Method of Splitting Tsunami (MOST) tsunami generation and propagation model has been coupled with the impact Simplified Arbitrary Lagragean Eulerian (iSALE) code, which can be used for modeling landslide-based tsunamigenic mechanisms. MOST also is being used to investigate the impact of the seismic tsunamigenic sources identified and characterized by the USGS. As an example, Figure 8.3 shows computed maximum tsunami wave amplitude using the MOST forecast model for the pacific basin for the 11 March 2011 Tohoku, Japan, earthquake.

The current phase of research includes development of probabilistic methods to evaluate landslide-based tsunami sources, analyses of typical sources in selected areas with the potential to impact existing and proposed power plants using probabilistic methods, implementation of NOAA's tsunami warning tools within the NRC, and development of a NUREG/CR describing acceptable tsunami modeling tools.

New Regulatory Guidance

Regulatory Guide 1.59, "Design-basis Floods for Nuclear Power Plants," Revision 2, dated August 1977, briefly discussed tsunami as a source of flooding. The NRC is in the final stages of updating this regulatory guide, taking into account the lessons learned from the 2011 Fukushima accident in Japan. However, the update of this guide will not include tsunami-induced flooding. The NRC staff currently is preparing a new regulatory guide focused on tsunami hazard assessment and risk.

The NRC also participated in the development of the International Atomic Energy Agency Safety Standard Series No. SSG-18, "Meteorological and Hydrological Hazards in Site Evaluation for Nuclear Installations," published in 2011. During this project, the NRC learned best practices of other countries for requirements for hazard assessment modeling that will benefit the hazard assessments for seismic events and flooding, including tsunami-induced flooding. Operating U.S nuclear power plants are currently undergoing a reassessment of flooding and seismic hazards as part of the Fukushima lessons learned initiative.

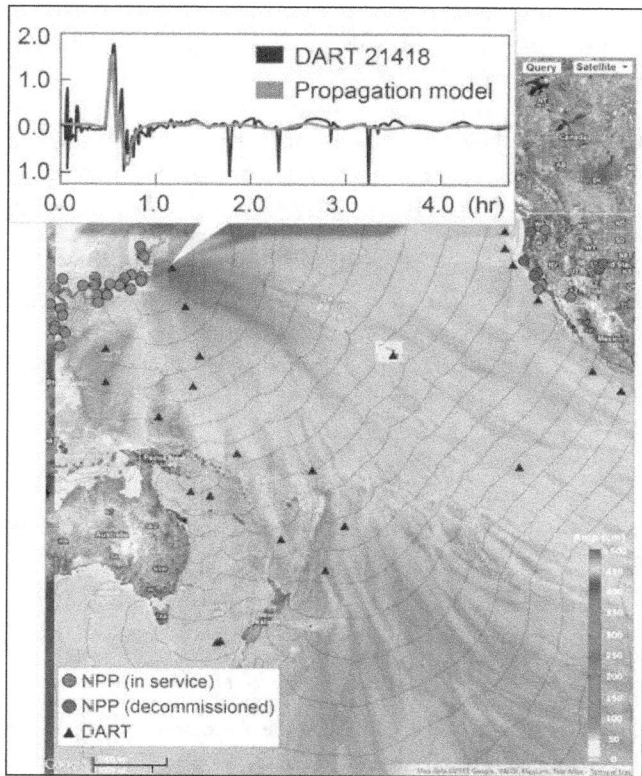

Figure 8.3 Computed maximum tsunami wave amplitude as calculated by MOST, NOAA's tsunami forecast system, for the Pacific Basin during the 11 March 2011 Tohoku event. DART (Deep-ocean Assessment and Reporting of Tsunamis) sensor locations are indicated by black triangles, and the global power plant locations are indicated by red circles. The inset shows the comparison between the observed and computed wave amplitudes at a DART station

For More Information

Contact Rasool Anooshehpoor, RES/DE, at
Rasool.Anooshehpoor@nrc.gov.

Seismic Isolation Technology Regulatory Research

Background

Seismic isolation technologies (also called base isolation technologies) are components and systems that isolate a structure from the motion of the ground during an earthquake. Modern seismic isolation devices and components were commercially developed in the 1970s and 1980s, and thousands of conventional buildings, industrial structures, and bridges have been seismically isolated in the United States and abroad. Seismic isolation has been used to design and construct nuclear facility structures in France and South Africa. The licensing and construction of new nuclear power plants are leading to an exploration of the use of seismic isolation technologies in U.S. nuclear facilities. Several new advanced reactor designs are expected to include seismic isolation systems. To prepare for the possible use of these technologies in nuclear plant design, the U.S. Nuclear Regulatory Commission (NRC) has initiated a program to identify and investigate these technical areas.

Approach

Development of NUREG on the Use of Seismic Isolation Systems in Nuclear Power Plants

The NRC, working with Lawrence Berkeley National Laboratory and the State University of New York, University at Buffalo, is addressing a range of technical considerations for analysis and design of safety-related nuclear facility structures using seismic isolation. An associated NUREG under development is intended to serve as a reference for engineers engaged in the design of structures using seismic isolation systems, as well as NRC staff charged with reviewing applications using these technologies. Typically, the seismic isolation components are treated as a civil or structural subsystem of a nuclear power plant whose risk-informed design is governed by specific performance objectives. Figure 8.4 shows a seismically isolated nuclear structure.

The NUREG will discuss the behavior, mechanical properties, modeling, structural response analysis, design issues, and performance criteria for seismic isolation design using the most commonly used seismic isolation devices. Testing, construction, and operational issues also will be discussed in the NUREG.

Ongoing Investigation of Nuclear Plant-Specific Issues

Further research is required in a number of technical areas. Some of these issues, such as the response to vertical excitation and soil-structure interaction, already are considered for non-isolated nuclear power plant designs such that current guidance could be applicable.

Currently, the NRC, in collaboration with the University of Nevada and E-Defense Laboratory in Japan, is undertaking a project of a large-scale simulation of a base-isolated structure subjected to multidirectional excitation of beyond-design-basis events (extreme earthquakes). This project will investigate a range of technical issues for analysis and design of safety-related nuclear facility structures using a combination lead rubber and cross linear bearing (LRB/CLB) seismic isolation system. These studies are focusing on the following:

- Investigation of the suitability of an elastomeric isolation system to meet the seismic design objectives for a nuclear facility.

- Evaluation of the relationship between bearing axial force and the lateral force-displacement loops at lateral displacements capacity.

- Evaluation of the vertical-lateral accelerations coupling.

- Investigation of the stability of the elastomeric bearings at large lateral displacements.

- Evaluation of the likelihood and possible consequences of rocking of the isolated superstructure.

- Development of test results database for the verification and validation of computer simulation models of base isolation devices.

An important conclusion from the work is that seismic isolation may be a viable technology for use in nuclear power plants. Research to investigate these critical areas is ongoing.

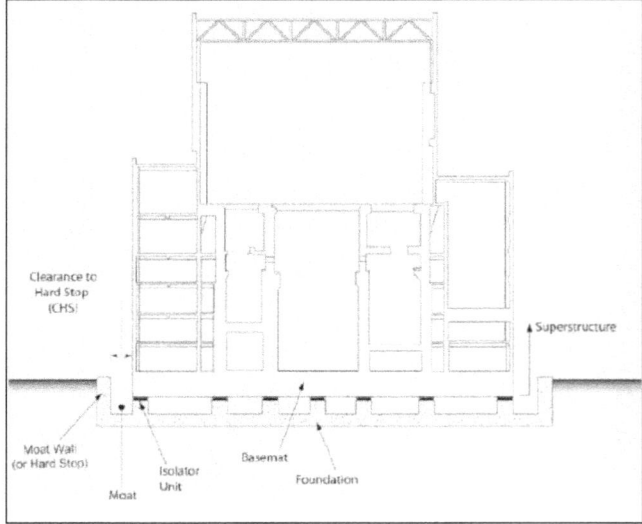

Figure 8.4 Seismically Isolated Nuclear Structure

Additional Plant-Specific Issues

Additional topics of interest include the following:

- Evaluation of the effect of differences among the mechanical properties of seismic isolation devices.

- Evaluation of the consequences of impact of the structure against sidewalls during horizontal motion or impact from isolator uplift.

For More Information
Contact Richard Rivera-Lugo, RES/DE, at Richard.Rivera-Lugo@nrc.gov, or Annie Kammerer, RES/DE, at Annie.Kammerer@nrc.gov.

Risk-Informed Assessment of Containment Degradation

Background

Over time, degradation has been observed in the containment vessels of a number of operating nuclear power plants in the United States. Forms of degradation include corrosion of the steel shell or liner, corrosion of reinforcing bars, loss of prestressing, and corrosion of bellows. The containment vessel serves as the ultimate barrier against the release of radioactive material into the environment. Because of this role, compromising the containment could increase the risk of a large release in the unlikely event of an accident. Previous work in this area assessed the effects of degradation on the pressure retaining capacity of the containment vessel through structural analyses that account for degradation. These analyses provided useful information about the effects of the degradation on the structural capacity of the containment in both deterministic and probabilistic fashions. However, additional studies still are required to identify adequate metrics and related methods that can be used to examine the effects of degradation in specific cases.

Approach

The U.S. Nuclear Regulatory Commission (NRC) is sponsoring research at Sandia National Laboratories (SNL) to assess the effects of containment vessel degradation in containment vessels in a risk-informed manner. Goals for the research being conducted at SNL include supporting license renewal reviews and inspections by providing methods to examine, on a casebycase basis, potential degradation effects from aging and repairs. Initially, the study evaluated the effects of degradation on several types of containments with respect to the guidelines given in Regulatory Guide 1.174, "An Approach for Using Probabilistic Risk Assessment in Risk-Informed Decisions on Plant-Specific Changes to the Licensing Basis." The study integrated fragility curves developed for nondegraded and postulated degraded conditions using structural analysis with preexisting probabilistic risk assessment models used in NUREG-1150, "Severe Accident Risks: An Assessment for Five U.S. Nuclear Power Plants." That phase of the study concluded that several cases of postulated degradation involving corrosion of the liner (see Figure 8.5) or shell showed small increases, no increases, or even decreases in the large early release frequency (LERF). Rather than leading to a containment rupture, the postulated liner degradation causes the containment to fail by leakage, with an increase in small early release frequency (SERF).

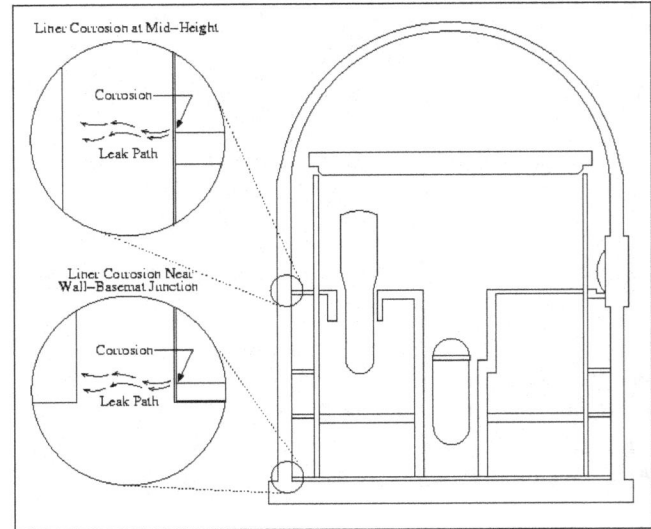

Figure 8.5 Example of reinforced concrete containment leak paths for postulated corrosion degradation (NUREG/CR-6920)

Since Regulatory Guide 1.174 does not provide guidance on the limits of SERF, additional deterministic analyses were performed to assess the effects of degradation on consequences to evaluate the feasibility of using metrics other than LERF. The study is continuing to assess the extent of corrosion, other containment types, and other degradation modes. Because most U.S. power plants have unique designs, a research goal is to develop results, approaches, and metrics that can be used for casebycase examination of degradation effects.

For More Information
Contact Thomas Herrity, RES/DE, at Thomas.Herrity@nrc.gov.

Structural Analyses in Regulatory Applications

Background

The U.S. Nuclear Regulatory Commission's (NRC's) Office of Nuclear Regulatory Research (RES) maintains a state-of-the-art capability in nonlinear structural analysis that supports multiple offices within the agency. These nonlinear structural analyses provide, for example, insights on the safety margins of structures in operating reactors and new reactor designs, as well as information on the performance of spent nuclear fuel transportation (SNFT) casks under regulatory accident conditions. Nonlinear structural analyses also provide structural performance information used to assess (1) beyond design-basis accidents initiated by internal or external hazards, and (2) security threats.

RES uses commercial finite-element codes, such as the nonlinear, dynamic explicit finite-element code LSDYNA (Livermore Software Technology Corporation), mechanical and structural codes from ANSYS Inc., and the ABAQUS/SIMULIA structural analysis codes. Expansion to coupled computational fluid dynamics and finite-element codes, specifically to the U.S. Navy's DYSMAS software, is currently underway. The office maintains a cluster of multiple-processor workstations that provide the capability needed to execute three-dimensional analyses using models that include critical reinforcement, connections, liner, and other structural details that affect structural performance. RES supplements these activities, as needed, through contracts to national laboratories and commercial entities, grants to universities, interagency agreements with other government agencies, and international collaborative research agreements.

Approach and Example Applications

RES uses nonlinear structural analyses to verify, validate, benchmark, and compile various modeling approaches and techniques. These results and information support staff assessments of analyses submitted by licensees and applicants and confirmatory analyses. Those results also guide structural evaluations of beyond-design-basis accidents or risk assessments performed by RES.

Finite-element Simulation of SNFT Drop Tests

Under an international cooperative research agreement between the NRC and Germany's Federal Institute for Materials Research and Testing (BAM), RES staff analyzes data from BAM's full-scale and scaled drop tests of SNFT casks (see Figure 8.6). Results from this study provide information on scale effects and insights on the effects of various modeling approaches that would inform guidance on modeling practices.

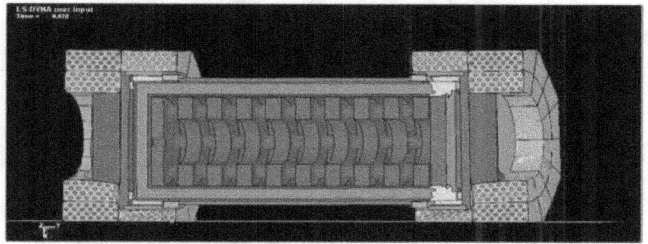

Figure 8.6 Finite-element analysis (LSDYNA) of 30-foot side drop test of SNFT cask

Containment and Spent Fuel Pool Liner Strains for Beyond-design-basis Seismic Events

RES staff uses commercial finite-element codes to analyze the load deformation response of structures, such as containment buildings to simulated seismic loads (see Figure 8.7). These analyses provide, for example, information on the magnitude of

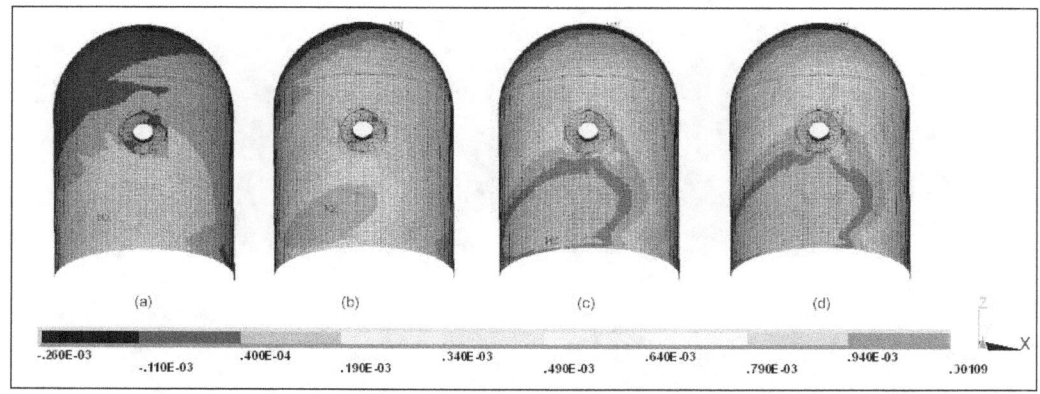

Figure 8.7 Maximum liner tensile strains for loads ranging from about two and half times (left) to eight times (right) a seismic design-basis load (calculated using ANSYS)

containment liner strains that may be induced by beyond-design-basis seismic events and their significance in severe accident consequence assessments. The results also provide insights into lateral load resisting mechanisms that are captured by various modeling approaches.

RES staff uses the nonlinear finite-element analysis code LSDYNA to calculate tensile strains in the liner of a spent fuel pool for beyond-design-basis seismic events (see Figure 8.8). Results of such finite-element analyses provide information on the load deformation behavior of spent fuel pool structures, which consist of thick reinforced concrete walls lined with leak-tight stainless steel liners. The analyses provide information on the condition of the spent fuel pool following beyond-design-basis earthquakes for subsequent safety analyses.

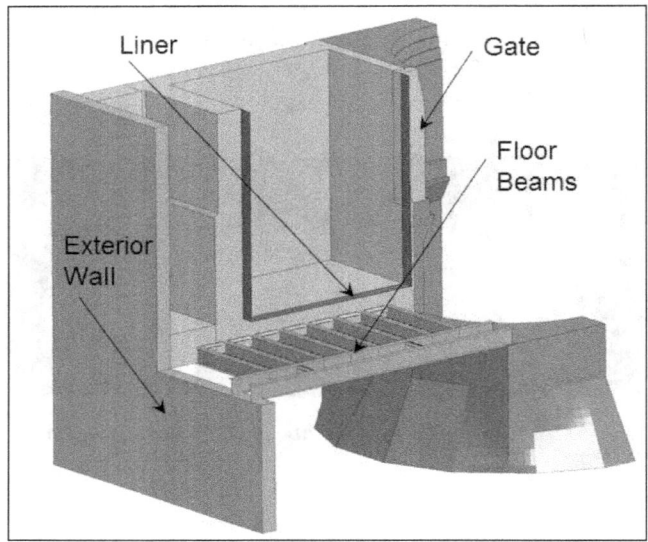

Figure 8.8 Cutout of the infinite-element model of a BWR spent fuel pool

For More Information
Contact Dr. Jose Pires, RES/DE, at <u>Jose.Pires@nrc.gov</u>.

Post-Tensioned Concrete Containment: Grouted Tendons vs. Ungrouted Tendons

Containment structures serve an important role in protecting the public and environment from potential radioactive releases during extreme or abnormal operating events. One way the strength of concrete containments can be increased is by placing steel tendons in concrete sections and post-tensioning them. Post-tensioning is a method where steel is tensioned and anchored against concrete to add compression to the concrete which improves strength and leak-tightness. The NRC participates in many international collaborative research efforts including a small-scale containment model tested at the Sandia National Laboratory (see Figure 8.9) and the International Standard Problem on containment integrity (ISP 48). The Sandia test and ISP 48 together supplied valuable information on the state-of-the-art for analyzing a post-tensioned containment. Over the past decade, post-tensioning technologies and computer codes have advanced remarkably. Therefore the NRC has initiated further studies on containment integrity to develop acceptable means of validating post-tensioning methods.

As part of an international collaborative research program with the Organisation for Economic Co-operation and Development's Nuclear Energy Agency (OECD/NEA), the NRC participates in an effort to investigate the comparative advantages and disadvantages of various post-tensioning techniques in reactor containments. Members of the OECD/NEA concrete working group are represented by knowledgeable staff from United States, Canada, Finland, United Kingdom, Sweden, Czech Republic, and France. The objective of the current investigation is to compare post-tensioned containments in a study of grouted vs. ungrouted tendons. The study is focused on 1) expected structural behavior and failure modes, 2) executing the post-tensioning and maintaining activities for whole expected life of the containment, and 3) corrosion protection and durability of the tendon material.

The outcome of this work will be valuable for modeling, analysis, durability, and inspection of post-tensioned concrete containments.

Another research project is underway to develop a technical basis for updating Regulatory Guides (RG) related to Inservice Inspection (ISI) of Post-tensioned Containments. The inspection and monitoring of post-tensioned concrete containment structures with grouted or ungrouted tendons are performed to ensure that the safety margins postulated in the design of containment structures are not reduced under

operating and environmental conditions. Maintaining the required prestressing force levels in post-tensioned concrete containments assure that the containment retains adequate margins with respect to structural and leak-tight integrity.

Figure 8.9 Prestressed concrete containment vessel model tendon sheaths at Sandia National Laboratories

This research benefits both new and existing reactors by developing the technical basis for testing guidance of the prestress levels in the future post-tensioned containments that may use new types of materials, as well as monitoring of these containments by instrumentation. The research findings will also be used to improve accuracy in predicting and measuring the prestress level of tendons in existing concrete containments.

The result of this project will be used to provide the technical basis for future revisions to Regulatory Guide (RG) 1.35, "Inservice Inspection of Ungrouted Tendons in Prestressed Concrete Containments," RG 1.35.1, "Determining Prestressing Forces for Inspection of Prestressed Concrete Containments," and RG 1.90, "Inservice Inspection of Prestressed Concrete Containment Structures with Grouted Tendons."

For More Information Contact
Madhumita Sircar, RES/DE, at Madhumita.Sircar@nrc.gov.

Concrete Degradation Issues

Background

By the year 2015, more than 65 percent of the 104 containments in the United States will be more than 20 years old (70 percent of the containments are made of reinforced or prestressed concrete). To date, the U.S. Nuclear Regulatory Commission (NRC) has issued renewed operating licenses (i.e., 40–60 years) to 71 reactor units. In addition, since 2010, long-term dry cask storage systems that used concrete elements (concrete overpacks and pads) have been the subject of studies that assessed the knowledge gaps of various degradation mechanisms that can affect structural performance.

As nuclear plants and dry cask storage systems (DCSS) age and continue to operate, incidence of degradation of structures, systems, and components caused by aging have been and will continue to be identified. Examples of some incidents of concrete and steel degradation include containment delamination, leaching of concrete in tendon galleries, corrosion of steel liners attached to concrete containments, freeze-thaw cracking in concrete and alkali-silicate reaction cracking in concrete caused by long-term exposure to ground water and adverse chemical interactions between aggregates and cement (see Figure 8.10). With the passage of time, the experiences noted above have the potential to degrade strength and increase risk to public safety and health. Thus, methods and acceptance criteria are needed to monitor and evaluate aging effects.

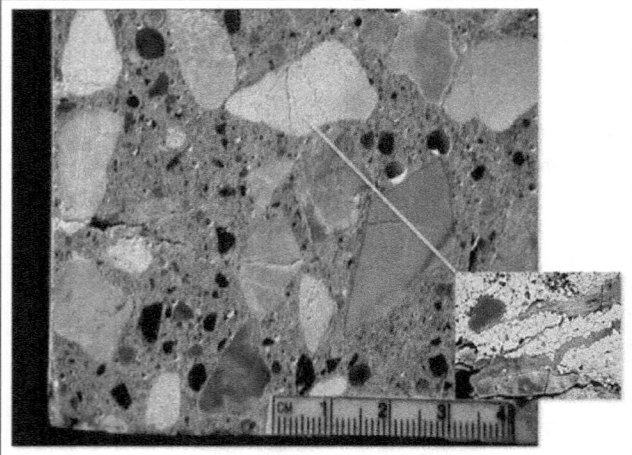

Figure 8.10 Alkali-Silica Reaction (ASR) effects on concrete. Inset is a magnified image of the location of concrete degradation .

Approach

Because concrete degradation has been observed in concrete containments and at DCSS, there is reason to suspect that degradation will continue to be found and that these conditions will have to be evaluated and addressed. There are three major issues involved in assuring adequate performance of concrete structures under accident or severe environmental loads for the operating life of the structures.

1. How confident is the NRC that, if degradation is in process, it will be found before the structure or component is challenged?

2. How well can the NRC assess the condition of the structure when an instance of degradation is found and, in the future, if the degradation mechanism at work continues? How can the NRC assess impact on plant safety and risk?

3. How effective are proposed plans for mitigation or repair likely to be?

The first issue is one of effectiveness of inspection. In the past, research programs have concentrated on evaluating the uncertainties associated with visually based inspection practices outlined in current American Concrete Institute (ACI) and American Society of Mechanical Engineer's (ASME) inservice inspection codes.

The major technical question underlying the second issue is confidence in ability to predict the difference in response of a structure to accident or severe environmental loads as a result of degradation.

For the last issue, NRC staff is involved in a Nuclear Energy Standards Coordination Collaborative task group effort to develop recommendations for the repair of concrete nuclear structures.

The objective of the NRC's research is to determine if current inspection and maintenance programs are sufficient in identifying and characterizing in a timely manner any degradation that may be present. Proper maintenance is essential to the safety of nuclear plant structures, and a clear link exists between effective maintenance and safety. Therefore, the aim of current and planned research is to review methods and establish acceptance criteria to monitor and evaluate aging effects. Recommendations for improved inspection and maintenance guidance and their bases will be addressed.

For More Information
Contact Bill Ott, RES/DRA, at <u>William.Ott@nrc.gov</u>.

Chapter 9: Materials Performance Research

Extremely Low Probability of Rupture

Research to Support Regulatory Decisions Related to
Second and Subsequent License Renewal Applications

Steam Generator Tube Integrity

Consequential Steam Generator Tube Rupture Program

Reactor Pressure Vessel Integrity

Degradation of Reactor Vessel Internals from Neutron
Irradiation

Primary Water Stress-Corrosion Cracking

Primary Water Stress-Corrosion Cracking Mitigation
Evaluations and Weld Residual Stress Validation Programs

Nondestructive Examination

Containment Liner Corrosion

Atmospheric Stress-Corrosion Cracking of Dry Cask
Storage Systems

High-Density Polyethylene Piping Research Program

Degradation of Neutron Absorbers in Spent Fuel Pools

Extended Storage and Transportation of Spent Nuclear Fuel

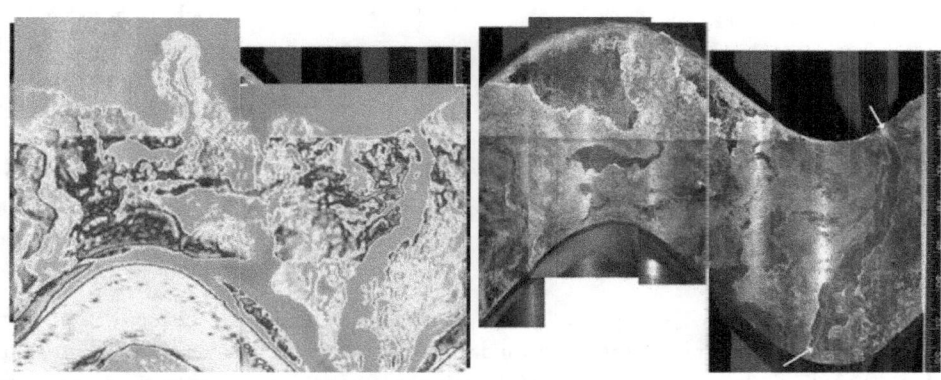

Destructive and nondestructive examination of nozzle leak path

Extremely Low Probability of Rupture

Background

In NUREG-0800, "Standard Review Plan for the Review of Safety Analysis Reports for Nuclear Power Plants: LWR Edition" (SRP), Chapter 3, "Design of Structure, Components, Equipment, and Systems," Section 3.6.3 "Leak-Before-Break (LBB) Evaluation Procedures," issued March 2007, the U.S. Nuclear Regulatory Commission (NRC) staff described acceptable analysis and assessment methodologies. Specifically, the SRP outlines a deterministic assessment procedure that can be used to demonstrate compliance with the requirement in General Design Criterion (GDC) 4, "Environmental and Dynamic Effects Design Bases," in Appendix A, "General Design Criteria for Nuclear Power Plants," to Title 10 of the *Code of Federal Regulations* (10 CFR) Part 50, "Domestic Licensing of Production and Utilization Facilities," for primary system pressure piping to exhibit an extremely low probability of rupture. SRP Section 3.6.3 does not allow for assessment of piping systems with active materials degradation mechanisms. However, primary water stress-corrosion cracking (PWSCC) is known to occur in systems that have been granted LBB exemptions to remove pipe-whip restraints and jet impingement shields.

To address this issue, the NRC has determined through a qualitative approach that these LBB-approved systems remain in compliance. (See NRC Regulatory Issue Summary 10-07, "Regulatory Requirements for Application of Weld Overlays and Other Migration Techniques in Piping Systems Approved for Leak-Before-Break," dated June 8, 2010.) This approach includes the following:

- As a qualitative rationale, the great majority of observed cracking is of limited extent and of shallow depth.

- These factors tend to mitigate the risk of piping rupture.

- PWSCC mitigation activities have been implemented (e.g., stress improvement and material replacement with overlays, mechanical stress improvement, inlays, and onlays).

Although such actions are prudent, timely, and warranted, they do not quantitatively address the issues of LBB-approved piping with active materials degradation, thus revealing a continued need for a new and comprehensive piping system assessment methodology. To address this need, an assessment tool that can be used to directly assess compliance with the probabilistic acceptance criterion of GDC 4 is necessary. This tool would properly model the effects of active degradation mechanisms, inservice inspection protocols, and associated mitigation activities. The probabilistic tool will be comprehensive with respect to known challenges, vetted with respect to the scientific adequacy of models and inputs, flexible enough to permit analysis of a variety of inservice situations, and sufficiently adaptable to accommodate evolving and improving knowledge and additional degradation modes.

Approach

As part of the effort for quantitatively ensuring the long-term extremely low probability of rupture in accordance with GDC 4, the Office of Nuclear Regulatory Research (RES) embarked on an effort to develop a modular-based computer code for the determination of the probability of failure for reactor coolant system components. In doing so, RES has sought the support of national laboratories and commercial contractors and communicates with the domestic nuclear industry under the auspice of the Electric Power Research Institute through an ongoing memorandum of understanding between the two organizations. This computer code will be capable of considering all degradation mechanisms that may contribute to low probability failure events while properly handling the uncertainty in the failure process. The code will be structured in a modular fashion so that, as additional operational experiences arise, additions or modifications can be easily incorporated without code restructuring. The first arm of the modular code that will be developed deals directly with primary piping integrity and is coined xLPR for "extremely low probability of rupture."

As part of a 2-year pilot study effort, RES developed a group of teams that determined the feasibility of developing a complex, comprehensive probabilistic computer code while properly accounting for uncertainties. As part of the pilot study, the team effort was focused on a particular problem (i.e., the failure of a pressurizer surge nozzle dissimilar metal weld as seen in Figure 9.1 with a circumferential crack due to PWSCC). The results of the pilot study effort include the following:

- The project team demonstrated that developing a modular-based probabilistic fracture mechanics code within a cooperative agreement while properly accounting for the problem uncertainties is feasible (Figure 9.2).

- The project team demonstrated that the cooperative management structure was promising and identified potential efficiency gains.

- The project team concluded the GoldSim commercial software is appropriate for future xLPR versions.

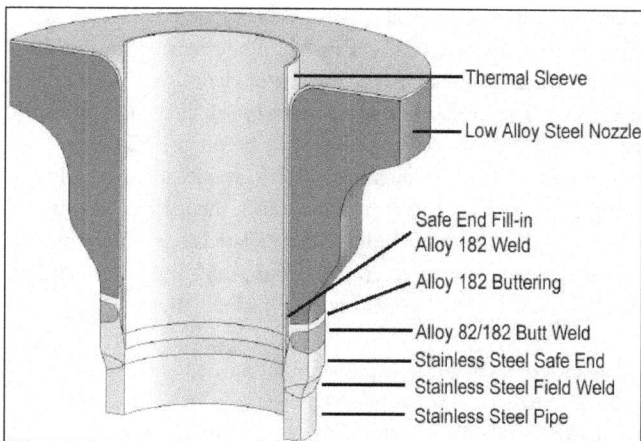

Figure 9.1 Pressurizer surge nozzle illustration

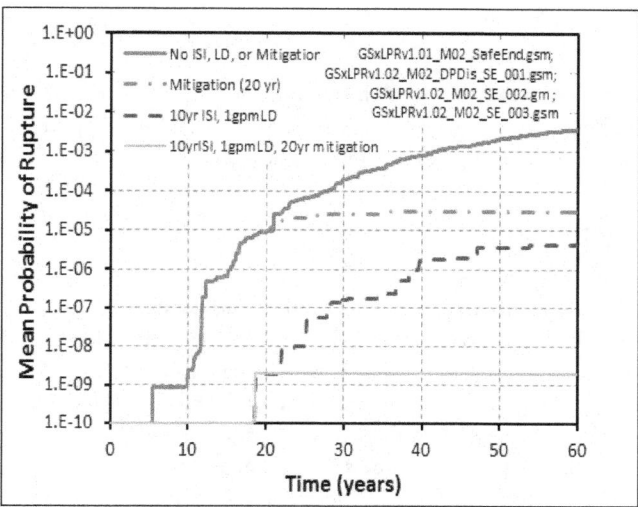

Figure 9.2 Mean probability of rupture for pressurizer surge nozzle with mitigation, leak detection (LD), and inspection (ISI)

Following the success of the xLPR pilot study, the project team is expanding the xLPR development from PWSCC in dissimilar metal welds (Version 1.0) to all relevant welds and material degradation in piping systems approved for LBB (Version 2.0). Using the project-specific quality assurance program, the project team will compile, code, and verify the modules needed for the stated purpose. These modules include loads (with weld residual stress), crack initiation, crack growth, crack stability, crack opening displacement, leakage, inspection, and mitigation. In addition, the project team will incorporate these self-contained modules into a computational framework that uses the GoldSim commercial software. The framework will control the time flow of the analyses while linking the modules and properly accounting for and propagating the problem uncertainty. After the xLPR Version 2.0 code development, the NRC staff will use the code in developing a technical basis and regulatory guidance for LBB.

Schedule

The planned schedule for the xLPR program is as follows:

- xLPR Version 2.0 complete—2014

For More Information
Contact Dr. David L. Rudland, RES/DE, at
David.Rudland@nrc.gov.

Research to Support Regulatory Decisions Related to Second and Subsequent License Renewal Applications

Background

Materials degradation phenomena, if not appropriately managed, have the potential to adversely affect the functionality and safety margins of nuclear power plant (NPP) systems, structures, and components (SSCs), especially as they continue to operate for longer periods. The Office of Nuclear Regulatory Research (RES) has an ongoing multiyear research program to develop an improved understanding of materials degradation failure mechanisms. This program will provide necessary technical data and enable the development of tools to better predict potential impacts on the long-term operability of NPP SSCs to support regulatory review of licensee's aging management programs (AMPs) to ensure continued safe plant operation.

As shown in Figure 9.3, to date about three-quarters of operating U.S. NPPs (73) have been granted a renewed license to operate from 40 to 60 years, which is known as the period of extended operation (PEO). The NRC is conducting a regulatory review of license renewal applications for 17 further units, with 10 more applications expected.

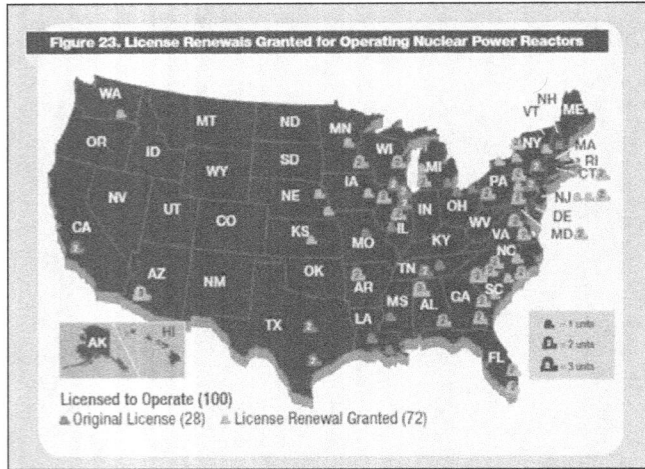

Figure 9.3 License Renewals Granted for Operating Nuclear Power Reactors

The U.S. commercial nuclear power industry has publicly informed the U.S. Nuclear Regulatory Commission (NRC) staff of its intentions to submit license renewal applications (LRAs) in the 2015–2019 timeframe for subsequent license renewal, which will allow plant operation up to 80 years. The agency permits subsequent license renewal requests under Title 10 of the *Code of Federal Regulations* (10 CFR) Part 54, "Requirements for Renewal of Operating Licenses for Nuclear Power Plants." However, the plants may need to resolve potential technical challenges from aging effects on passive long-lived SSCs before the NRC can approve LRAs for subsequent renewal. These challenges include aging effects on the reactor pressure vessel, the reactor pressure vessel internals, primary piping, safety-related secondary piping, buried and submerged structures, electric cable insulation, and concrete exposed to high temperature and radiation.

To ensure that the NRC is prepared for a timely review of possible LRAs for a subsequent renewal, research is needed to ensure the availability of the necessary technical information to support the agency's regulatory decision making,.

Objective

The objective of this research is to provide a sound technical basis and make guidance on the aging of SSCs available to support timely reviews of potential subsequent LRAs.

Approach

The NRC and industry have expended considerable resources over the last several decades to better understand the safety implications and risk associated with aging of SSCs. Key activities have included an assessment of the technical basis for an alternate pressurized thermal shock rule (10 CFR 50.61a, "Alternate Fracture Toughness Requirements for Protection against Pressurized Thermal Shock Events"), aging of electrical cables, and environmentally assisted cracking of materials. Furthermore, in February 2008 and again in February 2011, the NRC and the U.S. Department of Energy cosponsored a workshop on U.S. NPP life extension research and development, which requested stakeholder input into aging management research areas for "Life Beyond 60." (Summaries of the workshop proceedings can be found at Agencywide Documents Access and Management System [ADAMS] Accession No. ML080570419 for the 2008 workshop and at ADAMS Accession No. ML110630041 for the 2011 workshop.) Based on the results of these workshops and on the staff's long-term research plan, potential additional areas of focus for subsequent license renewal include aging management of reactor vessel and internal materials, cable insulation, buried and submerged structures, and concrete exposed to high temperature and radiation. The staff will be holding recurrent NRC/industry workshops on the status of operating experience from the initial renewal term and industry research activities to address aging management of technical issues for a subsequent license renewal term as a followup to the 2008 and 2011 workshops.

The NRC staff is presently updating the original NUREG/CR-6923, "Expert Panel Report on Proactive Materials Degradation Assessment," issued February 2007, to include longer timeframes (i.e., 80 or more years) and passive, long-lived SSCs beyond the primary piping and core internals, such as the pressure vessel, concrete containment building and cable insulation. This update will allow the staff to: (1) identify significant knowledge gaps and any new forms of degradation that may have emerged since the original proactive materials degradation assessment report was developed; (2) capture the current knowledge base on materials degradation mechanisms; and, (3) prioritize materials degradation research needs and directions for future efforts. The NRC staff is accomplishing this task through a collaborative effort with the U.S. Department of Energy's light-water reactor sustainability program (LWRS) and expects to complete the task in 2013.

In recent years, a variety of related research initiatives have emerged, including the creation of the Materials Aging Institute by the Electric Power Research Institute, Électricité de France, Tokyo Electric Power Company and others, and the development of expert networks and technical meetings focused on some elements of proactive management of materials degradation. The NRC, in cooperation with other national regulators and nongovernmental organizations, has implemented the "International Forum for Reactor Aging Management," which is a network of international experts to exchange technical information on operating experience, best practices, and emerging knowledge. These experts are working jointly to leverage the separate efforts of existing national programs into a coordinated research activity to support safe long-term operation. This coordination enables a pooling of technical expertise and avoids unnecessary redundant efforts by sharing responsibility, accountability, resources, and rewards from the activity. The three ongoing activities currently include: (1) the development of a handbook on reactor aging management issues; (2) the identification of aging management research issues; and, (3) the identification of technologies—existing and those needed—for the monitoring of degradation of reactor components .

RES also initiated an activity to collect results from the implementation of AMPs committed to by licensees for the initial license renewal period (i.e., the 40–60 year PEO), along with any information from other licensee aging management activities that will provide greater insights to materials aging phenomena in the PEO. This information and improved understanding will be used to identify any need for enhancements to AMPs for plant operation up to 80 years. The report from the first round of these audits is presently under final review, and discussions are underway with the industry for scheduling the second round of audits.

For More Information
Contact Makuteswara Srinivasan, RES/DE, at
Makuteswara.Srinivasan@nrc.gov.

Steam Generator Tube Integrity

Background

Steam generator (SG) tubes (Figure 9.4) are an integral part of the reactor coolant system (RCS) pressure boundary. They serve as a barrier to isolate the radiological fission products in the primary coolant from the secondary coolant and the environment. The understanding of SG tube degradation phenomena is continually evolving to keep pace with advances in SG designs and materials. To date, many modes of degradation have been observed in SG tubes, including bulk corrosion and wastage, crevice corrosion, pitting, denting, stress-corrosion cracking, and intergranular corrosion attack. Flaws have developed on both the primary and the secondary side of SG tubes. If such flaws go undetected or unmitigated, they can lead to tube rupture and possible radiological release to the environment.

Figure 9.4 Recirculating steam generator tube bundle

Overview

The main objective of this research program is to develop a technical basis for SG tube integrity evaluations. This basis is necessary to ensure that (1) SG tubes continue to be inspected appropriately, (2) flaw evaluations continue to be conducted correctly, and (3) repair or plugging criteria are implemented appropriately. To aid in regulatory decisions and to assess code applications, as depicted in Figure 9.5, this research program addresses the following areas:

- Assessment of inspection reliability.

- Evaluation of inservice inspection technology.

- Evaluation and experimental validation of tube integrity and integrity prediction modeling.

- Evaluation and experimental validation of degradation modes.

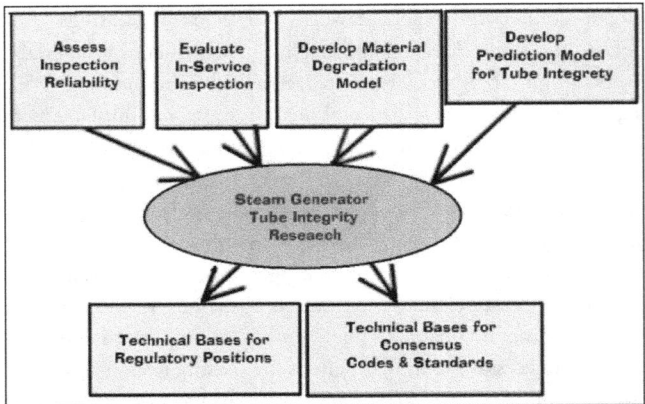

Figure 9.5 Tube integrity research schematic

Approach

The intent of the research is to formulate and document a comprehensive technical basis that will contribute directly to the safety, openness, and effectiveness of the U.S. Nuclear Regulatory Commission's (NRC's) regulatory actions related to SGs. The key elements of the program are best described by technical area.

Assessing Inspection Reliability

In this area, research aims to assess the reliability of current inspection methods based on the flaws observed in the field and to evaluate any new and emerging inspection methods as they arise. For example, one task in this area involves assessing the capabilities and limitations of automated eddy current analysis. The task will use the Argonne National Laboratory SG tube flaw mockup facility, which contains a variety of flaws typically found in the field. Results of automated eddy current analysis will be compared to a previous eddy current round robin test that studied the reliability of human analysts. In this way, the staff can assess the reliability of automated eddy current analysis techniques.

Inservice Inspection Technology

Advanced nondestructive examination (NDE) techniques are used to evaluate SG tube integrity. During inservice inspections, NDE is used to detect and characterize tube flaws. Research in this area aims to evaluate the reliability of NDE techniques for both original and repaired SG tubes. For eddy current inspection, this research will evaluate correlations of signal voltage to flaw morphology and structural integrity. A technical report on this research will present an evaluation of the differences and limitations between various eddy current methods, including bobbin coil, rotating pancake, and xprobe.

Research on Tube Integrity and Performance Modeling

When a flaw is detected in an SG tube, its potential for leaking or bursting must be assessed. Tube integrity is assessed using models that predict leak rates and burst pressures that a particular flaw might exhibit during normal operation or Design-Basis accidents. Although models exist to predict flaw behavior, they require that the complex flaw morphology be simplified. One means of simplifying a complex crack is to use a rectangular crack profile. Ongoing research will continue to assess the use of the rectangular crack method for estimating failure pressure and leak rate for complex crack geometries.

Research will also continue to examine the leak rate from postulated tube flaws in the region of the tubesheet under postulated severe accident conditions. Experimental tests will be conducted to calibrate and validate the leak models.

Another ongoing study examines the consequences of exposing RCS materials to high temperatures during severe accident scenarios. Such accidents may challenge the integrity of SG tubes; therefore, analyses are being conducted to determine whether certain RCS components may fail before SG tubes. Such a scenario would be preferable to an initial release through SG tubes because RCS leaks would leak into containment, whereas SG tube leaks could lead to a radiation release to the outside environment.

Research on Degradation Modes

Analytical models exist to predict potential degradation behavior in SG tubes during normal operating conditions. Research in this area seeks to evaluate and experimentally validate those models. Such research will require a better understanding of crevice conditions and stress-corrosion crack initiation,

evolution, and growth. The NRC has already conducted considerable research in these areas, which has established a better understanding of the nature of crevice behavior. A NUREG report will document the research.

International Cooperation

The NRC is currently administering the fourth 5-year term of the International Steam Generator Tube Integrity Program (ISG-TIP-4). In this program, regulators and researchers from member countries conduct and share research on tube integrity and inspection technologies. Current participants include organizations from Canada, France, Japan, Korea, and the United States.

For More Information
Contact Charles Harris, RES/DE, at Charles.Harris@nrc.gov.

Consequential Steam Generator Tube Rupture Program

Background

The U.S. Nuclear Regulatory Commission (NRC) and the nuclear power industry have expended considerable resources over the last two decades to better understand the safety implications and risk associated with consequential steam generator tube rupture (CSGTR) events (i.e., events in which steam generator (SG) tubes leak or fail as a consequence of the high differential pressures or SG tube temperatures, or both, predicted to occur in certain accident sequences). Key activities included an assessment of temperature-induced creep-rupture of the reactor coolant system (RCS) components in the study in NUREG-1150, "Severe Accident Risks: An Assessment for Five U.S. Nuclear Power Plants," issued December 1990; a representative analysis of the potential for induced containment bypass by an ad hoc NRC staff working group in NUREG-1570, "Risk Assessment of Severe Accident-Induced Steam Generator Tube Rupture" issued March 1998; and recent thermal-hydraulic (T/H) analyses and risk analyses as part of the steam generator action plan (SGAP). Severe accident analyses, performed as part of the State-of-the-Art Reactor Consequence Analyses project, provide additional insights into the likelihood and impact of subsequent failure of the reactor hot leg shortly following a CSGTR event.

Prior investigations of a Westinghouse plant concluded that the contribution of CSGTR events to the overall containment bypass frequency is, at best, of the same order of magnitude, if not lower than the containment bypass fraction associated with other internal events for most pressurized-water reactors. Thus, plant risk assessments should consider and monitor the risk associated with CSGTR in a manner commensurate with its expected importance at each plant. Although important conclusions were made, these investigations identified certain limitations of scope and a lack of thorough RCS component modeling with advanced simulation tools. Addressing these limitations to advance our understanding of associated risks and developing an enhanced risk assessment tool for CSGTR events are important activities.

Objectives

To close the technical gaps and to develop an enhanced risk assessment procedure for CSGTR, the current Office of Nuclear Regulatory Research (RES) program will attempt to fulfill the following objectives:

- Update computational fluid dynamics (CFD) and system code models for Combustion Engineering (CE) plants.

- Evaluate the impact of in-core instrument tube failures on natural circulation.

- Update SG flaw distributions.

- Complete structural analyses of CE and Westinghouse RCS components.

- Develop a user-friendly methodology for assessing the risk associated with consequential tube rupture and leakage in Design-Basis accidents and severe accident events.

- Conduct a reassessment of the conditional probabilities of CSGTR based on updated flaw distributions and updated T/H analyses.

- Compile and summarize key research, building upon NUREG-1570 (work performed as part of SGAP activities).

Approach

Combustion Engineering Thermal-Hydraulic and Severe Accident Analysis

The updated modeling approach and lessons learned from the most recent Westinghouse plant predictions will be applied to a CE plant model to improve the T/H predictions. This effort will update the hot-leg flow and mixing model and the hot-leg thermal radiation modeling.

The CE CFD model will be updated to include a simplified upper plenum, hot leg, surge line, and the SG primary side. This model will be used to predict hot-leg and inlet plenum mixing rates and the variations in temperature of the flow entering the hottest tubes in the SG.

The system code modeling effort will include the development of a MELCOR CE plant model that incorporates all of the lessons learned from the recent Westinghouse predictions completed in support of the SGAP. The modeling will also incorporate the updated CE CFD model predictions.

Assess Impact of In-Core Instrument Tube Failures

RES completed a study on the impact of the consequences of instrumentation tube failure during severe accidents, which is detailed in ERI/NRC-09-206, "Analysis of the Impact of Instrumentation Tube Failure on Natural Circulation During Severe Accidents." This work assesses the impact of instrumentation tube failures for Three Mile Island, Unit 2 (a Babcock & Wilcox design with a once-through SG) and Zion Nuclear Power Station (a four-loop Westinghouse design with a U-tube SG).

Updated Steam Generator Flaw Distributions

To assess the probability of an induced SGTR, detailed knowledge of the characteristics of SG tube flaws is needed with the tube temperature and stress profile during postulated accidents. For statistical analysis, flaw density distribution data as a function of size, shape, orientation, location, and type are needed. The potential for failure depends primarily on the upper tail of the size distribution (i.e., the most severe flaws) for a given flaw type and location. A verification process will also be used to confirm that the flaw distributions are consistent with operating experience for observed leakage rates.

By means of an existing memorandum of understanding addendum between the NRC and Electric Power Research Institute, RES will work with the industry to update flaw distributions originally developed in the mid-1990s. This update will include (1) evaluating the effect of improved inspection techniques on flaw density distributions, (2) developing distributions for both crack-like and wear-like defects, (3) accounting for flaws in the SG tube within the tubesheet regions, and (4) identifying any changes in flaw distribution caused by new tube materials, new SG designs, or new inspection techniques.

Structural Analysis of Combustion Engineering and Westinghouse Reactor Coolant System Components for Prediction of Reactor Coolant System Piping Failure

RES structural analyses will build upon the latest T/H and severe accident analyses to include specific RCS components for Westinghouse and CE plants (e.g., hot-leg nozzle and hot-leg to surge line nozzle). The failure analysis will consider uncertainty resulting from the shape, size, and location of potential flaws in the RCS components.

RES plans to identify, characterize, and model relevant RCS nozzles to assess their potential for failure during severe accidents for Westinghouse and CE plants. Two-dimensional axisymmetric and three-dimensional models will be developed to address variables such as nozzle geometries and configurations, boundary conditions, loading conditions, fabrication effects, stress-corrosion cracking mitigations, and degraded conditions. These models will be used to determine the time to failure for each analyzed component and the associated sensitivities to loadings and flaw geometry. Because of the importance of incorporating uncertainty, RES will develop a semi-empirical methodology based on numerical experiments to predict failure of critical RCS components. The NRC expects the resulting methodology to be more conducive to the procedure adopted in the CSGTR risk assessment method developed as part of the program.

Simplified Method for Assessing the Risk Associated with C-SGTR

In March 2009, RES provided the NRC's Office of Nuclear Reactor Regulation with a report that describes a method for assessing CSGTR risk. RES intends to extend the methods described in this previous report to incorporate a number of enhancements. These enhancements will include consideration of the updated T/H conditions, SG flaw distribution, and RCS component analyses. Additionally, CSGTR risk assessment methods described in previous reports by the NRC, Electric Power Research Institute, and industry will be reviewed to identify useful insights and modeling approaches for use with the new simplified method. RES anticipates that the level of analysis in the new approach will be comparable to that of the previous RES CSGTR risk report and the earlier NUREG-1570 study. Consistent with previous CSGTR risk assessment work, the new simplified method will consider both pressure-induced and thermally-induced SG tube failures.

Reevaluation of C-SGTR Conditional Probabilities

In support of SGAP, RES previously developed an SG tube failure probability calculator tool. RES plans to extend the framework and modeling approaches used in this tool, including pressure- and temperature-induced challenges. Consequently, this program will focus on further validation of the detailed modeling used in the calculator, extension of calculator capabilities, updates to basic data and parameters (including provisions for future data updates), improvements in calculator usability, and development of supporting documentation.

Status

In addition to the progress made on the development of a computational tool to assess risk associated with CSGTR with updated flaw distributions and on the assessment of in-core instrument failures, RES has identified appropriate "hold points" and decision criteria toward achieving the research goals to allow for potentially redirecting research toward the highest regulatory needs or for terminating research early if sufficient insights can be gathered from early phases to support a regulatory decision. These hold points would also provide a decision-making framework for the ongoing research activities to better define key milestones and regulatory end products.

Deliverables

The following deliverables are anticipated at the completion of the CSGTR program:

- Probabilistic risk assessment report.

- Risk assessment tool.

- draft regulatory guidance on risk-informed decision-making concerning CSGTR.

- Draft "Risk Assessment Standardization Project" handbook section on the assessment of CSGTR suitable to support revisions to the appendices to Inspection Manual Chapter 0609, "Significance Determination Process," dated June 2, 2011.

- Summary report compiling key research results.

For More Information
Contact Dr. Raj Mohan Iyengar, RES/DE, at
Raj.Iyengar@nrc.gov.

Reactor Pressure Vessel Integrity

Background

One aspect to the safe operation of a nuclear power plant is maintaining the structural integrity of the reactor pressure vessel during both routine operations (i.e., heat up, cool down, and hydro test) and during postulated accident scenarios (e.g., pressurized thermal shock [PTS]). To do this, procedures are needed to estimate and compare the driving force for structural failure to the resistance of the structure to this driving force (and the effect of radiation on this resistance). Current statutory procedures for these estimates appear in Title 10 of the *Code of Federal Regulations* (10 CFR) Part 50.61, "Fracture Toughness Requirements for Protection against Pressurized Thermal Shock Events" (hereafter referred to as the PTS rule); Appendix G, "Fracture Toughness Requirements," to 10 CFR Part 50, "Domestic Licensing of Production and Utilization Facilities"; Appendix H, "Reactor Vessel Material Surveillance Program Requirements," to 10 CFR Part 50; Regulatory Guide 1.99, "Radiation Embrittlement of Reactor Vessel Materials"; and Regulatory Guide 1.161, "Evaluation of Reactor Pressure Vessels with Charpy Upper Shelf Energy Less Than 50 ft-lb." Although these methods generally depend on empirically based engineering methods, they are known to incorporate large implicit conservatisms adopted to address state of knowledge deficiencies that existed at the time of their issuance. When coupled with the deterministic basis of current regulations, these conservatisms may unnecessarily reduce the possibility for continued operation and potential license renewals.

Objectives

The following two objectives apply:

1. Integration of the advances in the state of knowledge, empirical data, and computational power that has occurred in the 20 or more years since the adoption of the current regulatory requirements to develop the technical bases for state-of-the-science and risk-informed revisions to the statutory procedures that regulate the structural integrity of the current operational boiling and pressurized-water reactor fleets.

2. Use of the advances in the state of knowledge and empirical data that have accumulated over 20 or more years of structural materials research by the nuclear community to develop, validate, and refine physically based predictive models of material deformation and failure behavior to include the effects of radiation embrittlement.

Approach

RES has recently completed a multiyear study conducted in cooperation with Oak Ridge National Laboratory, other national laboratories and Government contractors, and the domestic nuclear power industry under the auspices of the Electric Power Research Institute's materials reliability project to develop the technical basis for a risk-informed revision to the PTS rule. The Office of Nuclear Reactor Regulation has used this technical basis to develop a voluntary alternative to the PTS rule, which uses improved knowledge to address many of the conservatisms in the current rule without affecting the public health and safety. The U.S. Nuclear Regulatory Commission completed this voluntary alternative rule in 2010.

Also in the coming years, RES may publish and make available for public comment a revised version of Regulatory Guide 1.99, along with its technical basis. This revision is based on over five times the quantity of empirical data used to develop the current regulatory guide. The insights gained from these activities provide a large part of the work needed as the technical bases to support revisions to Appendix G and Appendix H to 10 CFR Part 50.

In the next 5 to 10 years, RES will pursue the following two major initiatives to ensure the structural integrity of the pressurized nuclear power plant components in the existing fleet during the period of license extension and in the new reactor fleet:

- Development and validation of a method capable of identifying embrittlement mechanisms in reactor materials before they occur in commercial reactor service.

- Development and validation of a modular computational tool to perform probabilistic structural integrity assessments of passive primary reactor pressure boundary components.

For More Information
Contact Dr. Mark Kirk, RES/DE, at Mark.Kirk@nrc.gov.

Degradation of Reactor Vessel Internals from Neutron Irradiation

Background

The internal components of light-water reactor (LWR) pressure vessels are fabricated primarily with austenitic stainless steels because of their relatively high strength, ductility, and fracture toughness in the unirradiated state. During normal reactor operating conditions, the internal components are exposed to high-energy neutron irradiation and high-temperature reactor coolant. Prolonged exposure to neutron irradiation changes both the microstructure and microchemistry of these stainless steel components, increasing their strength and decreasing their ductility and fracture toughness. Neutron irradiation exposure also increases their susceptibility to irradiation-assisted degradation (IAD). Cracks caused by IAD have been found in a number of internal components in LWRs, including control rod blades, core shrouds, and bolts (Figure 9.6).

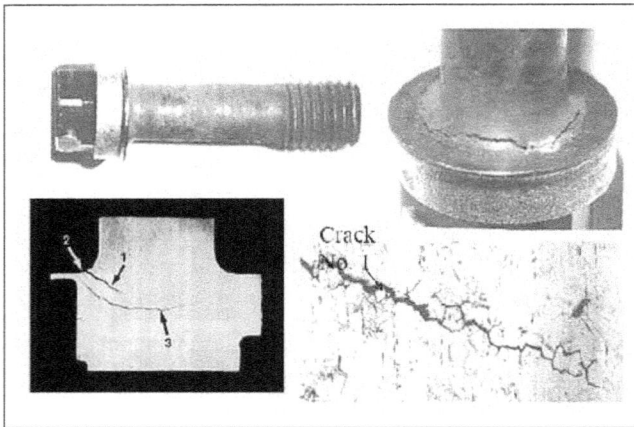

Figure 9.6 Cracking of a baffle bolt in a pressurized-water reactor (PWR)

As nuclear power plants age and as neutron irradiation dose increases, the degradation of the vessel internals becomes more likely and potentially more severe. Preliminary data suggest that the significance of the degradation of LWR vessel internals could increase during both the license extension period (i.e., 40 to 60 years) and during even longer term operation of nuclear power plants.

Objective

The U.S. Nuclear Regulatory Commission (NRC) has developed a broad research plan to address the degradation of reactor vessel internals from neutron irradiation. The results of the research will be used to provide insights into the causes and mechanisms of IAD in boiling-water reactors (BWRs) and PWRs and to inform regulatory decisions on the reliability of reactor vessel internals during long-term operation.

Approach

The NRC is doing the following in the conduct of research to characterize and evaluate IAD:

- Define a threshold neutron dose above which irradiation begins to affect material properties.

- Evaluate the adequacy of data used to estimate cyclic fatigue and crack growth rates for both BWR and PWR vessel internal materials.

- Assess the significance of void swelling and irradiation stress relaxation/creep on the structural and functional integrity of PWR internal components.

Test specimens have been and will be irradiated over a broad range of prototypical exposure levels to evaluate the expected performance of plant materials. Presently, a systematic research effort is underway to determine the causes of IAD, to establish a fracture toughness degradation threshold, and to investigate saturation effects in BWR and PWR internals. The Halden Nuclear Reactor facility in Norway is irradiating representative reactor internal materials, and Argonne National Laboratory is carrying out experimental testing. Specifically, within BWR environments, the effects on IAD from the hydrogen concentration in the reactor coolant and the concentration of light elements, such as sulfur and oxygen, within the steels were evaluated; this portion of the work has been completed, and the results have been documented. The effects of neutron dose on IAD and fracture toughness and the synergistic effects of neutron and thermal embrittlement on fracture toughness are currently being investigated for PWR environments.

Longer term research will focus on the effects expected during plant operation beyond 60 years. As previously indicated, the NRC staff is completing a multiyear study of the effect of BWR environments on IAD of austenitic stainless steel vessel internals. The products of this program have been used to evaluate licensee submittals related to managing degradation of these components and to inform other aspects of the regulatory process, such as inspection requirements and responses to relief requests. The results of this program have led to the resolution of regulatory issues and to the development, validation, and improvement of regulations and regulatory guidelines. In addition, the research plan includes the harvesting of internal structural materials from decommissioned nuclear reactors, such as the Zorita reactor in Spain. Materials from the Zorita reactor have higher levels of radiation exposure than experimental samples and would provide information on the expected behavior of domestic BWR and PWR components during long-term operation. The plan also provides for participation in other collaborative research efforts that (1) will leverage resources, (2) will extend knowledge acquired from previous research, and (3) will use unique testing facilities within the international community.

For More Information
Contact Dr. Appajosula S. Rao, RES/DE, at Appajosula.Rao@nrc.gov.

Primary Water Stress-Corrosion Cracking

Background

Primary water stress-corrosion cracking (PWSCC) in primary pressure boundary components composed of nickel-based alloys is a degradation mechanism that can affect the operational safety of pressurized-water reactors (PWRs).

Figure 9.7 shows PWSCC cracks that were found in control rod drive mechanism nozzle J-groove welds at North Anna Power Station, Unit 2. The narrow cracks are often located in complex structures either within or adjacent to welds and are difficult to detect and characterize. Undetected PWSCC led to reactor pressure boundary leaks and subsequent boric acid corrosion of the low-alloy steel reactor pressure vessel head at Davis-Besse Nuclear Power Station in 2002 (Figure 9.8).

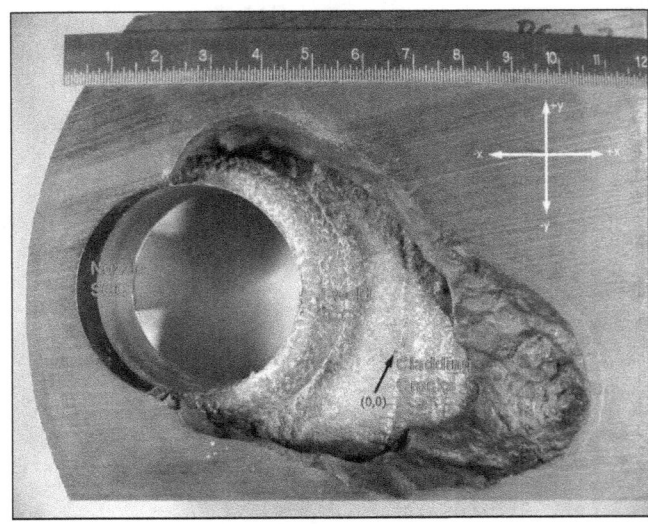

Figure 9.8 Photograph showing extensive boric acid corrosion in the low-alloy steel Davis-Besse reactor pressure vessel head. Reactor coolant leaked from PWSCC cracks in the Alloy 600 control rod drive mechanism nozzle and the nozzle J-groove weld.

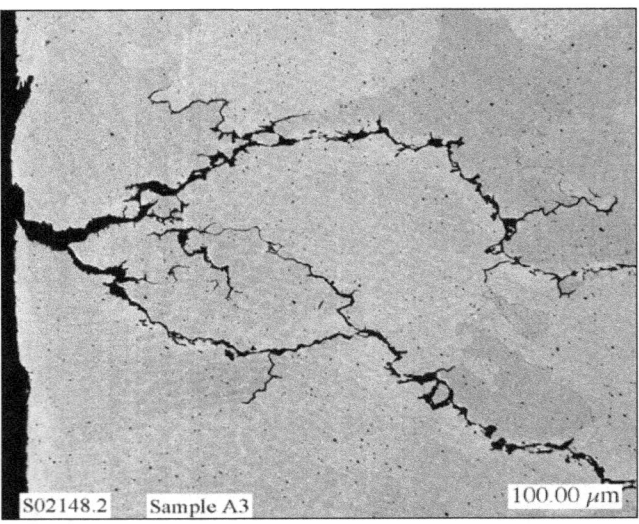

Figure 9.7 PWSCC cracks in the Alloy 182 J-groove weld in North Anna-2 Nozzle 31

Alloy 690 and associated weld metals Alloy 52 and Alloy 152, which have nominal chromium concentrations of 30 percent, have been used in replacement components, including steam generators, PWR replacement heads, reactor coolant system piping, nozzles, and instrument penetrations. PWSCC mitigation of the more susceptible alloys that remain in service (Alloy 600 and weld metals Alloy 82 and Alloy 182) has been done using Alloy 52 weld overlays.

Objective

The objectives of this program are to evaluate the PWSCC susceptibility of high-chromium Alloy 690, its weld metals Alloy 152 and Alloy 52, and dilution zones of dissimilar metal welds with low alloy steel or stainless steel to determine the relationship between PWSCC susceptibility and metallurgical characteristics of the chromium-containing nickel-based alloys used in replacement and new construction components. The work will also provide valuable information to assess potential mitigation methods for the lower chromium nickel-based alloys (600/182/82) that were originally used in PWRs and that were known to be susceptible to PWSCC.

Information obtained will be used to develop regulatory guidance and to establish inservice inspection requirements necessary to ensure the continued safe operation of PWRs.

Approach

The U.S. Nuclear Regulatory Commission (NRC) is sponsoring confirmatory research comprising crack growth rate measurements on nickel-based alloys in simulated PWR environments and microstructural and fracture surface analyses of test materials. The NRC is also participating in an international cooperative effort to evaluate factors that influence the PWSCC susceptibility of nickel-based alloys.

NRC-Sponsored Research

The NRC has ongoing research activities on the PWSCC susceptibility of nickel-based alloys. Specific tests are being conducted to evaluate the importance of the following:

- Fabrication processes and thermal treatments on Alloy 690.

- Shielded metal arc welding and gas tungsten arc welding processes.

- Heat-affected zones adjacent to shielded metal arc welds and gas tungsten arc welds.

- Weld defects, including hot cracking and ductility dip cracking.

- Dilution zones in dissimilar metal welds.

Examination of test specimen fracture morphology, metallurgical analyses, and crack tip characterizations of test specimens and actual plant components that have been removed from service will provide data to determine how the microstructural features affect PWSCC growth rates.

Results obtained from the NRC-sponsored research have shown that possible combinations of cold work applied to Alloy 690 with specific thermal treatments can significantly affect PWSCC susceptibility. High-chromium weld filler alloys are generally more resistant to PWSCC; however, higher susceptibility of some welds and the dilution zone in dissimilar metal welds has been observed. Evaluation of the compositional and metallurgical variations that increase the PWSCC susceptibility in the welded materials is under investigation.

PWSCC International Cooperation

The NRC is also participating in an international cooperative effort that includes representatives from the Electric Power Research Institute, industry, and licensees. This cooperative effort has led to the development of PWSCC testing protocols and analysis methods, the evaluation of representative plant materials, and the testing of newly developed weld alloys. The cooperative effort provides a forum for the dissemination and discussion of research results to the benefit of all participants.

For More Information
Contact Darrell S. Dunn, RES/DE, at Darrell.Dunn@nrc.gov.

Primary Water Stress-Corrosion Cracking Mitigation Evaluations and Weld Residual Stress Validation Programs

Background

In pressurized-water reactor (PWR) coolant systems, nickel-based dissimilar metal (DM) welds are typically used to join carbon steel components, including the reactor pressure vessel, steam generators, and the pressurizer, to stainless steel piping. Figures 9.9 and 9.10 show a representative nozzle to piping connection cross-section, including the DM weld. The DM weld is fabricated by sequentially depositing weld beads as high-temperature molten metal that cools, solidifies, and contracts, retaining stresses that approach or potentially exceed the material's yield strength.

These DM welds are susceptible to primary water stress-corrosion cracking (PWSCC) as an active degradation mechanism that has led to reactor coolant system pressure boundary leakage. PWSCC is driven by tensile weld residual stresses (WRSs) and other applied loads within the susceptible DM weld material. Hence, proper assessment of these stresses is essential to accurately predict PWSCC flaw growth and to ensure component integrity.

The nuclear power industry has developed several PWSCC mitigation techniques for DM welds that are currently being implemented in the PWR fleet. Examples include the following:

- Full structural and optimized weld overlays, in which replacement material less susceptible to PWSCC is welded onto the outer diameter of the affected joint that also imparts a stress improvement to the susceptible joint.

- Weld inlays, in which a layer of replacement material less susceptible to PWSCC is welded to the inner diameter to act as a barrier between the corrosive reactor coolant and the DM weld material (e.g., similar to cladding).

- Mechanical stress improvement processes (MSIP), in which the pipe is squeezed using a large hydraulically driven clamp that imparts a stress improvement to the susceptible joint.

Weld overlays and MSIP reduce and, in some cases, reverse tensile residual stresses in DM welds, thus decreasing the driving force for crack growth. However, weld inlays have been shown to increase tensile WRSs, which can potentially increase PWSCC initiation and growth of the less susceptible replacement material.

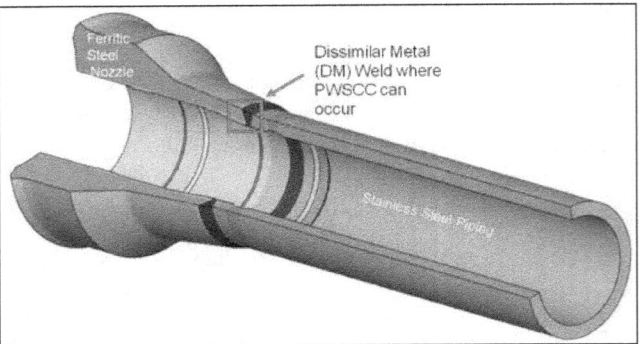

Figure 9.9 Cutaway view of a carbon steel nozzle DM weld and stainless steel piping that is typical in a light-water-cooled nuclear power plant

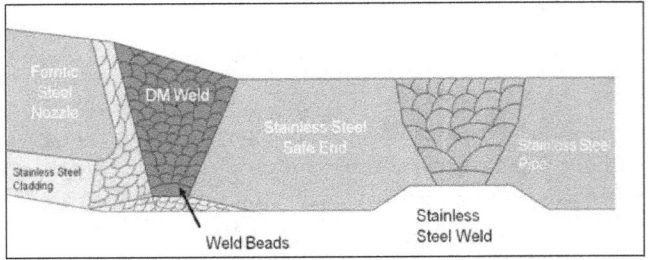

Figure 9.10 Cross-section of a nozzle to pipe weld highlighting the weld bead pattern

Validation Program

Recent improvements in computational capabilities have facilitated advances in WRS predictions using finite-element analysis (FEA). Although no universally accepted methodology exists to model WRSs using FEA, the Electric Power Research Institute (EPRI) has developed draft guidelines for streamlining these procedures. The assumptions and estimation techniques vary from analyst to analyst, which causes variability in the predicted WRS profiles.

The Office of Nuclear Regulatory Research (RES) is supporting the Office of Nuclear Reactor Regulation (NRR) in developing appropriate regulatory requirements to address PWSCC in reactor coolant piping systems. A portion of this effort includes the WRS validation program aimed at refining and validating the FEA procedures for modeling WRS and for characterizing the uncertainties in the resulting predictions. The U.S. Nuclear Regulatory Commission (NRC) is conducting the WRS validation program cooperatively with EPRI under a memorandum of understanding (MOU) addendum.

Figure 9.11 shows a typical WRS FEA performed using the ABAQUS software for a reactor pressure vessel to pipe nozzle DM weld. The distribution of stresses shows the area in which a flaw may initiate (typically on the inner diameter of the DM weld), propagate, and cause leakage or structural instability. Validation of the results of this analysis is being done by comparing predicted temperature, thermal strain, and residual stress fields to a variety of physical measurements performed on actual and representative plant components and mockups.

The WRS validation program has been successful in a number of areas to date, including the following:

- Evaluations of various PWSCC mitigation techniques (full structural weld overlays, optimized weld overlays, MSIP, and inlays).

- Safety evaluation report technical basis development provided to NRR for approving several PWSCC mitigation techniques for use by the PWR fleet.

- Input to the American Society of Mechanical Engineers Boiler and Pressure Vessel Code (ASME Code) case reviews.

- Multiple plant-specific PWSCC flaw evaluations for NRR review.

- Development of technical letter reports on the WRS validation program.

- Development of a draft NUREG report on the WRS validation program.

RES and its contractors in cooperation with the nuclear power industry through an NRC/EPRI MOU addendum have completed a multi-phase program to validate predictions of WRSs based on FEA. A major element of this program involved the international WRS round robin in which 15 organizations blindly and independently analyzed the WRSs in a representative pressurizer surge nozzle DM weld mockup (Figure 9.12). RES has completed a blind validation of this mockup by measuring WRSs and by comparing the measurements to blindly conducted FEA predictions.

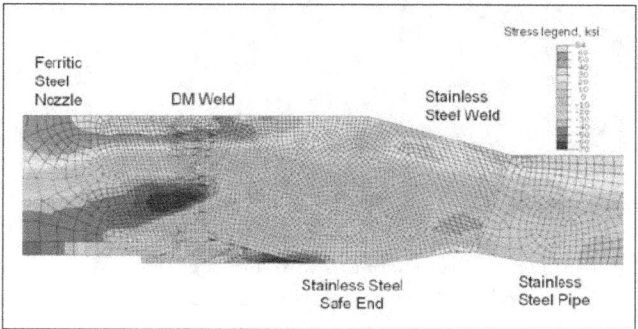

Figure 9.11 Stress magnitude distribution in a nozzle to pipe weld configuration

Figure 9.12 Pressurizer surge nozzle DM weld mockup being measured for WRSs

Remaining Work

Further validation of WRS predications will be performed using different weld geometries under the auspices of the NRC/EPRI MOU. The research performed for the WRS validation program will help improve the following activities:

- Development of appropriate acceptance criteria for validation of WRS predictions.

- Development of best practice guidelines for WRS prediction procedures.

- Incorporation of best practice guidelines into the ASME Code.

- Determination of estimates for the uncertainty and distribution of WRSs for use in the Extremely Low Probability of Rupture Code.

Figures 9.9 through 9.11 are courtesy of Dr. Lee Fredette of Battelle Memorial Institute, Columbus, OH.

For More Information
Contact Dr. Michael Benson, RES/DE, at Michael.Benson@nrc.gov or Dr. Shah Malik, RES/DE, at Shah.Malik@nrc.gov.

Nondestructive Examination

Background

In accordance with Title 10 of the *Code of Federal Regulations* (10 CFR) Part 50.55(a), "Codes and Standards," licensees must inspect structures, systems, and components to ensure that the requirements of the American Society of Mechanical Engineers Boiler and Pressure Vessel Code (ASME Code) are met and that structures, systems, and components can continue to perform their safety functions. Research on nondestructive examination (NDE) of light-water reactor (LWR) components and structures provides the technical basis for regulatory decisionmaking related to these requirements. For example, the Office of Nuclear Reactor Regulation and the Office of New Reactors will use the findings to evaluate licensees' alternatives to ASME Code requirements, new plant submittals, proposed changes to the ASME Code, and ASME Code cases for U.S. Nuclear Regulatory Commission (NRC) endorsement. In addition, results from the NDE of these components and structures are used to assess models developed to predict the effects of materials degradation mechanisms and are used as initial conditions for component-specific fracture mechanics calculations. Pacific Northwest National Laboratory (PNNL) is conducting this work.

Regulatory Needs

Areas of interest addressed by NDE research include the following:

- Evaluation of the accuracy and reliability of NDE techniques used for inservice inspection of LWR systems and components.

- Support of NRC rulemaking efforts in materials reliability, such as General Design Criterion (GDC) 4, "Environmental and Dynamic Effects Design Bases" of Appendix A, "General Design Criteria for Nuclear Power Plants," to 10 CFR Part 50, "Domestic Licensing of Production and Utilization Facilities," to demonstrate that the probability of fluid system piping rupture is extremely low.

- Assessment of inspection requirements, procedures, and inspector qualifications.

- Incorporation of modeling into NDE.

The four specific project areas highlighted below address these regulatory needs.

Approach

Evaluation of the Accuracy and Reliability of NDE Techniques

Research activities include evaluating the accuracy, effectiveness, and reliability of NDE, as currently practiced in the nuclear industry. The research objectives are (1) to determine the relationships among preservice inspection methods, inservice degradation (e.g., cracking and aging), and inservice inspection practice and results and (2) to evaluate the effectiveness, accuracy, and reliability of new techniques that the NRC expects licensees to apply in current, new, and advanced reactors. Many reported events over the past several years have revealed that there are issues regarding NDE, as it is currently being employed in the field, with respect to qualification and certification of inspectors, inspection methods, and inspection practices. In addition, certain materials and locations susceptible to degradation are difficult to inspect in the current fleet of reactors and will most likely remain challenging for new reactors. This NRC program is using fabricated mockups and components removed from reactors, including some canceled plants and some operating reactors, to determine the effectiveness of existing and emerging NDE techniques (Figure 9.13).

Figure 9.13 Components and material that have been removed from canceled plants

The NRC performs some of this work under cooperative agreements to help defray costs and to gain access to the expertise of other organizations. For example, the NRC program at PNNL is evaluating the ability to detect and characterize primary water stress-corrosion cracking (PWSCC) in LWR components and an NRC-initiated international project known as the Program for the Inspection of Nickel-Alloy Components and the Program To Assess the Reliability of Emerging Nondestructive Techniques.

Under its current program at PNNL, the NRC is directing research on the inspection of coarse-grained austenitic alloys and welds (Figure 9.14). NDE of these components is difficult because of signal attenuation and reflections. In these materials, grain boundaries and other microstructural features appear similar to cracks (Figure 9.15). Research findings will support appropriate inspection requirements for these components to ensure safety. The NRC is performing some of this work under cooperative agreements with the Electric Power Research Institute (EPRI) and the Institut de Radioprotection et de Surete Nucleaire (IRSN).

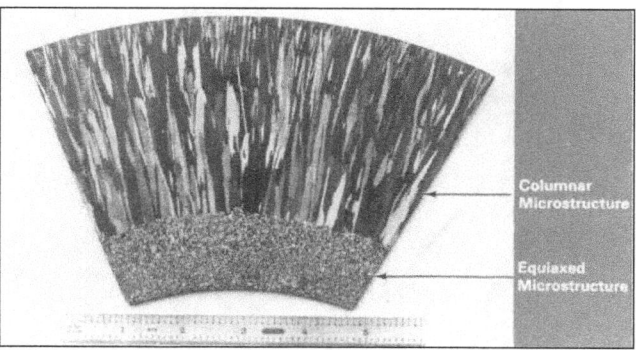

Figure 9.14 Sample illustrating the coarse grain microstructure of centrifugally cast stainless steel

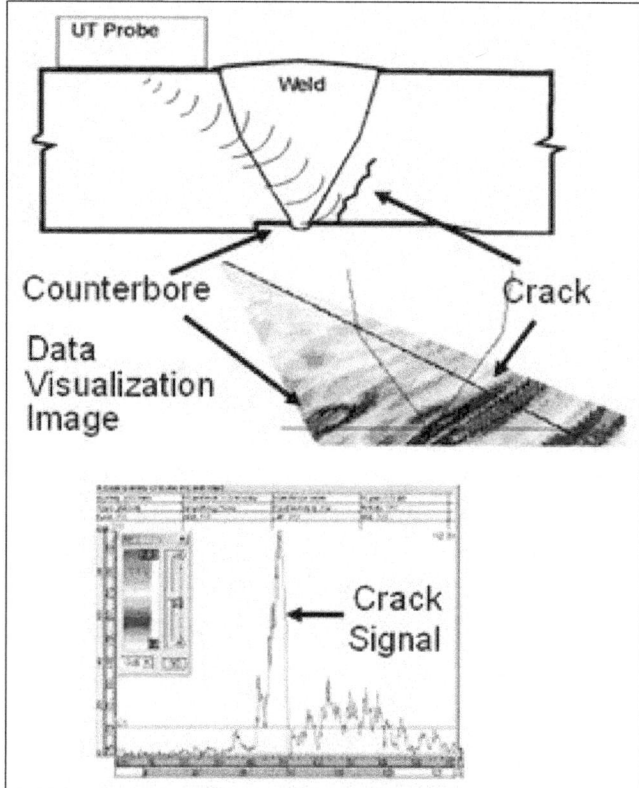

Figure 9.15 Schematic view of flaw detection at the far side of a weld using a phased array ultrasonic technique that improves flaw detection in coarse-grained metals and welds

Support for NRC Rulemaking Efforts General Design Criterion 4

The NRC is analyzing strategies for managing PWSCC in leak-before-break systems to ensure that the probability of fluid system piping rupture remains extremely low in accordance with the requirements in GDC 4 of Appendix A to 10 CFR Part 50. Management by inspection or monitoring is one of the strategies that the NRC is analyzing. To rely on inspections as a management technique, the inspections must be capable of detecting PWSCC before the probability of failure would no longer be considered extremely low. The NRC is analyzing the reliability of the inspections being performed under Section XI, "Rules for Inservice inspection of Nuclear Power Plant Components," of Appendix VIII, "Performance Demonstration for Ultrasonic Examination Systems," to the ASME Code for the range of dissimilar metal butt weld configurations that exist in leak-before-break piping to determine whether the inspections being performed are adequate to ensure that the probability of leakage is extremely low.

Assessment of Inspection Requirements, Procedures, and Inspector Qualifications

Several incidents have occurred over the past several years in which flaws in components were either mischaracterized or missed entirely. Some of the examinations that were conducted were examinations that were qualified through the Performance Demonstration Initiative. This raises questions with respect to the qualifications and training of inspectors, the procedures used, and the effectiveness and reliability of some of the techniques that are being used. A cooperative research program is being conducted with EPRI to assess the current process for optimizing and modifying essential variables and the design, fabrication, and use of site-specific mockups. In addition, NRC is participating on an ASME committee that is developing a new personnel qualification and certification process. The current process is applicable to any industry and is employer-based. The goal of this committee is to develop a standard process specific to the nuclear industry.

Incorporation of Modeling into NDE

Events over the past several years have revealed deficiencies with respect to the application of NDE in the field, especially with the use of conventional transducers. Modeling at PNNL has revealed issues such as poor transducer design, improper selection of transducer(s), inadequate insonification of the weld, and improper focal depth of beams (Figures 9.16 and 9.17). The modeling shows that such inattention to detail is likely to result in missed detections. A cooperative program with EPRI is being initiated to evaluate the incorporation of modeling software into NDE as applied to the inspection of nuclear components to address these issues.

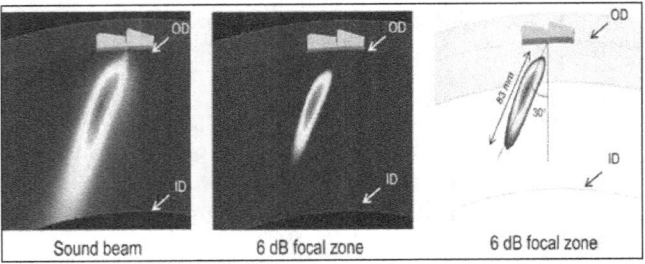

| Sound beam | 6 dB focal zone | 6 dB focal zone |

Figure 9.16 Sound field simulations showing the acoustic beam density generated by a probe (simulated sound field energy decreases from red to blue). Image on left is an unfiltered sound beam. Middle image has a –6 dB filter (50% of the maximum beam intensity is predicted at the ID surface emphasizing the insufficient extent of the –6 dB focal zone [or highest intensity sound field] to reach the specimen ID surface region). Image on right has pipe surfaces drawn (note that pipe layer above OD is an artifact of the simulation software).

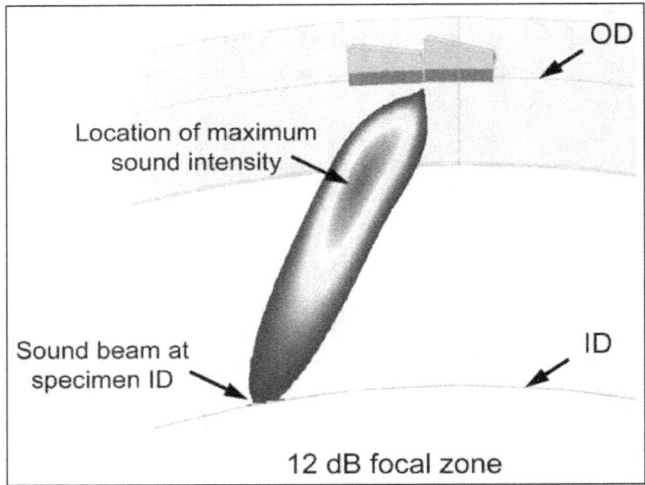

Figure 9.17 The same sound field simulated in Figure 9-16 was projected to the specimen ID. The sound field was substantially diminished in strength (–12 dB level). A drop of 12 dB from the maximum sound intensity is a ratio of 4:1 indicating that only about 25% of the maximum beam intensity is predicted at the ID surface. This is a best scenario given that a homogeneous, isotropic material was modeled versus the anisotropic nature of a weld microstructure. The reduction in signal level predicts a significantly reduced chance of detecting ID surface-connected flaws.

For More Information

Contact Wallace Norris, RES/DE, at Wallace.Norris@nrc.gov or Carol Nove, RES/DE, at Carol.Nove@nrc.gov.

Containment Liner Corrosion

Background

Commercial nuclear power plant containment buildings are designed to act as a barrier to prevent radioactive release under severe accident conditions. Many nuclear power plants have containment buildings constructed with either reinforced or post-tensioned concrete in contact with a thin steel containment liner, which serves as a leak-tight membrane. Although leak-rate tests and inspections required by the U.S. Nuclear Regulatory Commission (NRC) assess the integrity of containment liners, three instances of through-wall corrosion of the liner have occurred since 1999. In all cases, liner corrosion was associated with foreign material embedded in the concrete during original construction (Figure 9.18). Prior leak tests or inspections detected neither the foreign material nor the corrosion-related material loss. Active corrosion was identified only after penetration of the liner had occurred.

Figure 9.18 Photograph of the through-wall corrosion detected at Beaver Valley, 2009. A piece of wood embedded in the concrete during original construction was found behind the corroded area of the steel containment liner. The area was identified by a large paint blister which was filled with steel corrosion products

Objective

The objectives of this program were to evaluate historical information about liner corrosion events, determine the mechanisms for through-wall corrosion, and determine whether plant designs and construction practices influence the susceptibility to liner corrosion. Based on the historical information collected, modeling of the corrosion damage to the containment liner will be performed. The results of the program will be used to assess the current methods of inspecting the liner and possible methods to mitigate liner corrosion. Knowledge gained will also be applied to the effects of plant aging on the integrity of the containment structure and the steel liner.

Approach

Historical information on incidents of liner corrosion was gathered from several sources, including the following:

- Inservice inspection reports and leak-rate test results.
- NRC inspection reports.
- Licensee event reports.
- International operating experience.

Information on containment liner corrosion events was analyzed to determine the relationships between liner corrosion incidents and plant design, operational parameters, and the presence of construction defects. Initial calculations of containment liner corrosion showed that localized attack could result in liner penetration in a few decades. More extensive modeling is planned to evaluate the maximum liner corrosion penetration rate and size of the affected area in the liner. The analysis conducted will be used to identify whether additional regulatory action is needed.

Results

Review of historical information showed that containment liner corrosion initiated on the inside surface of containment liners as a result of degraded or damaged coatings and water collection behind moisture barriers occurs more frequently than corrosion at the liner-concrete interface. Although damage to moisture barriers and coating are more frequent, NRC-required inspections have resulted in early detection and mitigation of these incidents.

Operating experience indicates construction defects, such as fragments of wood present from the time of original construction, are a major contributor to liner corrosion at the concrete-liner interface. For containment structures designed so that the liner is in contact with the concrete (Figure 9.19), a foreign material in contact with the steel may retain moisture, promote crevice corrosion, and be the source of acidic decomposition products. Limited initial calculations indicate that when a foreign material is in contact with the liner, corrosion of the liner proceeds at a rate that is much faster than the passive corrosion rate expected for steel in contact with a typical alkaline concrete environment.

Future Efforts

The efficacy of current inspection methods and the value gained from augmented inspections will be assessed. More detailed modeling will be conducted to better understand the effects of construction defects on corrosion of the steel containment liner and the potential benefits of mitigation methods for concrete degradation and liner corrosion. Evaluation of the aging and degradation of these passive components and potential mitigation methods will be necessary as nuclear power plants enter extended operation periods beyond 40 years of service and potentially beyond 60 years of service.

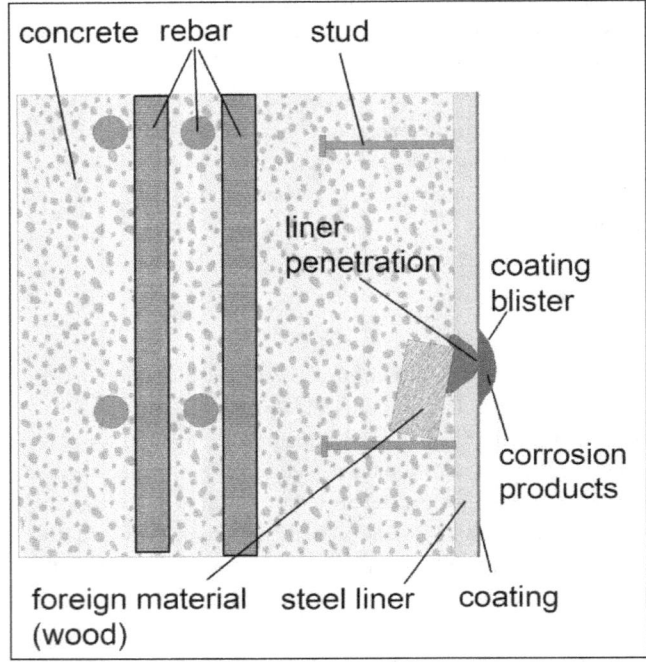

Figure 9.19 Schematic showing a cross-section of a reinforced concrete containment structure with embedded foreign material from original construction and corrosion penetration of the steel containment liner

For More Information
Contact Greg Oberson, RES/DE, at Greg.Oberson@nrc.gov, or Darrell S. Dunn, RES/DE, at Darrell.Dunn@nrc.gov.

Atmospheric Stress-Corrosion Cracking of Dry Cask Storage Systems

Background

Commercial nuclear power plants refuel every 18 to 24 months. Fuel removed from the core is placed in spent fuel pools for a minimum of 5 years. Independent spent fuel storage installations (ISFSIs), licensed under Title 10 of the *Code of Federal Regulations* Part 72, "Licensing Requirements for the Independent Storage of Spent Nuclear Fuel, High-level Radioactive Waste, and Reactor-Related Greater Than Class C Waste," are used when spent fuel pools have reached capacity. (See Figure 9.20 for map of ISFSI locations.) ISFSIs are initially licensed for up to 40 years, and license renewals for 40 years were recently completed for three ISFSI sites.

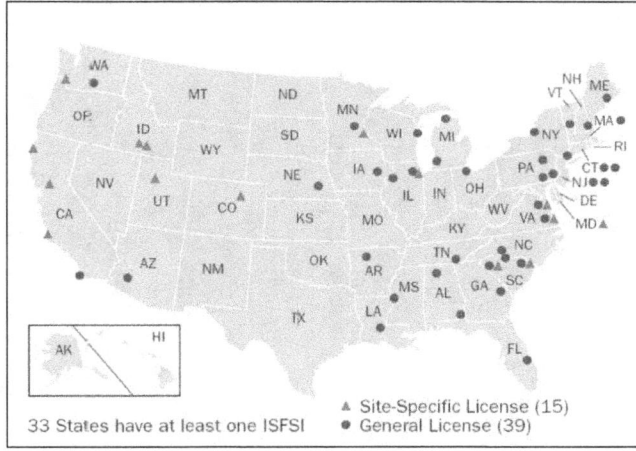

Figure 9.20 ISFSI locations

33 States have at least one ISFSI
▲ Site-Specific License (15)
● General License (39)

Dry storage systems at operating ISFSIs consist of canisters constructed using austenitic Type 304/304L/316/316L stainless steels (Figure 9.21). Some of the current and possibly future ISFSI sites are located in coastal atmospheres where chloride-containing salt as an airborne aerosol may deposit on the canister surfaces. A review of previous research provided little insight on the possible effects of salt accumulation over the expected range of operating temperatures for dry storage system canisters. Understanding the environmental conditions and material factors that influence atmospheric chloride stress-corrosion cracking (SCC) of austenitic stainless steel is necessary to evaluate the long-term operation of dry cask storage systems.

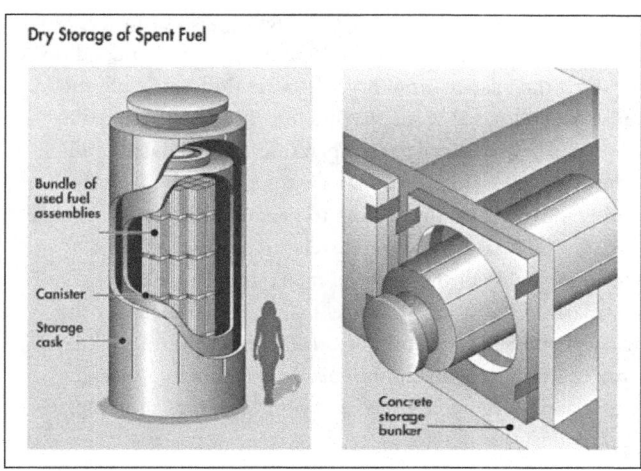

Figure 9.21 Dry storage system designs

Objective

The objective of this research is to evaluate the potential for canister degradation at ISFSIs, including the deposition and accumulation of sea salts and atmospheric deposits that may induce SCC. Evaluation of SCC susceptibility must consider the time-dependent changes in the environmental conditions on the surfaces of the stainless steel canisters, canister construction materials, and fabrication effects. Information obtained will help identify potential issues and regulatory requirements for long-term ISFSI operation in coastal atmospheres.

Approach

The U.S. Nuclear Regulatory Commission (NRC) previously sponsored research to evaluate the chloride SCC susceptibility of austenitic stainless steel dry storage systems exposed to coastal atmospheres. Accelerated laboratory tests showed that stainless steel Type 304, 304L, and 316L base metals and Type 304/308, 304L/308L, and 316L/316L gas tungsten arc welded alloys may be susceptible to SCC at temperatures and relative humidity values at which the deliquescence of sea salts could occur. Additional testing has been initiated to better understand the effects of temperature, relative humidity, salt deposit composition, and amount of salt present on the SCC of stainless steels used in the construction of spent fuel canisters.

Results

Ongoing test results indicate that SCC can occur on Type 304 stainless steel with sea salt surface concentrations as low as 1 gram per square meter. Systematic test results (Figure 9.22) show the relationship between temperature, relative humidity, and the deliquescence behavior of the major sea salt constituents.

SCC has been observed at temperatures of 35° Celsius (95° Fahrenheit) and 45° Celsius (113° Fahrenheit) under conditions in which the relative humidity values are sufficiently high to cause deliquescence of the magnesium chloride contained in the deposited sea salt. The temperature and relative humidity conditions are realistic and may be experienced by spent fuel canisters in service. In addition to producing sufficient residual stress for SCC, fabrication and welding processes used in the construction of the canisters may impart some degree of sensitization to the stainless steel materials that increases the atmospheric SCC susceptibility. Initial test results using deposits from other sources, including species that should be present in industrial environments, have not been shown to promote SCC of the canister materials

The NRC will continue to share the results of this research with industry as part of the ongoing cooperative efforts to address the safe long-term storage of spent fuel.

For More Information

Contact Greg Oberson, RES/DE, at Greg.Oberson@nrc.gov.

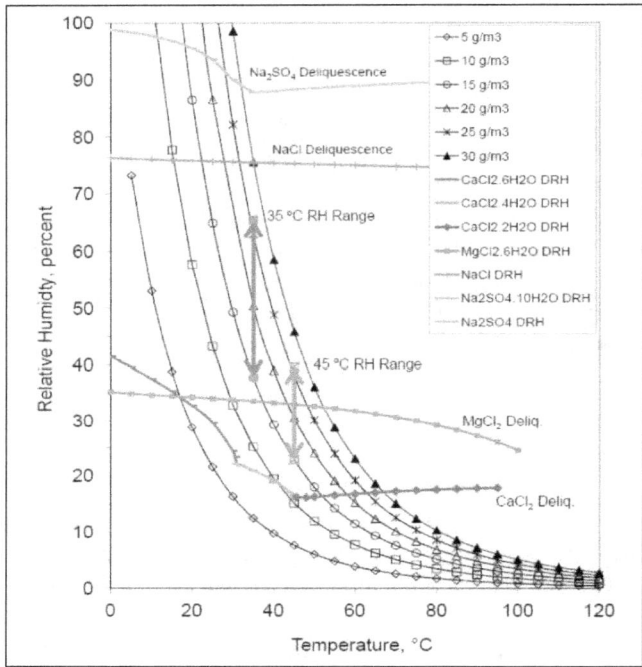

Figure 9.22 Relationship between temperature, relative humidity, and the deliquescence behavior of the major sea salt constituents

Summary and Future Work

Results of accelerated testing under representative atmospheric conditions indicate that deposited sea salts can form chloride-containing solutions at high relative humidity values and can promote SCC in austenitic stainless steels. Higher temperatures and lower relative humidity prevent the formation of chloride-containing solutions that can promote SCC. The implications of this research suggest that the SCC of ISFSI storage casks appears to be limited to a narrow range of conditions; however, SCC may be more likely to occur during extended operation as the storage canister surface temperatures decrease.

High-Density Polyethylene Piping Research Program

Background

As seen in Figure 9.23, carbon steel piping used for nuclear power plant Class 3 safety-related service water systems has experienced general corrosion, microbiologically induced corrosion, and biofouling resulting in leakage and flow restriction.

Figure 9.23 Corrosion and biofouling of carbon steel service water system piping

As a result, the nuclear power industry proposed to replace buried carbon steel piping service water systems with high-density polyethylene (HDPE) piping. HDPE piping is typically immune to general corrosion, microbiologically induced corrosion, and biofouling; is less costly to install; and has a potential service life that exceeds 50 years. HDPE piping is used in nuclear non-safety applications in U.S. nuclear power plants with great success. British Energy (part of EDF Energy) has installed HDPE piping in nuclear safety-related service water piping applications; however, lower allowable temperature requirements were imposed compared to proposed domestic nuclear applications.

Section III of the American Society of Mechanical Engineers Boiler and Pressure Vessel Code (ASME Code) governs the design and installation of Class 3 safety-related service water piping systems. However, the ASME Code does not include the design and installation of HDPE piping systems. The ASME Section III/XI Special Working Group on Polyethylene Piping developed Code Case N755 to provide rules for the design and installation of HDPE piping systems. Code Case N755 addresses many of the issues related to using HDPE piping in Class 3 safety-related buried piping systems; however,

the U.S. Nuclear Regulatory Commission (NRC) identified several issues related to the allowable service life conditions (i.e., temperature and stress), pipe fusion, and inspection that need resolution before the agency will allow its general use by licensees. ASME is working to resolve these issues.

Because the NRC has not approved Code Case N755, licensees submitted relief requests for the substitution of carbon steel piping with HDPE piping for Class 3 safety-related applications. The agency granted two such relief requests, which relied heavily on Code Case N755; however, the NRC imposed several additional requirements to ensure piping and fusion joint integrity. (See Figure 9.24, which shows HDPE piping installed at a nuclear power plant.)

Figure 9.24 Installation of underground Class 3 safety-related HDPE piping

Regulatory Needs

The objective of this program is to conduct confirmatory research to assess the service life, design, fabrication, and inspection requirements proposed in Code Case N755. Because HDPE is a new material for safety-related applications at nuclear power plants, data and analyses are needed to independently verify the requirements in Code Case N755 and its application to existing and new nuclear power plants.

Approach

The Office of Nuclear Regulatory Research (RES) is performing confirmatory testing and analyses on HDPE piping to evaluate the following:

- Allowable Service Life Conditions for Pipe and Fusion Joints. Slow crack growth is the most relevant failure mechanism for HDPE piping in service water applications and is strongly influenced by service temperature and stress. Therefore, the conduct of confirmatory testing needs to be done to verify the service life requirements proposed in Code Case N-755. Full-scale pipe testing and small-scale coupon testing are being performed on both parent materials (i.e., no joints) and on fusion joints to verify the resistance of PE4710 to slow crack growth.

- Fusion Procedure Qualification Requirements. HDPE pipes are joined together by heat fusion processes developed experimentally for small-diameter, thin-walled pipes used for natural gas applications. The essential variables and ranges used to qualify the processes for fusing small diameter pipes may not be applicable to large-diameter, thick-walled pipes used in nuclear service water applications. The Plastics Pipe Institute has revised the document that specifies the fusion procedure (TR33) to include large pipes. However, although the processes used to fuse small-diameter pipes may appear to work for large-diameter pipes, the soundness and long-term performance of large-diameter fusion joints needs to be verified to determine the effect of flaws and incomplete fusion on slow crack growth resistance. The NRC is using a combination of analytical modeling of the fusion procedure and long-term pipe testing of fusion joints to identify the critical variables that affect the service life of HDPE fusion joints to help specify fusion procedure qualification requirements.

- Nondestructive Testing Methods and Procedure Qualification Requirements. Currently, no rules exist for volumetric inspections of HDPE piping in Code Case N755. Although industry is working to develop methods for detecting volumetric flaws in HDPE parent pipes and fusion joints, the NRC is performing research to confirm the capability, effectiveness, and reliability of the proposed NDE methods. Specifically, the NRC will perform research to assess (1) the acceptance criteria, (2) the effectiveness of various proposed NDE technologies for inspecting a variety of product forms, and (3) the probability of detection and uncertainty associated with the characterization of flaws.

RES is active in ASME Code activities related to HDPE piping and coordinates HDPE piping issues with the Office of Nuclear Reactor Regulation and the Office of New Reactors for eventual ASME resolution.

For More Information
Contact Eric Focht, RES/DE, at Eric.Focht@nrc.gov.

Degradation of Neutron Absorbers in Spent Fuel Pools

Background

In spent fuel pools, a stainless steel rack structure aligns and supports spent fuel assemblies. Assemblies are spaced closely together in such a manner that means other than distance alone are required to maintain subcriticality in the pool. Therefore, subcriticality assurance is often credited to neutron absorber panels containing boron that are placed within the rack walls. In the past 15 years, neutron absorber materials, especially Boral and Boraflex, have shown various types of degradation, such as blistering or matrix degradation (Figures 9.25 and 9.26). Information Notice 09-26, "Degradation of Neutron-Absorbing Materials in the Spent Fuel Pool," dated October 28, 2009, summarizes specific incidents of excessive degradation. Degradation of credited neutron absorber panels may affect criticality calculations and challenge the subcriticality requirement of keff < 0.95 in Title 10 of the *Code of Federal Regulations* (10 CFR) 50.68, "Criticality Accident Requirements."

Currently, plants detect and manage neutron absorber aging and degradation through surveillance programs, such as sample coupons, In-Situ BADGER testing, and RACKLIFE modeling. If significant degradation occurs to the degree at which absorber panels no longer provide sufficient neutron absorption, plants may credit additional sources of negative reactivity, such as storage spacing patterns, based on burnup credit or, in pressurized-water reactors only, soluble boron. However, as extended plant operations produce increasing numbers of spent fuel assemblies that require storage in previously unoccupied cells and as neutron absorber materials continue to age and degrade, these strategies may no longer be sufficient to maintain pool subcriticality, especially in boiling-water reactor pools that do not contain soluble boron.

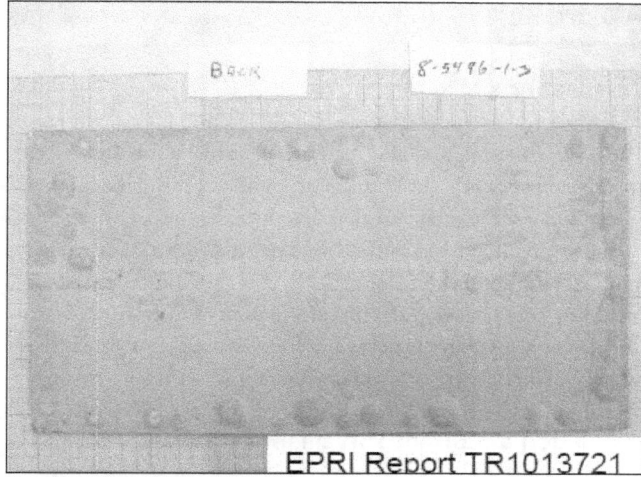

Figure 9.25 Blistering on the aluminum cladding of a Boral neutron absorber

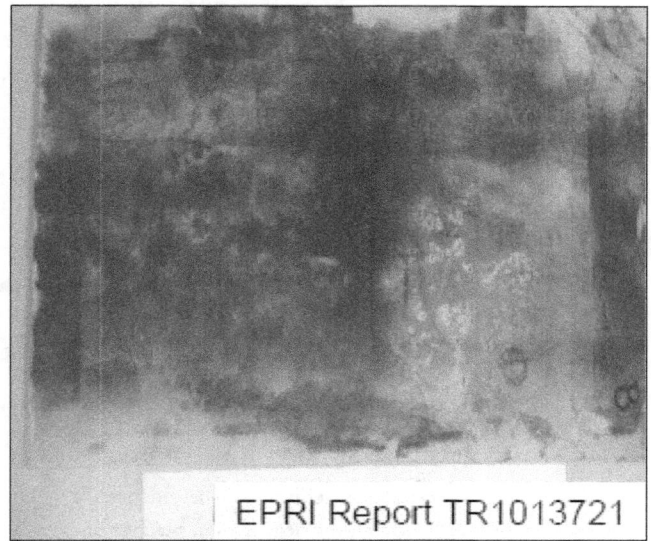

Figure 9.26 Degradation of the composite matrix in a Boraflex neutron absorber

Objective

In light of these new concerns, the U.S. Nuclear Regulatory Commission (NRC) has cataloged the current strategy for pools to maintain subcriticality. The NRC has also initiated a program to evaluate the efficacy of current surveillance methods. Results of this program will be used to guide future regulatory decisions on spent fuel pools.

Approach

Compilation of Existing Data

The NRC collected available data on spent fuel pools and absorbers from both public and non-public licensing documents to develop a ready reference that describes the means by which each pool maintains subcriticality. The NRC has also developed a more complete database of neutron absorber information, including surveillance program information, which will assist in identifying degradation trends as a function of factors, such as panel age, fluence, or pool environment.

Evaluation of Current Surveillance Methods and Programs

Over the past 2 years, the NRC has prepared two reports that assess the uncertainties associated with current methods of surveillance. The first report entitled, "Boraflex, RACKLIFE, and BADGER: Description and Uncertainties," focuses on the degradation of Boraflex and assesses the uncertainties in the RACKLIFE degradation program. Figure 9.27 shows an example of uncertainty in RACKLIFE results as a function of temperature variation. The second report entitled, "Assessment of Uncertainties Associated with the BADGER Methodology," describes uncertainties in BADGER blackness testing. Figure 9.28 shows head misalignment as one example of a source of uncertainty. The NRC is currently initiating a research program to assess the surveillance adequacy of coupon programs.

Neutron Absorber Degradation Mechanisms

To support the assessment of the efficacy of surveillance methods, the NRC is studying the degradation mechanisms and rates of neutron absorber panels composed of materials, such as Boraflex, Boral, and phenolic resin-based matrix absorbers. An understanding of these degradation mechanisms is necessary to determine whether surveillance methods and programs can detect the loss of neutron absorption ability in a timely manner.

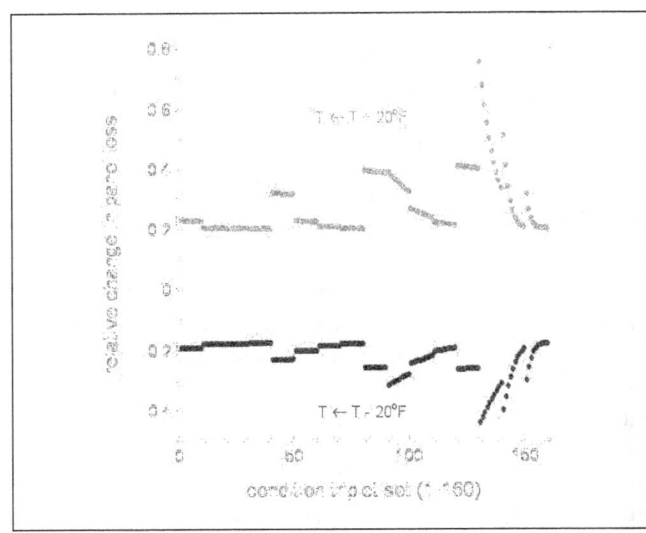

Figure 9.27 Model of uncertainty in RACKLIFE-calculated panel loss as a function of temperature variation

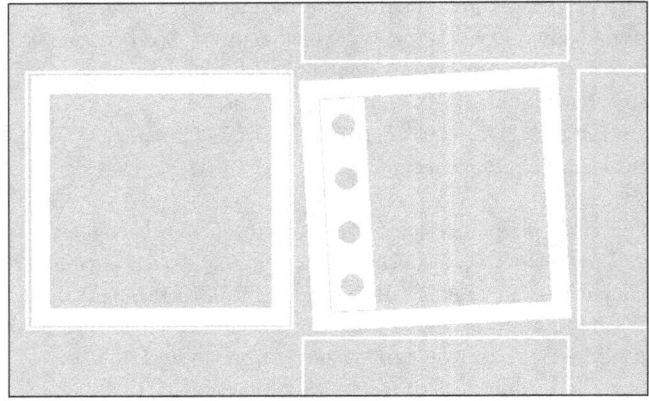

Figure 9.28. Head misalignment in the BADGER neutron detection instrument

For More Information
Contact April Pulvirenti, RES/DE, at April.Pulvirenti@nrc.gov.

Extended Storage and Transportation of Spent Nuclear Fuel

Background

Commercial nuclear power plants use independent spent fuel storage installations (ISFSIs), licensed under Title 10 of the *Code of Federal Regulations* (10 CFR) Part 72, "Licensing Requirements for the Independent Storage of Spent Nuclear Fuel, High-level Radioactive Waste, and Reactor-Related Greater Than Class C Waste," when spent fuel pools have reached capacity. ISFSIs are initially licensed for 40 years, and the U.S. Nuclear Regulatory Commission (NRC) recently completed license renewals for 40 years for three ISFSI sites. Because of the lack of a permanent solution for spent fuel disposal, extended storage at current or future ISFSI locations will be necessary before the subsequent transportation of spent fuel to a consolidated storage site, processing facility, or future repository. Safety during the extended storage and transportation (EST) of spent fuel requires a review of the technical basis and requirements for operation of dry cask storage systems.

Objective

The objective of this work is to develop the technical bases for the EST of spent fuel. The development of the technical bases necessitates an enhanced understanding of the time dependencies and environmental conditions that affect the possible degradation modes of safety significant structures, systems, and components in the dry cask storage systems. Significant operational parameters may include fuel burnup, material composition, dry cask design, thermal loading, ISFSI location, and the age of the systems. The NRC will use the information obtained in this program to evaluate ISFSI license renewals and to assess the benefits of additional inspections or monitoring of the condition of dry cask storage systems.

Approach

The NRC developed a multitask approach to identify the technical information needs, assemble relevant information from ongoing efforts, review the consequence and risk information needs, and conduct focused investigations on significant technical issues. An assessment of aging and degradation phenomena that affect dry cask structures, systems, and components was done to identify areas for which additional technical information was required. The staff reviewed ongoing industry and international research and the integration of existing agency efforts to identify activities that can provide

information to address the technical information needs for the EST program. A reassessment of radionuclide release parameters and the probability and consequences of events that may lead to damage of the dry cask systems considered the likely extended period of operation of existing ISFSIs. The NRC developed research plans and initiated scoping studies on previously identified emergent technical issues, such as the possibility of stress-corrosion cracking (SCC) of stainless steel canisters in marine environments.

Results

The draft EST Technical Information Needs Report, issued for public comment in May 2012, considers degradation processes identified in previous assessments by the Nuclear Waste Technical Review Board, the U.S. Department of Energy (DOE), the Electric Power Research Institute (EPRI), and Savannah River National Laboratory under contract to the NRC. The report identifies several high-priority information needs, including SCC of stainless steel canisters, fuel pellet swelling and fuel rod pressurization stresses on cladding.

The staff reviewed ongoing industry and international research and existing agency efforts to identify activities that are likely to provide data necessary to address the identified technical information needs. Ongoing work led by EPRI in the Extended Storage Collaboration Program will provide information on the condition of canisters in marine environments, the application of nondestructive examination methods for dry cask inspection, and possibly data on the storage of high-burnup fuel. Existing agency efforts will provide needed information on hydrogen effects in cladding, pellet cladding interactions, and fuel performance. International research may provide valuable data on cask and seal performance, cladding integrity, and advancements in monitoring and inspection technologies for dry cask storage systems.

The staff's review of the factors that affect release fractions for low and high-burnup fuel has identified several significant parameters, including the fraction of fuel rods that fail in an event, the fraction of fuel converted to respirable fines, and the filtering and deposition of respirable fines. The staff is assessing risk information needs by reviewing previous hazard identification studies conducted by the NRC, EPRI, and DOE. This effort will examine the validity and completeness of the existing hazard assessments for longer than the originally anticipated storage times.

Emergent technical issues identified include the corrosion of stainless steel casks, concrete degradation, and the need for improved cladding and dry cask temperature profiles. Research plans have been developed for concrete degradation and temperature profiles during EST. Follow-on work on the corrosion of stainless steel casks is focusing on the range of conditions for which SCC is possible. The results of this study will be compared to the results obtained by EPRI that are designed to measure the range of actual conditions on stainless steel canisters. The NRC will use the results to support the basis for any additional inspection requirements or proposed mitigation actions.

Future Efforts

Based on the assessment of technical information needs, future efforts will comprise refined testing and analyses of the SCC susceptibility of dry cask storage materials and the assessment of methods for inservice nondestructive evaluation of dry cask storage systems. In addition, efforts will focus on realistically modeling the thermal behavior of fuel in dry cask storage systems to assess the effects of temperature on cladding embrittlement and to accurately predict the initiation of corrosion of the dry cask materials. The staff will analyze the effects of residual moisture after drying for fuel and internals and will conduct scoping studies to evaluate the available methods for functional monitoring of dry cask storage systems.

For More Information
Contact Darrell Dunn, RES/DE, at Darrell.Dunn@nrc.gov.

Chapter 10: Digital Instrumentation and Control and Electrical Research

Digital Instrumentation and Control

Digital Instrumentation and Control Probabilistic Risk Assessment

Analytical Assessment of Digital Instrumentation and Control Systems

Susceptibility of Nuclear Stations to External Faults

Evaluation of Equipment Qualification Margins to Extend Service Life

Battery Testing Program

Digital Instrumentation and Control

Background

The digital instrumentation and control (I&C) area continues to evolve as the technology changes and the U.S. Nuclear Regulatory Commission (NRC) continues to refine its regulatory approach. Many current control rooms are dominated by analog equipment, such as electromechanical switches, annunciators, chart recorders, and panel-mounted meters. However, as operating nuclear power plants (NPPs) upgrade their control rooms, analog equipment is being replaced with modern digital equipment, including flat screen operator interfaces and soft controls. Future plants will have highly integrated control rooms similar to those in Figure 10.1. The NRC has seen a substantial increase in the proposed use of digital systems for new reactors and retrofits in operating reactors. As a result, the NRC continues to update applicable licensing criteria and regulatory guidance and perform research to support licensing these new digital I&C systems.

In the 1990s, the NRC developed guidance to support the review of digital systems in NPPs. Since that time, the NRC has been effectively using the current licensing guidance for review of applications of digital technology in operating reactors and in certification of new reactor designs. In an effort to continually improve the licensing process, the NRC commissioned the National Research Council of the National Academy of Sciences to review issues associated with the use of digital systems. The National Research Council issued its report, "Digital Instrumentation and Control Systems in Nuclear Power Plants," and made several recommendations, which included a recommendation to update the NRC research program to balance short-term regulatory needs and long-term anticipatory research needs. The Advisory Committee on Reactor Safeguards (ACRS) also has encouraged research in the digital I&C area to keep pace with the ever-changing technology.

Overview

In 2005, the Office of Nuclear Regulatory Research (RES) developed a comprehensive 5-year Digital System Research Program Plan, which defined the I&C research programs to support the regulatory needs of the agency. In 2007, the NRC formed a Digital I&C Steering Committee and seven task working groups (TWGs) to work with the nuclear industry in improving regulatory guidance for digital I&C system upgrades in operating reactors, support design certification submittals for new reactors, and support review of digital I&C systems in fuel cycle facilities. The TWGs issued seven new interim staff

guidance documents to address specific digital I&C regulatory issues. In 2011, the Digital I&C Steering Committee was sunset, with further regulatory improvements managed by the NRC regulatory offices.

In 2010, the agency developed an updated Digital System Research Plan with input from several sources, including the National Research Council's report on digital I&C systems at nuclear power plants, ACRS, external stakeholders, and the NRC staff. The updated research plan consists of five research program areas: (1) safety aspects of digital systems, (2) security aspects of digital systems, (3) advanced nuclear power concepts, (4) knowledge management, and (5) carryover projects. The products of these research programs include technical review guidance, information to support regulatory-based acceptance criteria, assessment tools and methods, standardization, and knowledge management initiatives.

Figure 10.1 Highly integrated control room

Research Program

RES currently is conducting research in several key technical areas that support licensing of operating reactors, new reactors, and advanced reactors.

Work in the area of safety aspects of digital systems includes analytical assessment research to support safety evaluations of digital I&C systems. Ongoing research is developing an inventory and classification for NPP digital systems, exploring the state-of-the-art in analysis of safety-critical digital systems, performing failure mode and operational experience analysis, and examining the need for new regulatory review tools, such as the use of system hazard analysis, a safety demonstration framework, and guidance for review of software tools. This research will improve the understanding of how digital systems may fail and will support the development of the commensurate criteria to ensure that these systems will not compromise their safety functions and will not affect NPP safety adversely. Other research projects are investigating fault-tolerant testing techniques and advanced diagnostics and prognostics. The NRC and the industry are interested in risk-informing digital safety system licensing reviews. One of the major challenges to risk-informing digital system reviews is developing an acceptable method for modeling digital system reliability. The staff examined a number of reliability and risk methods that have been developed in other industries, such as aerospace, defense, and telecommunications. Based on its review of these techniques and available failure data, the staff performed benchmark studies of digital system modeling methods, including traditional event-tree, fault-tree, and dynamic methods. Internal staff and ACRS reviews of the studies challenged the viability of the methods and availability of data needed. Further research on the failure modes of digital systems and quantitative software reliability is being pursued.

With respect to the security aspects of digital systems, the staff developed a new Regulatory Guide 5.71, "Cyber Security Programs for Nuclear Facilities," in support of Title 10 of the *Code of Federal Regulations* (10 CFR) 73.54, "Protection of Digital Computer and Communication Systems and Networks." The staff has completed research to explore cyber vulnerabilities in digital systems and networks, including wireless networks that are expected to be deployed in NPPs. This research validated the need for the new regulatory guidance and cyber security programs required under 10 CFR 73.54.

The staff also is staying abreast of advanced nuclear power concepts in the digital systems area. In support of the proposed license applications for small modular reactors, research projects to investigate unique regulatory aspects for advanced I&C are underway.

In the knowledge management area, collaborative research efforts in the United States and internationally support sharing regulatory standards and research data for digital systems. There are ongoing efforts to share operational experience data and analysis techniques with industry through the Electric Power Research Institute, with other Government agencies, such as the National Aeronautics and Space Administration, and with research organizations in other countries. Research supports international NPP digital system standards harmonization and NRC knowledge management and regulatory efficiency improvements.

For More Information
Contact Russell Sydnor, RES/DE, at Russell.Sydnor@nrc.gov.

Digital Instrumentation and Control Probabilistic Risk Assessment

Background

Nuclear power plants have traditionally relied on analog systems for their monitoring, control, and protection functions. With a shift in technology to digital systems and their functional advantages, existing plants have begun to replace current analog systems, while new plant designs fully incorporate digital systems. Since digital instrumentation and control (I&C) systems are expected to play an increasingly important role in nuclear power plant safety, the U.S. Nuclear Regulatory Commission (NRC) has developed a digital I&C research plan that defines a coherent set of research programs to support its regulatory needs.

The current licensing process for digital I&C systems is based on deterministic engineering criteria. In its 1995 policy statement on probabilistic risk assessment (PRA), the Commission encouraged the use of PRA technology in all regulatory matters to the extent supported by the state-of-the-art in PRA methods and data (Federal Register (FR) 60 FR 42622). One of the programs included in the NRC digital I&C research plan addresses risk assessment methods and data for digital I&C systems since, at present, no consensus methods exist for quantifying the reliability of digital I&C systems.

Objective

The objective of this research is to identify and develop methods, analytical tools, and regulatory guidance for (1) including models of digital systems in nuclear power plant PRAs, and (2) incorporating digital systems in the NRC's risk-informed licensing and oversight activities.

Approach

Previous and current Office of Nuclear Regulatory Research (RES) projects have identified a set of desirable characteristics for reliability models of digital systems and have applied various probabilistic reliability modeling methods to an example digital system (i.e., a digital feedwater control system). Figure 10.2 provides an illustration of one of these modeling methods. Several NUREG/CR reports, which have received extensive internal and external stakeholder review, document this work. The results of these benchmark studies have been compared to the set of desirable characteristics to identify areas where additional research might improve the capabilities of

the methods. One specific area currently being pursued by RES is the quantification of software reliability. To examine the substantial differences in PRA modeling of software (versus conventional nuclear power plant components), RES convened a workshop in May 2009 involving a team of experts with collective knowledge of software reliability or nuclear power plant PRA. At the workshop, the experts established a philosophical basis for modeling software failures in a reliability model. RES subsequently reviewed quantitative software reliability methods and selected two methods to apply to an example software-based protection system in a proof-of-concept study. The two methods currently being pursued are the Bayesian Belief Network approach and the statistical testing method. These methods are being applied to the Loop Operating Control System of the Idaho National Laboratory Advanced Testing Reactor.

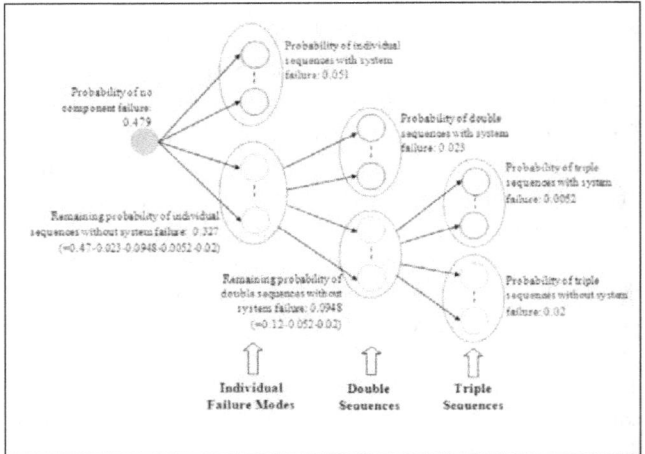

Figure 10.2 Condensed Markov state transition model for quantifying DFWCS failure frequency from hardware failures

The results of the benchmark studies also have highlighted the following areas for which enhancement in the state-of-the-art for PRA modeling of digital systems is needed:

- Approaches for defining and identifying failure modes of digital systems and determining the effects of their combinations on the system.

- Methods and parameter data for modeling self-diagnostics, reconfiguration, and surveillance, including using other components to detect failures.

- Better data on hardware failures of digital components, including addressing the potential issue of double-crediting fault-tolerant features, such as self-diagnostics.

- Better data on the common-cause failures (CCFs) of digital components.

- Methods for modeling software CCF across system boundaries (e.g., when there is common support software).

- Methods for addressing modeling uncertainties in modeling digital systems.

- Methods for human reliability analysis associated with digital systems.

- Process for determining if and when a dynamic model of controlled plant processes is necessary in developing a reliability model of a digital system.

Even if an acceptable method is established for modeling digital systems in a PRA and progress is made in the above areas, (1) the level of effort and expertise required to develop and quantify the models will need to be practical for vendors and licensees, and (2) the level of uncertainty associated with the quantitative results will need to be sufficiently constrained so that the results are useful for regulatory applications.

International Collaboration

In October 2008, RES staff led a technical meeting on digital I&C risk modeling for the working group on risk (WGRisk) of the Organisation for Economic Co-operation and Development, Nuclear Energy Agency (NEA), Committee on the Safety of Nuclear Installations. The objectives of this meeting were to make recommendations regarding current methods and information sources used for quantitative evaluation of the reliability of digital I&C systems for PRAs of nuclear power plants, and identify, where appropriate, the near- and long-term developments necessary to improve modeling and evaluation of the reliability of these systems. While the meeting did not produce specific recommendations of the methods or information sources that should be used for quantitative evaluation of the reliability of digital I&C systems for PRAs of nuclear power plants, it did provide a useful forum for the participants to share and discuss their experience with modeling these systems. The report documenting the meeting is available on the NEA Web site at http://www.nea.fr/nsd/docs/2009/csni-r2009-18.pdf. A follow-on WGRisk activity now is underway and is focused on development of hardware and software failure mode taxonomies for digital I&C systems for use when incorporating these systems into PRAs of nuclear power plants.

References

NUREG/CR-6962, "Traditional Probabilistic Risk Assessment Methods for Digital Systems," October 2008.

NUREG/CR-6997, "Modeling a Digital Feedwater Control System Using Traditional Probabilistic Risk Assessment Methods," September 2009.

Chu, T.L., et al., "Workshop on Philosophical Basis for Incorporating Software Failures into a Probabilistic Risk Assessment," Brookhaven National Laboratory, Technical Report, BNL90571-2009-IR, November 2009 (Agencywide Documents Access and Management System [ADAMS] Accession No. ML092780607).

Chu, T.L., et al., "Review Of Quantitative Software Reliability Methods," Brookhaven National Laboratory, Technical Report, BNL-94047-2010, September 2010 (ADAMS Accession No. ML102240566).

For More Information
Contact Ming Ling, RES/DRA, at Ming.Li@nrg.gov or Alan Kuritzky, RES/DRA, at Alan.Kuritzky@nrc.gov.

Analytical Assessment of Digital Instrumentation and Control Systems

Background

New and proposed digital instrumentation and control (I&C) systems in nuclear power plants (NPPs) pervade and affect nearly all plant equipment, with increasing interdependencies (e.g., through interconnections, resource sharing, and data exchanges) and increasing complexity. These interdependencies are becoming increasingly difficult to identify, analyze, and understand. Configurations of these networked systems tend to have plant-specific differences, such that no two systems are identical. The operating history of such systems is relatively limited and, by the very nature of the systems and expected changes, failure data from these systems is not likely to become statistically significant. In addition, unanticipated failure modes could create very confusing situations that might place the plant, or lead operators to place the plant, in unexpected or unanalyzed configurations. Under these conditions, evaluation for licensing has become very challenging. The Advisory Committee on Reactor Safeguards also has raised similar concerns.

Overview

This research project, which addresses these concerns, is driven, in part, by the Commission's Staff Requirements Memoranda (SRM)-M070607, "Meeting with Advisory Committee on Reactor Safeguards (ACRS), Thursday, June 7, 2007," and SRM-M080605B, "Meeting with Advisory Committee on Reactor Safeguards (ACRS), Thursday, June 5, 2008." It is also driven by the needs expressed by the regulatory offices through the fiscal year (FY) 2010–FY 2014 U.S. Nuclear Regulatory Commission (NRC) Digital Systems Research Plan (Agencywide Documents Access and Management System [ADAMS] Accession No. ML100541484).

Using existing theoretical knowledge in the fields of software and systems engineering for high-confidence, real-time control systems, this research will develop a framework of knowledge about how and why digital I&C systems may fail. The framework will allow continuous enrichment with new knowledge gained from operating experience and other research inside and outside the NPP application domain.

Objectives

In support of the safety evaluation of digital I&C systems, this research will improve the understanding of how systems may fail and develop the commensurate criteria to ensure that these systems will not compromise their safety functions and affect NPP safety adversely.

Knowledge in the form of a causality framework will be useful for improving root-cause analysis of operating experience and will serve to inform companion research in probabilistic risk assessment. Knowledge about modes of degradation also will inform research concerning the effects of degraded I&C on human performance.

Approach

Based on an inventory of current and future digital I&C devices and systems in NPPs, the three pre-approved digital I&C platforms, and emerging trends in digital technology, this research will characterize the NPP application domain. For this bounded domain, the research will identify credible failure and fault modes and analyze their effects, including the operating crew, the plant, and other affected systems. Since the limited failure modes of mature technology hardware components are relatively well understood and the practice of their application is relatively mature, the scope of this research focuses on understanding the failure modes of systems, the causes of such failures, and the criteria or conditions to avoid or prevent such failures. Of particular interest are the system failures caused by complex logic, whether implemented in the form of software, field-programmable gate arrays, or complex programmable logic devices.

To acquire relevant knowledge outside the NPP industry, the NRC reached out to the world's leading researchers in safety-critical software and systems engineering and pursued an elicitation process culminating in a two-day clinic. The results from this expert elicitation activity have shaped the direction of some of the research described below.

Inventory and Classification of Digital Instrumentation and Controls Systems

This study will establish an inventory of current and future digital I&C devices and systems in NPPs. The purpose is to understand the scope and nature of the systems on which safety assessment research should be focused. The inventory will include enough information to allow characterization of the domains of applications in NPPs (the digital I&C devices and systems and their relationships to their environments) and the ability to cluster the inventoried items into classes of similarities. Example elements of information include: (1) the role or NPP function in which the item is applied, (2) whether the item stands alone or is interconnected, (3) various aspects and indicators of the complexity of the item, (4) the degree of verification and qualification, (5) properties of its architecture, and (6) properties of its development process to the extent that these elements or information are available. The intent of such characterization is to support operational experience analysis and facilitate the understanding of possible adverse behaviors and approaches to ensure freedom from adverse behaviors.

Digital Instrumentation and Control Failure and Fault Modes Research

This study will establish an analytical framework for organizing knowledge about how and why digital I&C systems may fail. The scope of the study will be limited to the domain of digital I&C devices and systems (e.g., classes of devices and systems represented in the existing inventory, NRC-approved platforms, and trends in new licensing applications). The scope includes an analysis of systems with tightly coupled integration of traditionally decoupled or loosely coupled functions, applications (e.g., reactor trip system, engineered safety features actuation system), signals, and infrastructural services, as exemplified in new licensing applications.

Knowledge about failures, faults, and their causes will be organized in a reusable manner. Coupled with causal knowledge will be research on criteria or conditions to avoid or prevent such faults (e.g., constraints on the architecture and the development process).

Knowledge Elicitation from Experts

To acquire relevant knowledge outside the NPP industry, the NRC reached out to the world's leading researchers in safety-critical software and systems engineering and pursued an elicitation process culminating in a two-day clinic held in January 2010 to identify the following:

- Current limitations in the assurance of complex logic and areas of uncertainties.

- Evidence needed for effective assurance.

- Areas in need of research and development.

The pool of experts represented a broad diversity in cultural backgrounds, application domains, and thought processes. Countries of origin included the United Kingdom, Sweden, Germany, Australia, New Zealand, Canada, and the United States. Application domains included defense, space flight, commercial aviation, medical devices, automobiles, telecommunications, and railways.

Through a chain of referrals by the experts, the NRC built a candidate pool of over 75 experts, of which more than 30 experts were available for individual interviews. Based on common

patterns emerging from the collective interviews, the NRC drafted a reference position to confirm with the experts the areas of general agreement and to identify areas for deeper discussion. While certain findings confirmed NRC staff positions, other findings revealed opportunities to improve the rigor and depth of NRC reviews. For example, the experts confirmed that the safety assessment of a digital I&C system will continue to require high-caliber judgment from a diverse team, commensurate with the complexity of the system and its development process and environment, such as in systems containing complex software or other manifestations of complex logic. To exercise reasonable judgment, the review team will require a variety of complementary types of evidence, integrated with reasoning to demonstrate that the remaining uncertainties will not affect system safety adversely. The experts recommended that, in the absence of such demonstration, there should be diverse defensive measures independent from digital safety systems using complex software or other implementations of complex logic or products of software-intensive tools.

Safety Assessment Techniques

In accordance with recommendations from the experts, as mentioned above, the NRC is examining the need for new regulatory review tools, such as guidance for review of software tools, the use of system hazard analysis, and a safety demonstration framework. Other regulatory agencies have used or considered the application of the evidence-argument-claim structure (variously known as an assurance case or a safety case) to systematize the safety evaluation of a complex digital I&C system (see Figure 10.3). Although the NRC has a comprehensive regulatory guidance framework, certification or licensing applications submitted for review tend to address the various requirements and guidelines separately, rather than in a safety-goal-oriented integrative manner. This research will explore mapping the NRC's regulatory guidance framework into a safety-goal-oriented, evidence-argument-claim framework.

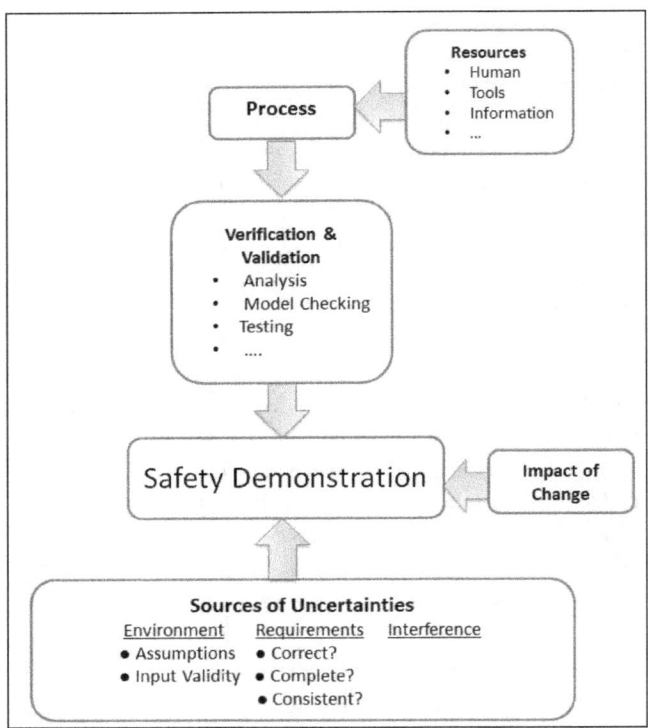

Figure 10.3 Integrating different types of evidence to demonstrate that a system is safe

For More Information

Contact Luis Betancourt, RES/DE, at Luis.Betancourt@nrc.gov.

Susceptibility of Nuclear Stations to External Faults

Background

Offsite power is considered to be the most reliable electrical source for safe operation and accident mitigation in nuclear power plants (NPPs). It is also the preferred source of power for normal and emergency NPP shutdown. When offsite power is lost, emergency diesel generators provide onsite power. Consequently, if both power sources are lost, a total loss of alternating current power could occur, resulting in station blackout, which is one of the significant contributors to core damage frequency.

In August 2003, an electrical power disturbance in the northeastern part of the United States caused nine NPPs to experience a loss-of-offsite-power (LOOP) event. This event, which was initiated by an overgrown tree touching electrical transmission lines, resulted in cascading outages, caused trips of NPP stations, and disabled offsite power supplies.

At Catawba Nuclear Station on May 20, 2006, both units tripped automatically from 100 percent power following a LOOP event. (See the licensee event report for Event Number 41322006001, "Loss-of-offsite-power Event Resulted in Reactor Trip of Both Catawba Units from 100% Power.") That event began when a fault occurred within a current transformer associated with one of the switchyard power circuit breakers. A second current transformer failure, along with the actuation of differential relays associated with both switchyard buses, de-energized both buses and separated the units from the grid. These events illustrated that design and maintenance practices for NPP switchyard protection systems can affect the reliability and availability of the plants' offsite power sources.

Since deregulation of the electric power industry, NPP switchyards may have become more vulnerable to external faults because most of those switchyards are no longer owned, operated, or maintained by companies that have an ownership interest in the NPPs. Instead, the switchyards are maintained by local transmission and distribution companies, which may not fully appreciate the issues associated with NPP safety and security. Maintenance practices may also be inconsistent among these companies. In addition, circuit breaker components (i.e., relays, contacts, and opening and closing mechanisms) have begun to show age related degradation. Improper maintenance of these components could affect the detection and mitigation of faults, which could, in turn, delay protective actions at NPPs.

Objective

The U.S. Nuclear Regulatory Commission (NRC) staff initiated a research project to develop a better understanding of the current power system protection in electrical switchyards and identify the system vulnerabilities that contribute to electrical fault propagation into nuclear facilities.

Approach

This research project comprises multiple tasks. First, the operation of electrical protection systems associated with events that have resulted in plant trips and LOCPs will be reviewed. The study will then identify the root causes of the propagation of external electrical faults into the NPP switchyards and assess the level of protection of current NPP switchyard breaker arrangements and relay schemes used for protection. The NRC will coordinate this study with the North American Electric Reliability Corporation (NERC) and Federal Energy Regulatory Commission (FERC) and their assessments of switchyard protection. Lastly, the study will illustrate through analysis and modeling how an actual fault outside an NPP switchyard will affect an operating NPP station. The study will compare existing NPP switchyard designs with modeling and analysis of the settings and identify the desirable level of protection offered for responding to external faults.

Products

Upon completion of this research project, the NRC may develop a NUREG series report to provide an assessment of NPP switchyard protection designs in its response to external electrical faults. The agency also will consider publishing a regulatory guide, in coordination with NERC and FERC, to address the desirable level of protection acceptable for NPP switchyards. The study results will be published in 2013.

For More Information
Contact Darrell Murdock, RES/DE, at
Darrell.Murdock@nrc.gov.

Evaluation of Equipment Qualification Margins to Extend Service Life

Background

Title 10 of the *Code of Federal Regulations* (10 CFR) 50.49, "Environmental Qualification of Electric Equipment Important to Safety for Nuclear Power Plants," requires that Class 1E electrical equipment located in a harsh environment be environmentally qualified to perform its safety-related function during and following a Design-Basis event such as a loss-of-coolant accident.

In particular, 10 CFR 50.49, the Environmental Qualification (EQ) Rule, states that "margins must be applied to account for unquantified uncertainty, such as effects of product variations and inaccuracies in test instruments."

The Institute of Electrical and Electronics Engineers (IEEE) Standard (Std.) 323-1974, "IEEE Standard for Qualifying Class 1E Equipment for Nuclear Power Generating Stations," defines margin as the difference between the most severe specified service conditions of the plant and the conditions used in type testing.

Furthermore, Section 6.3.1.5 of IEEE Std. 323-1974 lists suggested factors for licensees and equipment manufacturers to apply to service conditions for type testing, including temperature, pressure, radiation, voltage, frequency, time, vibration, and environmental transients.

The margins, as indicated in IEEE Std. 323-1974, are used in the test profiles to determine the qualified life of equipment. However, the margins are expected to account for the following:

- Manufacturing tolerances and measurement uncertainties.

- Lack of sufficient oxygen in the test chamber.

- Lack of simultaneous age conditioning (temperature and radiation).

- High dose rate for radiation aging.

- License renewal to extend the life of equipment to 60 years.

- Inconsistencies in activation energy values used in the Arrhenius equation[1] for thermal aging.

---

[1]   The Arrenhius equation is a methodology for addressing time-temperature aging effects, where the key assumption is that material thermal degradation is dominated by a single chemical reaction whose rate is determined by the temperature of the material and a material constant called the activation energy.

Since manufacturing tolerances and measurement uncertainties cannot be readily quantified when establishing qualified life, margins are added to ensure that the equipment can perform its safety function. The existing margin is applied to extend the life of equipment to 60 years for license renewal, but when an imprecise activation energy is used, the impact on the time needed for thermal aging can be affected. Therefore, to correct for any errors in activation energy, margins are added. As a result, the margins are used to account for a variety of factors, as opposed to only production variations.

For example, the lack of oxygen in the test chamber during accelerated aging could impact the qualified life since the equipment could have greater degradation because of the oxidation effects. As a result, equipment testing does not consider the effects of oxygen, and the margins account for this phenomena.

Furthermore, recent data have shown that simultaneous aging (radiation and thermal) may produce synergistic effects that reduce the qualified life when compared to sequential aging. The same margins also are used to account for any variations between sequential and simultaneous aging. Using a smaller dose rate for the radiation aging of cables would more adequately result in showing radiation degradation effects, but the margin also is credited for the use of a high radiation dose rate.

Because licensees are pursuing license renewal, EQ out to 60 years is now a concern. Industry guidance on methods for assessing cable aging include: (1) mechanical condition indicators (e.g., elongation, indenter methods, recovery time), (2) dielectric condition indicators (e.g., insulation resistance, dielectric loss, time domain reflectometry, line resonance analysis, partial discharge), and (3) chemical indicators (e.g., oxidation time/temperature, Fourier transform infrared spectroscopy).

The purpose of this testing is to confirm (1) that cables are acceptable for 60 years of life, (2) the industry recommended condition-monitoring tests adequately track degradation (i.e., aging), and (3) that the industry recommended condition-monitoring methods are applicable for the common insulation materials.

The regulatory use for this research is to establish the technical basis for assessing the qualified life of electrical cables in light of the uncertainties identified following the initial qualification testing. As a followup to this project, confirmatory testing on a variety of cables will be performed to assess and evaluate condition monitoring methods on electrical cables subjected to aging under normal operating conditions and design-basis-event (accident) conditions. Various insulation types, low and medium voltage cables, power and instrumentation and control cables, naturally aged cables, and new cables will be tested. Further, aged cable samples from the decommissioned Zion nuclear plant will be used.

Furthermore, the staff will coordinate research efforts with the international community in this subject area. For example, with the Organisation for Economic Co-operation and Development/ Nuclear Energy Agency, the staff is participating in the Cable Aging Data and Knowledge Project to evaluate the qualification of cables, inspection methods used, and condition monitoring techniques applied. The outcome will be a database providing an up-to-date encyclopedic source to monitor and predict the performance of numerous unique applications of cables and a commendable practices report that will aid regulators and operators to enhance aging management.

Approach

Through this research, the staff aims to (1) confirm whether EQ requirements for electrical equipment are being met throughout the current and renewed license periods of operating reactors, (2) quantify the margin, and (3) verify its adequacy to address the uncertainties discussed above. This research will assess the existing margins and evaluate their adequacy in light of known problems. Researchers will perform a background literature search, including a review of several key reports on the aging of cables to help inform the test program.

This confirmatory testing of cables will comprise two-phases. The first phase of the project will focus on assessing condition-monitoring techniques during normal operational aging, including submergence. Thus, cables should be subjected to normal operating conditions (temperature, radiation, humidity) in both mild and harsh environments. The second phase will focus on cables subject to accident conditions and located in harsh environments. These cables should be exposed to a simulated accident (temperature, pressure, humidity, radiation,and chemical or steam spray). The condition-monitoring techniques should be evaluated for the capability to predict proper operation of cables during and after the accident (post-accident period).

Upon the conclusion of this research, the NRC will publish NUREG/CR reports that outline the margin available to address the known uncertainties when qualifying electrical equipment and assessing condition-monitoring methods for common insulation materials.

For More Information
Contact Sheila Ray, RES/DE, at Sheila.Ray@nrc.gov.

Battery Testing Program

Background

The U.S. Nuclear Regulatory Commission (NRC) is sponsoring confirmatory battery testing and extended battery operation testing research to determine if charging current is a suitable indicator of a fully charged condition for lead-calcium batteries and to evaluate a commercial nuclear power plant batteries' response to station blackout (SBO) events outside the scope of the current SBO rule. The research program will validate whether the batteries generally used in the nuclear industry remain in a fully charged condition and operational readiness while in standby mode. The research also will determine battery ultimate performance capabilities for an extended duration.

Approach

Traditionally, the typical plant technical specifications required the measurement of specific gravity to determine whether a battery was fully charged. However, newer battery types and the need to know the state-of-charge when the battery is not fully charged prompted the necessity for a change from measuring specific gravity to measuring battery current. Also, older nuclear plant designs relied heavily on diesel generators and batteries, while in most of the newest plant designs the batteries have replaced the diesel generators as the only source of standby power. Lastly, there is a need to better understand the capability of station batteries for an extended duration to bring the plant to a safe shutdown condition.

To ensure that a station battery has the capability to perform its design function, the staff initiated research and arranged the testing of batteries to be performed in three phases: (1) evaluation of charging current as a monitoring technique, (2) evaluation of the use of charging current to monitor battery capacity, and (3) evaluation of the criteria for selecting the point when a battery can be returned to service and meets its design requirements.

To evaluate a battery response to SBO events, the staff will test batteries to the existing and revised SBO load sequence profiles. These profiles would address the condition where all alternating current power is recognized to be lost for a prolonged period. The staff has partnered with other stakeholders to obtain the profiles.

The approach for the confirmatory battery testing involved the testing of lead-acid batteries from three different types of vendors to obtain a good sample of what is currently being used at nuclear power plants. The batteries were installed in a configuration similar to that used in the plants, and subjected to deep discharge and charge cycles to simulate an expected service life for the batteries. All testing was performed in accordance with Institute of Electrical and Electronics Engineers (IEEE) Standard 450-2002, "IEEE Recommended Practice for Maintenance, Testing, and Replacement of Vented Lead-Acid Batteries for Stationary Applications," along with a quality assurance plan developed specifically to meet the needs of the project to ensure an acceptable level of quality for the test results.

Status

The test has been completed, and a NUREG-series report will be issued to document the assessment of whether charging current is an appropriate means of determining battery compliance with technical specification requirements and whether a tested battery can be returned to service. Also, there will be a revision to Regulatory Guide 1.129, Revision 2, "Maintenance, Testing, and Replacement of Vented Lead-Acid Storage Batteries for Nuclear Power Plants," (ML063490110) that will address the major findings and observations of the test. Figure 10.4 shows NRC staff reviewing the first set of batteries that the contractor received before commencing confirmatory battery testing.

The approach for the extended battery operation testing comprises two sequences using the same battery setup configuration from the confirmatory testing. The first sequence will verify how long the batteries can operate at reduced discharge rates and provide data that shows the state of health of the batteries. The second sequence will test the batteries to the revised SBO profiles to determine how long the batteries can support maintaining the plant in a safe shutdown condition.

Figure 10.4 NRC staff reviewing the first set of batteries that the contractor has received before commencing confirmatory battery testing

For More Information
Contact Liliana Ramadan, RES/DE, at
Liliana.Ramadan@nrc.gov.

Chapter 11: Fukushima Activities

Overview of the Office of Nuclear Regulatory Research Followup Activities Related to the Fukushima Dai-ichi Accident

Containment Venting Systems Analysis

Near-Term Task Force Recommendation 3 Potential Enhancements to the Capability to Prevent or Mitigate Seismically Induced Fires and Floods

Hydrogen Control and Mitigation Inside Containment and Other Buildings

Consequence Study of a Beyond-Design-Basis Earthquake Affecting the Spent Fuel Pool for a U.S. Mark I Boiling-Water Reactor

Fukushima Dai-ichi Accident Study with MELCOR 2.1

Fukushima Units 1, 2, 3, and 4 after the accident showing extensive damage to the reactor buildings

Overview of the Office of Nuclear Regulatory Research Followup Activities Related to the Fukushima Dai-ichi Accident

On Friday, March 11, 2011, a 9.0-magnitude earthquake struck Japan and was soon followed by a tsunami that was estimated to have exceeded 45 feet (14 meters) in height. The incident resulted in extensive damage to the six nuclear power reactors at the Fukushima Dai-ichi site.

Since that time, the U.S. Nuclear Regulatory Commission (NRC) has been working to understand the events in Japan and to relay important information to U.S. nuclear power plants (NPPs). In particular, the NRC established a Near-Term Task Force (NTTF) of senior agency experts to determine lessons learned from the accident and to initiate a review of NRC regulations to determine whether the agency needs to take additional measures to ensure the safety of U.S. plants. The NTTF issued its report entitled, "Recommendations for Enhancing Reactor Safety in the 21st Century," on July 12, 2011 (Agencywide Documents Access and Management System Accession No. ML111861807), which concluded that continued operation and licensing activities pose no imminent risk. The report also concluded that enhancements to safety and emergency preparedness are necessary, and presented a dozen recommendations for the Commission's consideration. The NRC subsequently prioritized and expanded the NTTF recommendations (SECY-11-0137, "Prioritization of Recommended Actions To Be Taken in Response to Fukushima Lessons Learned," dated October 3, 2011 (ML11272A111), and it continues to make additions and modifications, as appropriate. The recommendations were divided into three tiers based on the urgency of the issues, as described in SECY-11-0137.

On March 12, 2012, the NRC issued the first regulatory requirements for the nation's 104 operating reactors based on lessons learned from Fukushima Dai-ichi and on the prioritized NTTF recommendations. The NRC continues to evaluate and act on the lessons learned to ensure that appropriate safety enhancements are implemented at NPPs in the United States. The NRC established the Japan Lessons Learned Project Directorate (JLD) to focus exclusively on implementing the lessons learned.

The Office of Nuclear Regulatory Research (RES) has been supporting the agency's lessons learned effort through participation in the JLD Steering Committee and several JLD working groups and through followup research. In addition to the brief overview of support that RES provides the JLD described below, this chapter contains multiple summary sheets that describe the ongoing followup research.

Tier 1—Seismic Hazard Reevaluations (Recommendation 2.1, Seismic)

The NRC has requested licensees perform seismic hazard reevaluations consistent with NTTF Recommendation 2.1 (seismic and flood hazard reevaluations) in the following two-phases:

1. In Phase 1, the NRC issued requests for information under Title 10 of the *Code of Federal Regulations* (10 CFR) 50.54(f) to all licensees (1) to request that they reevaluate the seismic hazards at their sites using updated seismic hazard information and present-day regulatory guidance and methodologies and (2) to ask them to perform a risk evaluation, if necessary (ML12053A340).

2. In Phase 2, the NRC staff will determine, based on the results of Phase 1, whether additional regulatory actions are necessary to provide additional protection against the updated hazards (e.g., update the design-basis and upgrade structures, systems, and components important to safety).

RES supported the Office of Nuclear Reactor Regulation in defining acceptable approaches for responding to these information requests. Attachment 7 to SECY-12-0025, "Proposed Orders and Requests for Information in Response to Lessons Learned from Japan's March 11, 2011, Great Tohoku Earthquake and Tsunami," dated February 17, 2012, (ML12039A103) provided those acceptable approaches. The agency will consider alternate approaches with appropriate justification. Currently, RES supports the NRC's review of the guidance under development by the industry on responding to the information requests. RES is also supporting the development of technical bases and guidance for use by the staff in its review of the plant-specific industry responses.

RES has been especially involved with the development and review of the acceptable approaches to, and guidance on, site-specific probabilistic seismic hazard analysis and on the development of the related up-to-date site-specific ground motion response spectra that account for local site amplification effects. These aspects of the reevaluations relate directly to the comparison of the up-to-date ground motion response spectra to the seismic design spectra for existing plants. This comparison is an essential element of information for making the decision on whether a risk evaluation is necessary for a particular response to the information request.

Tier 1—Flood Hazard Reevaluations and Walkdowns (Recommendations 2.3 and 2.1, Flooding)

JLD Flooding Work Group 2.3 (2.3 Flooding WG) is addressing NTTF Recommendation 2.3 for flooding protection walkdowns. The NRC has asked licensees to perform flooding protection walkdowns to identify and address plant-specific degraded, nonconforming, or unanalyzed conditions, to assess available physical margin, and to verify the adequacy of monitoring and maintenance procedures. The 2.3 Flooding WG consulted with licensees and external stakeholders and endorsed an industry-developed methodology and acceptance criteria for flooding walkdowns. The 2.3 Flooding WG will review the licensee flooding walkdown reports and NRC resident inspector reports and will conduct plant audits.

RES supported the 2.3 Flooding WG in drafting the information request letter, defining the scope and acceptable approaches for performing the flooding protection walkdowns, and reviewing the industry guidance document on the topic. RES staff will help review the licensee submittals and will participate in the site audits.

JLD Flooding Work Group 2.1 is addressing NTTF Recommendation 2.1 for flooding hazard reevaluations. The NRC has asked licensees to reevaluate flooding hazards at NPP sites using updated flooding hazard information and present-day regulatory guidance and methodologies. The NRC has also asked licensees to compare the reevaluated hazard to the current design-basis at the site for each potential flood mechanism. If the reevaluated flood hazard at a site is not bounded by the current design-basis, the NRC has asked licensees to perform an integrated assessment. The integrated assessment will evaluate the total plant response to the flood hazard and will consider multiple and diverse capabilities, such as physical barriers, temporary protective measures, and operational procedures.

RES has been supporting Flooding Work Group 2.1 in developing the information request letter; drafting interim staff guidance on how to perform the integrated assessment; and contributing to interim staff guidance on estimating flooding hazards caused by a storm surge, seiche, and tsunami. RES is also supporting the review of an industry-developed white paper on evaluating flood hazards caused by dam failure. RES staff will also support the review of licensee flooding hazard reevaluation reports.

Tier 1—Station Blackout Advanced Notice of Proposed Rulemaking (Recommendation 4.1)

RES is supporting the rulemaking team and is developing a revision to Regulatory Guide 1.155, "Station Blackout," which was last updated in 1988. Additionally, the RES staff is leading an effort at Brookhaven National Laboratory (BNL) under a memorandum of understanding among RES, the Electric Power Research Institute, and the U.S. Department of Energy to study extended battery operation. This study will evaluate the response of typical commercial NPP batteries to station blackout (SBO) events beyond the scope of the current SBO rule (i.e., extended SBO events). BNL will test batteries to the modified SBO profiles to determine the capability of station batteries for extended operation (i.e., beyond 4 hours). The goal is to gain an improved understanding of station battery performance under select extended SBO load profiles and to publish the results in an NRC technical report (i.e., NUREG/CR). This effort will support the development of new guidance on extending battery performance during a prolonged SBO event. For more information on the NRC's battery testing program, see Chapter 10, page 212.

Tier 3—Enhanced Reactor and Containment Instrumentation

The Advisory Committee on Reactor Safeguards recommended supplementation of the NTTF recommendations by enhancing selected reactor and containment instrumentation to withstand beyond-design-basis accident conditions. The NRC accepted this recommendation, and RES staff members were assigned the lead for developing an action plan. A project team was formed, and the team subsequently developed the following three recommendations:

1. Coordinate with NTTF recommendations related to seismic and flood protection (Recommendation 2.3), loss of alternating current power (Recommendation 4.1), and integration of onsite emergency response capabilities (Recommendation 8). Additionally, ensure the consideration of instrumentation needs in the agency's followup activities related to NRC Order EA-12-049 (ML12054A736) on mitigation strategies for beyond-design-basis external events and Order EA-12-051 (ML12054A682) on reliable spent fuel pool instrumentation.

2. Obtain and review information and insights from previous and ongoing research and coordinate with international and domestic efforts to identify enhanced instrumentation needs.

3. Evaluate Tier 1 action information and information obtained from internal, domestic, and international research to recommend regulatory framework changes, if any, for enhanced reactor and containment and instrumentation.

The team is following NTTF Tier 1 items, identifying potential collaborative research activities, and participating in an International Atomic Energy Agency activity to develop improved international standards for severe accident instrumentation.

Follow-on Research Related to Fukushima Dai-ichi

The following major RES efforts resulting from the Fukushima Dai-ichi accident at the time of publication are described in more detail in subsequent sections of this chapter:

- Containment overpressure mitigation systems analysis.

- Potential enhancements to the capability to prevent or mitigate seismically induced fires and floods.

- Hydrogen control and mitigation inside containment and other buildings.

- Consequence Study of a Beyond-Design-Basis Earthquake Affecting the Spent Fuel Pool for a U.S. Mark I Boiling-Water Reactor.

- Fukushima Dai-ichi accident study with MELCOR 2.1.

Containment Venting Systems Analysis

Background

In SECY-11-0137, "Prioritization of Recommended Actions To Be Taken in Response to Fukushima Lessons Learned," dated October 3, 2011, the U.S. Nuclear Regulatory Commission (NRC) staff described its proposals for the regulatory actions that must be taken to address the recommendations of the Fukushima Near-Term Task Force. One of the immediate (Tier 1) actions proposed by the NRC staff for containment overpressure mitigation was the issuance of orders requiring reliable hardened containment vents to those licensees with boiling-water reactor (BWR) facilities with Mark I (see Figure 11.1) and Mark II containment designs. The NRC subsequently issued orders requiring reliable hardened vents for these plants on March 12, 2012. In SECY-11-0137, the NRC staff also identified an additional longer term issue related to the possible use of filters on the containment vents to limit the release of radioactive materials if the venting systems were used after significant core damage had occurred. In the staff requirements memorandum for SECY-11-0137, dated October 19, 2011, the Commission directed the NRC staff to remove the filtered containment vents issue from the "additional issues" category and, instead, merge it with the Tier 1 issue of hardened vents for Mark I and Mark II containments. In response to the staff requirements memorandum, the staff included plans to address the filtered venting issue for Mark I and Mark II containments in SECY-12-0025, "Proposed Orders and Requests for Information in Response to Lessons Learned from Japan's March 11, 2011, Great Tohoku Earthquake and Tsunami," dated February 17, 2012. These plans included accident progression and consequence analyses of various containment overpressure mitigation strategies.

Objective

The objective of this study is to evaluate various post-accident containment overpressure mitigation and fission product retention strategies by performing MELCOR and MACCS calculations to provide a technical basis for regulatory analysis of these strategies.

Approach

The MELCOR calculations considered a set of accident prevention and mitigation measures. The staff selected the evaluated measures based on the events that occurred during the Fukushima accident, the accident management alternatives previously developed by the industry, the current state of

knowledge of severe accident progression in a BWR, and the insights gained from the State-of-the-Art Reactor Consequence Analyses (SOARCA) study. Specifically, the calculations assessed the impacts of containment venting characteristics, such as vent location (drywell or wetwell), vent filtration, reactor core isolation cooling operating time, and water injection to the core or drywell, on the timing of the severe accident progression and associated source terms. This analysis used the MELCOR and MACCS2 input decks developed for the Peach Bottom Atomic Power Station and its surroundings from the SOARCA project with a few minor changes (Figured 11.2).

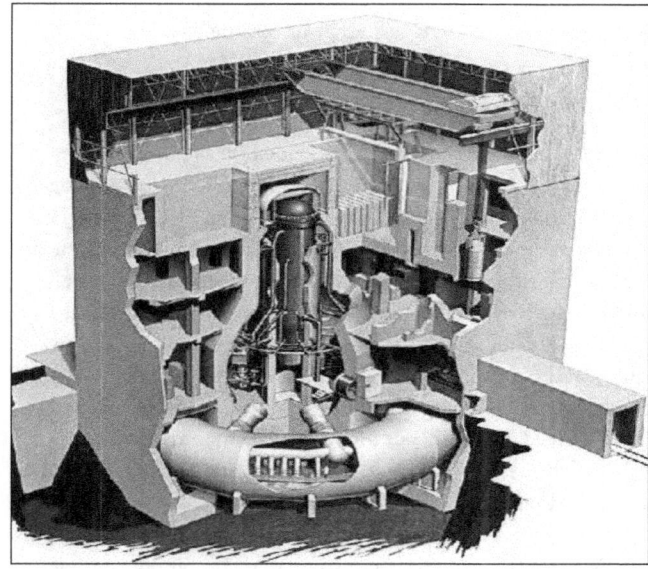

Figure 11.1 Schematic of a Boiling Water Reactor with Mark I containment

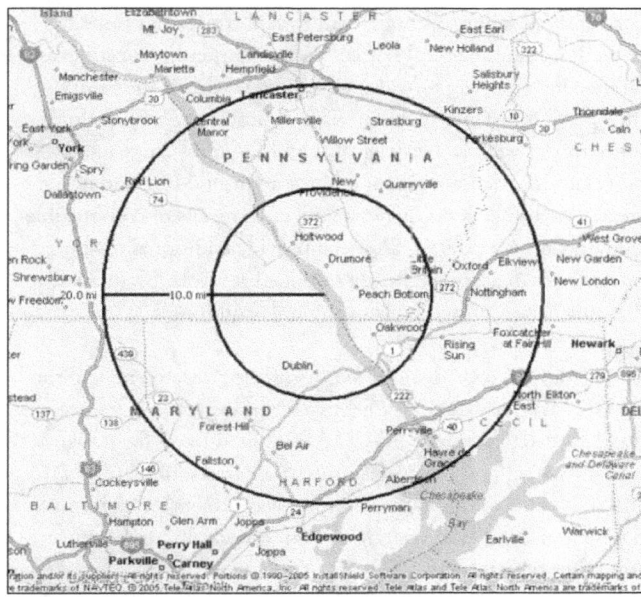

Figure 11.2 Peach Bottom Atomic Power Station and its surroundings

The analysis considered both long-term and short-term station blackout scenarios with various venting and water addition mitigation methods. The following three possible release modes from containment were calculated, depending on the combinations of mitigation measures chosen:

1. Overpressure failure with release from the drywell head to the refueling floor and then out to the environment.

2. Drywell liner melt-through with release into the lower portion of the reactor building and then to the environment.

3. Intact containment as a result of venting to the environment.

The magnitudes of the releases depended on the mitigation measures assumed. The highest releases resulted from venting through the drywell without an external filter. The next highest releases resulted from drywell liner melt-through with no water addition. Still lower releases resulted from combining water addition and venting through the wetwell, thus taking advantage of fission product retention in the plant from containment sprays and suppression pool scrubbing. The lowest releases resulted from assuming the presence of external filters with the vents. MELCOR does not model the decontamination factor (DF) of the external filters in any mechanistic manner; instead, a prescribed value of DF is assigned to the filter.

The MELCOR analyses demonstrate that combining venting and spraying (or any mitigation action that includes water on the drywell floor) is an effective strategy for mitigating radiological releases. Venting alone or spraying alone does not prevent containment failure, although either action provides some reduction in radiological releases. Venting prevents overpressure failure, and spraying provides water to cool debris in the drywell, thus preventing liner melt-through. Combining both actions prevents drywell head leakage and buildup of hydrogen and other non-condensable gases in the reactor building and other areas, which thereby provides an effective means of combustible gas control. An external filter can provide additional fission product attenuation of already scrubbed aerosols (by spray or flooding action or by suppression pool scrubbing).

The MACCS2 code determined the offsite consequences of the releases. In those cases in which venting was present, release fractions calculated by MELCOR were used to perform two sets of MACCS2 calculations (one using a prescribed filter DF and the other assuming that no filter was present) to determine population dose, land contamination, economic consequences, and other relevant quantities.

The MACCS consequence analyses show a clear benefit from applying an external filter to either the wetwell or drywell vent paths. Applying an external filter to either a wetwell or drywell vent path (with DFs of ≥ 10 or ≥ 1,000, respectively) results in a lower conditional latent cancer fatality risk (i.e., a 40- to 95-percent reduction) when compared to the unfiltered cases. The population dose is also lowered (i.e., a 50- to 95-percent reduction) when compared to the unfiltered cases.

Unlike the latent cancer fatality risk calculations, the population dose includes public doses from the ingestion pathway and doses to offsite decontamination workers. All the filtered cases with an external filtered vent path result in a reduction by several orders of magnitude in cesium-137 land contamination. For all cases considered, the conditional prompt fatality risk is either zero or essentially zero.

For the cases considered, a DF ≥ 10 for all wetwell venting filtered cases and a DF ≥ 1,000 for all drywell venting filtered cases result in lower economic costs (i.e., > 60 percent to orders of magnitude reduction) than the costs for their respective unfiltered cases.

The results from MELCOR and MACCS2 analyses are documented in SECY-12-0157, "Consideration of Additional Requirements for Containment Venting Systems for Boiling Water Reactors with Mark I and Mark II Containments."

Conclusions and Additional Activities

Combining MELCOR and MACCS2 analyses provides an effective methodology for establishing the technical basis for strategies to mitigate radiological consequences of severe accidents in BWR Mark I containments. These strategies include combining venting and water addition actions (including spraying), supplemented further by the installation of an external filter.

The Commission issued a Staff Requirements Memorandum on March 19, 2013, in which it directed the staff to modify, as a near-term action, the current order, EA-12-050, on reliable hardened vents to provide additional capability to remain functional under severe accident conditions. The Commission also directed the staff to develop technical bases and rulemaking for filtering strategies as a longer-term action. The follow-on work based on the Commission direction will likely include a more detailed evaluation of accident management strategies and an examination of plant-specific features.

For More Information
Contact Sudhamay Basu, RES/DSA, at Sudhamay.Basu@nrc.gov or Edward Fuller, RES/DSA, at Edward.Fuller@nrc.gov.

Near-Term Task Force Recommendation 3 Potential Enhancements to the Capability to Prevent or Mitigate Seismically Induced Fires and Floods

Background

Seismically induced fires have the potential to cause multiple failures of safety-related structures, systems, or components (SSCs) and induce separate fires in multiple locations at the site. Events, such as pipe ruptures (and subsequent flooding), could also cause such problems in multiple locations simultaneously. Additionally, seismic events could degrade the capability of plant SSCs intended to mitigate the effects of fires and floods. Although the generic issues program has examined these issues to a limited degree (e.g., Generic Safety Issue 172, "Multiple Systems Responses Program," closed out in 2001) and responses to Supplement 5, "Individual Plant Examination of External Events (IPEEE) for Severe Accident Vulnerabilities," to Generic Letter 88-20, "Individual Plant Examination for Severe Accident Vulnerabilities," dated November 23, 1988, the U.S. Nuclear Regulatory Commission's (NRC's) Near-Term Task Force (NTTF) concluded that the staff should reevaluate the potential for common-mode failures of plant safety-related SSCs as the result of seismically induced fires and floods. The NTTF identified this issue as Recommendation 3, "Evaluate Potential Enhancements to the Capability To Prevent or Mitigate Seismically-Induced Fires and Floods." SECY-11-0137, "Prioritization of Recommended Actions To Be Taken in Response to Fukushima Lessons Learned," dated October 3, 2011, prioritizes this issue as a longer term Tier 3 item because longer term staff evaluation was required to support a decision on the need for regulatory action. Although the staff believes that the use of traditional deterministic design-basis methods can enhance the capability to prevent seismically induced fires and floods, accident sequences and complex dependencies needed to evaluate the mitigation of these events can be done more systematically through Probabilistic Risk Assessments (PRAs). In the staff requirements memorandum to SECY-11-0137, the Commission directed the staff to initiate the development of a PRA methodology to evaluate potential enhancements to the capability to prevent or mitigate seismically induced fires and floods as part of Tier 1 activities. Therefore, the Commission indicated that the staff should start the prerequisite activity to initiate the development of an appropriate PRA methodology to support this issue without unnecessary delay while other aspects of this activity remain prioritized as Tier 3.

Approach

Certain activities that are being conducted over the near-term would provide valuable information for the ultimate resolution of Recommendation 3. For example, ongoing efforts to address seismic and flood hazard and mitigation strategies should provide a more complete understanding of plant-specific hazards, vulnerabilities, and mitigation capabilities. The staff plans to monitor the progress of these Tier 1 areas before dedicating substantial resources to the evaluation of seismically induced fires and floods. Therefore, the staff plans to do the following during fiscal years 2012–2016 to address NTTF Recommendation 3, as augmented by Commission direction:

- Initiate the development of a PRA methodology for addressing seismically induced fires and floods. As initially described in SECY-12-0025, "Proposed Orders and Requests for Information in Response to Lessons Learned from Japan's March 11, 2011, Great Tohoku Earthquake and Tsunami," dated February 17, 2012, the staff has completed a detailed plan for developing this method (Agencywide Documents Access and Management System Accession No. ML121450222). The staff plans to focus method development activities under the following two tasks:

1. Coordinate with standards development organizations (e.g., American Society of Mechanical Engineers and American Nuclear Society) and develop more generalized approaches for assessing concurrent hazards. These activities will help identify the technical elements and the associated high-level and supporting requirements for a suitable PRA method and will suggest specific areas for which detailed guidance is necessary.

2. Perform a feasibility scoping study to identify issues associated with the risk assessment of multiple concurrent hazards and evaluation of available PRA methods within this context. This study would provide information on the capabilities of traditional and advanced risk assessment methods (e.g., linked event tree and fault tree and dynamic simulation-based approaches) for accident scenarios in which issues, such as event timing, dependencies, and concurrency, can influence risk significance. This study would also include an evaluation of the current state-of-the-art for addressing seismically induced fires and floods and, more generally, concurrent hazards.

- Once the staff has obtained sufficient information from the Tier 1 activities related to seismic and flooding hazard evaluations and mitigation strategies for beyond-design-basis external events (e.g., activities under Recommendations 2.1, 2.3, and 4.2), the staff will reevaluate NTTF Recommendation 3. This evaluation will be based on experience gained in developing a PRA methodology for seismically induced fires and floods and insights derived from activities under Recommendations 2.1, 2.3, and 4.2. The staff expects that this reevaluation will result in one of the following outcomes (along with the supporting technical and regulatory basis):

  - A recommendation for regulatory action (e.g., rulemaking and order).

  - A recommendation for no regulatory action.

  - A recommendation for further research to support future regulatory decisionmaking.

For More Information

Contact Kevin Coyne, RES/DRA, at <u>Kevin.Coyne@nrc.gov</u>.

Hydrogen Control and Mitigation Inside Containment and Other Buildings

Background

The physical damage to Fukushima reactor buildings will perhaps be the most enduring visible image of plant damage initiated by the earthquake and tsunami in Japan on March 11, 2011. The apparent cause was combustion of hydrogen that was generated from the high-temperature oxidation of fuel cladding. Extensive cladding oxidation and core material melting had occurred in Fukushima, Units 1, 2 and 3, although the timelines for core damage differed in each unit because of the differences in equipment availability and operator response.

In SECY-11-0137, "Prioritization of Recommended Actions To Be Taken in Response to Fukushima Lessons Learned," dated October 3, 2011, the U.S. Nuclear Regulatory Commission (NRC) staff described its proposals for the regulatory actions that will be taken to address the recommendations of the Fukushima Near-Term Task Force (NTTF). For Recommendation 6, the NTTF recommended, as part of the longer term review, that the NRC identify insights about hydrogen control and mitigation inside containment or in other buildings as further study of the Fukushima Dai-ichi accident reveals additional information. SECY-11-0137 prioritizes this recommendation as Tier 3 because longer term staff evaluation was required to support a decision on the need for regulatory action. In the staff requirements memorandum to SECY-11-0137, the Commission agreed with the Tier 3 prioritization of Recommendation 6.

Objective

In accordance with Title 10 of the *Code of Federal Regulations* (10 CFR) 50.44, "Combustible Gas Control for Nuclear Power Reactors," licensees are required to use various hydrogen control and mitigation schemes inside containment buildings, depending on their unique design characteristics. As a result of insights and continued post-accident analyses of the Fukushima events, the NRC will reassess the hydrogen control rule as it relates to the various containment designs. In addition, the agency will evaluate connected buildings for the potential of combustible gas ingress and will determine what design enhancements will be necessary.

Approach

The NRC will reassess hydrogen control while recognizing the various interrelated operating aspects and conditions. For example, the Fukushima accident revealed that the primary containment pressure in boiling-water reactor (BWR) Mark I and II containments significantly exceeded its design limit, particularly as a result of hydrogen gas generated by severe core damage and relocation along with steam buildup. Licensees' severe accident management guidelines (SAMGs) address containment pressure control. However, damage to the equipment and other factors hampered the timely mitigation of increasing pressures in the Fukushima containments. As a result, copious amounts of hydrogen leaked into the associated reactor buildings. Therefore, pressure and hydrogen control for severe accidents in Mark I and II containments should now consider the effect of leakage into the reactor buildings. The availability of containment sprays, as shown in Figure 11.3, to reduce containment pressure and temperature will also influence the plant damage state.

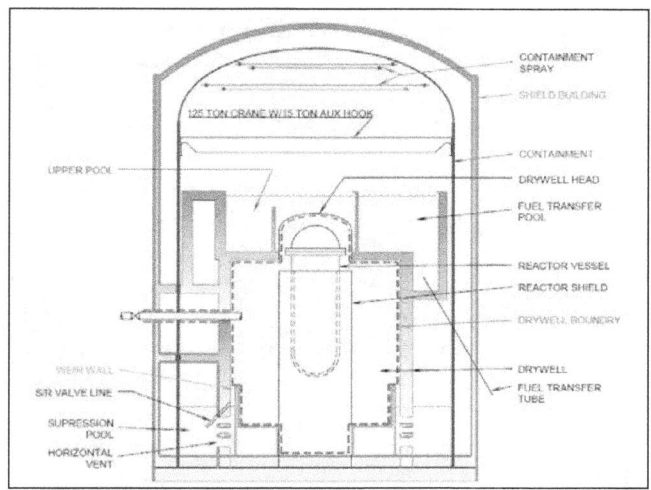

Figure 11.3 Mark III containment

Consequently, the NRC is reevaluating the integration of reliable containment venting strategies into the SAMGs under NTTF Recommendation 5.1, which states, "Order licensees to include a reliable hardened vent in BWR Mark I and Mark II containments,"and under Recommendation 5.2, which states, "Reevaluate the need for hardened vents for other containment designs, considering the insights from the Fukushima accident. Depending on the outcome of the reevaluation, appropriate regulatory action should be taken for any containment designs requiring hardened vents." (The recommended action under Recommendation 5.1 should include performance objectives for the design of hardened vents to ensure reliable operation and ease of use [both opening and closing] during a prolonged station blackout.) Therefore, the connection between Recommendations 5 and 6 is that, during postulated severe accidents, the generation of hydrogen is likely to contribute to an increased containment pressure. Figure 11.4 is a pictorial overview that shows the relationship of containment venting and hydrogen control for differing containment designs. Because of the smaller primary containment relative to other designs, pressure control and venting are more strongly coupled to hydrogen control in the Mark I and Mark II containments.

Currently, the NRC is participating in an Organisation for Economic Co-operation and Development/Nuclear Energy Agency benchmark study of the accident at Fukushima. This effort will place particular emphasis on hydrogen generation from all sources and will compare the information derived to the current understanding used as the basis for existing hydrogen control and mitigation schemes. Moreover, the NRC will pursue

the assessment of potential or, if possible, any identifiable leakage paths from the primary containments into the reactor buildings as a result of Fukushima accident forensic studies.

As needed, the NRC will perform accident progression studies using the MELCOR code to contrast the plant performance of the different containment types (e.g., BWR Mark III, pressurized-water reactor ice condenser, and large dry containments). These studies will focus on containment performance and the potential adverse consequences on adjacent buildings.

For More Information
Contact Allen Notafrancesco, RES/DSA, at Allen.Notafrancesco@nrc.gov.

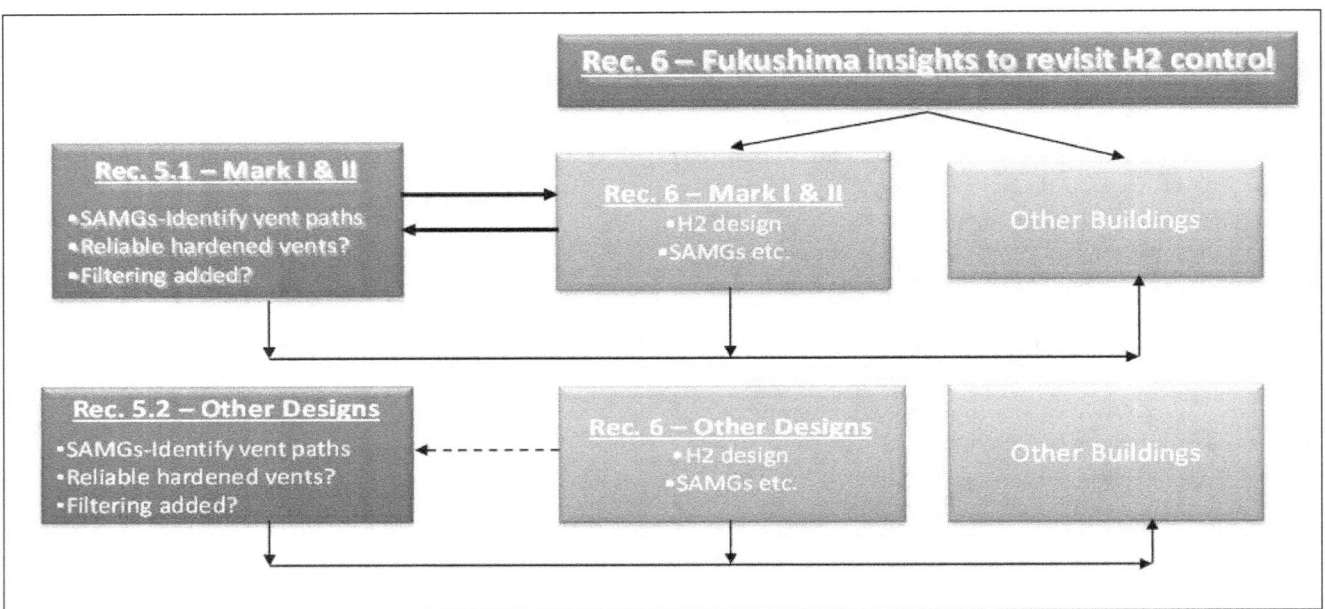

Figure 11.4 Nexus of Recommendations 5 and 6

Consequence Study of a Beyond-Design-Basis Earthquake Affecting the Spent Fuel Pool for a U.S. Mark I Boiling-Water Reactor

Background

All U.S. nuclear power plants store spent nuclear fuel in spent fuel pools (SFPs). These pools are robust structures made of reinforced concrete several feet thick with steel liners. The water is typically about 40 feet deep and serves to shield the radiation and to cool the rods. As the pools near capacity, utilities move some of the older spent fuel into dry cask storage. Fuel is typically cooled at least 5 years in the pool before its transfer to dry cask storage. The U.S. Nuclear Regulatory Commission (NRC) believes SFPs and dry casks both provide adequate protection of the public health and safety and the environment. Although the NRC has a rich regulatory basis for its current position on spent fuel storage, a number of events (e.g., the Fukushima accident) have motivated a reassessment of the underlying knowledge base. To launch this reassessment, the Office of Nuclear Regulatory Research has undertaken a study to produce updated consequence estimates for scenarios of interest related to a SFP. Results from this study will inform a regulatory decisionmaking process guided by the Fukushima lessons-learned Tier 3 issue of whether spent fuel should be transferred to dry cask storage earlier than currently planned by the nuclear power plant licensees.

The events at Fukushima Dai-ichi demonstrated that SFPs are robust structures and that the fuel remained adequately cooled through addition of water to compensate for boiling. These observations are in agreement with NRC conclusions from past studies that both SFP and dry casks provide adequate protection of public health and safety and the environment and that the likelihood of an accident involving significant radiological release from the SFP remains small.

Objective

The objective of the Consequence Study of a Beyond-Design-Basis Earthquake Affecting the Spent Fuel Pool for a U.S. Mark I Boiling-Water Reactor (also called the Spent Fuel Pool Study or SFPS) is to reexamine the potential impacts on SFP safety in the event of a beyond-design-basis seismic event.

Approach

This SFPS evaluates the consequences associated with a large seismic event and its impact on the SFP. Prior studies have concluded that a beyond-design-basis earthquake accident scenario is the principle contributor to SFP risk. The SFPS considers ground motion associated with a rare but credible seismic event and uses structural analysis methods to determine the potential damage states, including some damage states that affect SFP integrity, for a boiling-water reactor with a Mark I containment design (Figure 11.5). The SFPS uses detailed modeling of the event progression to determine the consequences of any resulting fission product release for various cases that reflect the effect of changes in configuration during the operating cycle, the loading of the SFP (Figure 11.6), and the deployment of mitigation capabilities. Insights from the SFPS have already identified other areas in which additional research would inform the staff's decisionmaking process.

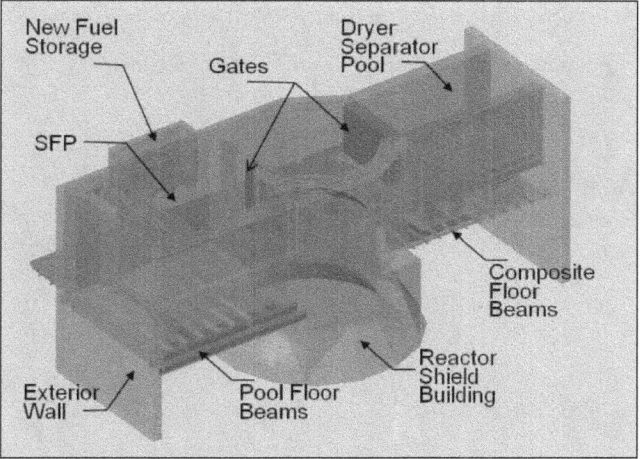

Figure 11.5 Model used to generate three-dimensional finite-element models of the SFP structure and its supports

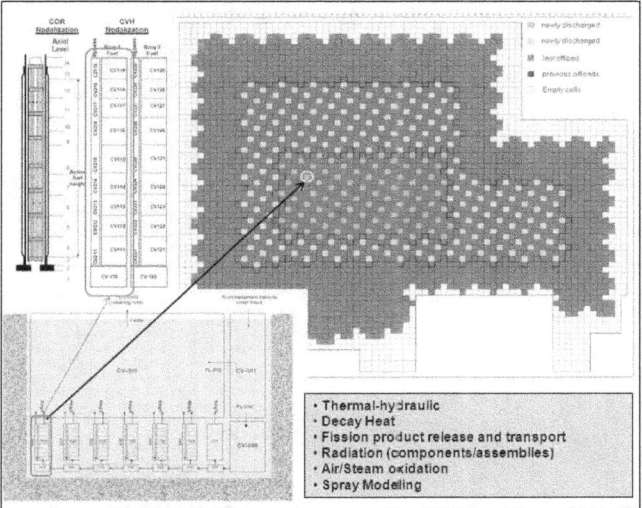

Figure 11.6 SFP MELCOR model

A draft report titled "Consequence Study of a Beyond-Design-Basis Earthquake Affecting the Spent Fuel Pool for a U.S. Mark I Boiling-Water Reactor" was released for public comment in June 2013 and is available in the Agency Document and Management System (ADAMS), ML13133A132. The final report is expected to be published in late 2013.

The technical elements of the study include the following:

- Seismic and structural assessments based on available information to define initial and boundary conditions.

- SCALE analysis of reactor building dose rates.

- MELCOR accident progression analysis (e.g., effectiveness of mitigation and fission product release).

- Emergency planning assessment.

- MACCS2 offsite consequence analysis (e.g., land contamination and health effects).

- Probabilistic considerations.

- Human reliability analysis of mitigation measures.

- Regulatory Analysis.

For More Information
Contact Don Algama, RES/DSA, at Don.Algama@nrc.gov.

Fukushima Dai-ichi Accident Study with MELCOR 2.1

Background

On March 11, 2011, the Tohuku earthquake struck near the Fukushima Dai-ichi power station, causing a regional loss of electric power and the operating reactors (Units 1, 2, and 3) to scram. The emergency onsite diesel-powered generators started as designed and supplied power to the emergency cooling systems needed to keep the reactors cool. Several tsunami waves produced by the earthquake reached the Fukushima Dai-ichi site roughly an hour later, resulting in the loss of emergency diesel-powered alternating current generators. Consequently, each of the three units suffered core damage of varying degrees. Without adequate containment cooling, the pressure suppression pools began to boil, which produced rising pressures in the containment vessels that eventually exceeded their design pressures. Containment venting was attempted; however, because of difficulties in accessing and manually operating the vent valves, venting was either unsuccessful or delayed. Ultimately the containment systems leaked, failed, or were intentionally vented, resulting in the release of radioactivity to the reactor buildings and the environment. Combustible gases produced from the damaged cores and perhaps molten core-concrete interaction accumulated in the reactor buildings and caused explosions and destruction of portions of the buildings (Figure 11.7).

In response to the accident at the Fukushima Dai-ichi nuclear power station in Japan, the U.S. Nuclear Regulatory Commission (NRC) and U.S. Department of Energy agreed to jointly sponsor an accident reconstruction study as a means of assessing the severe accident modeling capability of the MELCOR code. MELCOR is the state-of-the-art system-level severe accident analysis code used by the NRC.

Objective

The objectives of the study were to (1) collect, verify, and document data on the accidents by developing an information portal system, (2) reconstruct the accident progressions using computer models and accident data, and (3) assess the MELCOR code and the Fukushima models and suggest potential future data needs.

Approach

The study is focused on using the MELCOR code and known initial and boundary conditions surrounding the accidents in the Fukushima reactors to "reconstruct" the accidents and to predict, as well as possible, plant state parameters (e.g., reactor and containment pressures); the implied core damage and fission product release; and, if possible, the circumstances leading up to the reactor building explosions during the first four days of the accidents. Sandia National Laboratories developed MELCOR 2.1 models of the Fukushima Dai-ichi Units 1, 2, and 3 reactors and the Unit 4 spent fuel pool. Oak Ridge National Laboratory developed a MELCOR 1.8.5 model of the Unit 3 reactor. The MELCOR boiling-water reactor (BWR) models used in this analysis were previously developed in the NRC's state-of-the-art reactor consequence analysis project. The preliminary results of the study (Figure 11.8) are documented and available at http://melcor.sandia.gov/docs/Fukushima_SAND_Report_final.pdf.

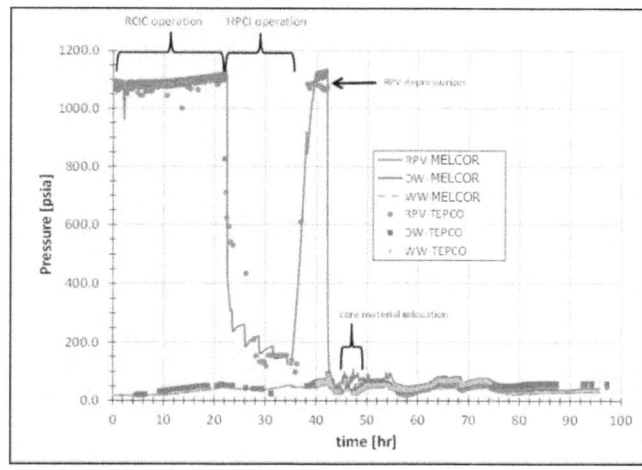

Figure 11.8 MELCOR-predicted reactor and containment pressures compared to TEPCO data (Unit 3)

For More Information

Contact Richard Lee, RES/DSA, at Richard.Lee@nrc.gov.

Figure 11.7 Fukushima Units 1, 2, 3, and 4 after the accident showing extensive damage to the reactor buildings

Chapter 12: International Cooperative and Long-Term Research

Cooperative International Research Activities and Agreements

The Organisation for Economic Co-operation and Development Halden Reactor Project

The Organisation for Economic Co-operation and Development/Nuclear Energy Agency PKL2 Project

The Organisation for Economic Co-operation and Development ROSA-2 Program

Studsvik Cladding Integrity Project

Zirconium Fire Research

International Nondestructive Examination Round Robin Testing

International Cooperative Research on Impact Testing

Round Robin Analysis of Containment Performance Under Severe Accidents

Collaborative Research with Japan on Seismic Issues

Agency Forward-Looking and Long-Term Research

Completed prestressed concrete containment vessel ¼-scale model at Sandia National Laboratories

Cooperative International Research Activities and Agreements

Cooperative Research Agreements

The U.S. Nuclear Regulatory Commission's (NRC's) Office of Nuclear Regulatory Research (RES) has implemented over 100 bilateral or multilateral agreements with more than 30 countries and the Organisation for Economic Co-operation and Development (OECD). These agreements cover a wide range of activities and technical disciplines, including severe accidents, thermal-hydraulic code assessment and application, digital instrumentation and control, nuclear fuels analysis, seismic safety, fire protection, human reliability, and more.

Bilateral, Multilateral, and Code User Groups

Many of the agreements are established bilaterally with a foreign regulator or research institution for participation in one of the two largest nuclear safety computer code sharing programs. The Code Applications and Maintenance Program includes thermal-hydraulic code analysts from more than 20 member nations. The Cooperative Severe Accident Research Program also includes more than 20 member nations that focus on the analysis of severe accidents using the MELCOR code. Both programs include user group meetings at which participants share experience with the NRC codes, identify code errors, perform code assessments, and identify areas for additional improvement, experiments, and model development.

The OECD's Nuclear Energy Agency (NEA) coordinates most of the NRC's multilateral research agreements. A few examples show how diverse the agreements can be. Large-scale experiments include the Halden Reactor Project (HRP), based in Norway, and the domestically based Sandia Fuel Pool project. The OECD Piping Failure Data Exchange Project database is an example of a different sort of shared resource for participants. RES applies a set of established criteria when considering the cooperative research program proposals it receives. Considerations include cost, benefit, timeliness of expected results for current and expected regulatory uses, and more.

NRC participation in these agreements allows broader sharing of data obtained from physical facilities not available in the United States. As a result, NRC tools, data, and safety knowledge stay current and are state-of-the-art. This enhances the NRC's ability to soundly make realistic regulatory and safety decisions based on worldwide scientific knowledge and promotes the effective and efficient use of agency resources. Data obtained are used to develop new analytical models; to validate NRC safety codes; to enhance assessments of plant risk, including decisionmaking, fire, and human performance and reliability; and to develop risk-informed approaches to regulation.

NEA Activities

The NRC plays a very active role at the OECD/NEA, with RES maintaining leadership roles in the Committee on the Safety of Nuclear Installations (CSNI) (including CSNI's seven working groups and joint research projects) and the Committee on Radiation Protection and Public Health. The RES Director is the Chairman of CSNI, and RES senior management represents the NRC on the Halden Reactor Project's Board of Management.

International Atomic Energy Agency Activities

RES also serves as the agency lead on codes and standards. By acting as the agency lead in the International Atomic Energy Agency's (IAEA's) Nuclear Safety Standards Committee, RES coordinates NRC contributions to the many IAEA safety standards guides. RES also participates in two "extra-budgetary programs" within IAEA entitled, "Protection against Tsunamis and Post Earthquake Consideration in the External Zone," and "Seismic Safety of Existing Nuclear Power Plants," which feeds into IAEA's International Seismic Safety Center.

Bilateral Information Exchange

RES actively seeks international cooperation to obtain technical information on safety issues that require test facilities not available domestically and would require substantial resources to duplicate in the United States. RES often will propose modifications to a project sponsor so that the proposed project can better meet the NRC's needs. In addition, the NRC may propose to sponsor cooperative international participation in research projects conducted by the NRC. Bilateral exchanges with counterparts multiply the amount of information available to RES staff. As an example, RES has developed an extremely beneficial relationship with the Canadian Nuclear Safety Commission in the area of environmental modeling, groundwater monitoring, and more. The relationship allows the NRC to apply lessons learned at and around Canadian reactors to domestic reactors. Similarly, the NRC's relationship with the French Institute of Radiation Protection and Nuclear Safety (IRSN) is multifaceted and mutually beneficial. Each organization has developed expertise in areas the other can learn from. The NRC and IRSN cooperate in dozens of technical areas.

RES has long been a leader in the area of enhancing domestic resources with international knowledge, skills, and use of foreign facilities. The staff has worked, and continues to work, to ensure that the international activities in which it participates have direct relevance to the NRC's regulatory program.

For More Information
Contact Donna-Marie Sangimino, RES/IPT, at
Donna-Marie.Sangimino@nrc.gov.

The Organisation for Economic Co-operation and Development Halden Reactor Project

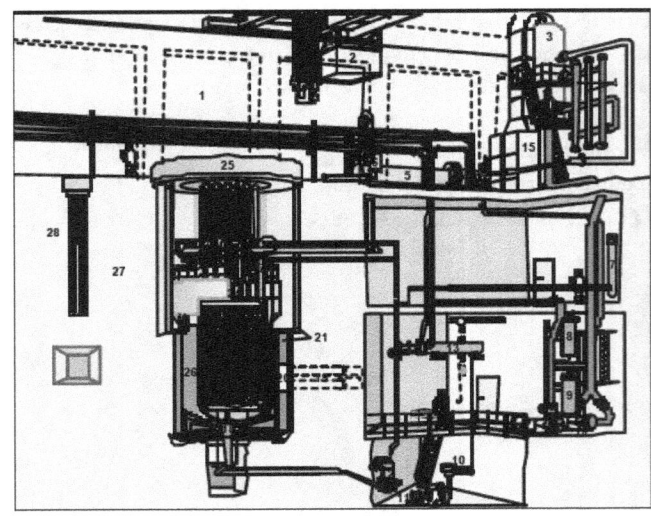

Figure 12.1 HBWR test reactor

Background

The U.S. Nuclear Regulatory Commission (NRC) and its predecessor, the U.S. Atomic Energy Commission, have been participating in the Organisation for Economic Co-operation and Development (OECD) Nuclear Energy Agency Halden Reactor Project (HRP) since its inception in 1958. During this period, the NRC has used numerous research products from this internationally funded cooperative effort, which is located in Halden, Norway, and managed by the Norwegian Institute for Energy Technology (Institutt for Energiteknikk [IFE]). For example, Halden tests on high-burnup fuel under loss-of-coolant accident (LOCA) conditions contributed to the technical basis for a rulemaking underway related to fuel cladding embrittlement. As another example, Halden's human factors research has supported regulatory guidance in areas such as alarm systems, hybrid control rooms, display navigation, and guidance for the review of proposed staffing configurations in computer-based control rooms. The HRP operates on a 3-year research cycle, and the current program plan runs from 2012–2014.

Facilities and Activities

Fuels and Materials Research

The Halden boiling-water reactor (see Figure 12.1), which currently operates at 18 to 20 megawatts, is fully dedicated to instrumented in-reactor testing of fuel and reactor materials. Since its initial startup, the reactor facility has been progressively updated and is now one of the most versatile test reactors in the world. The HRP fuels and materials program focuses on the performance of fuel and structural materials under normal or accident conditions using the numerous experimental channels in the core that are capable of handling many test rigs simultaneously.

Recent NRC reviews of industry fuel behavior codes have directly employed data from the HRP fuels program. These data are also essential for updating the NRC's fuel codes and materials properties library, which are used to audit industry analyses. Currently, the NRC is particularly interested in the previously mentioned LOCA tests, which are investigating such phenomena as axial gas flow, maintaining or breaking fuel-to-cladding bonding, fuel axial relocation, and fuel fragment spillage through cladding burst opening.

Regarding the HRP's nuclear reactor materials testing program, the HRP has, over the years, provided fundamental technical information to support the understanding of the performance of irradiated reactor pressure vessel materials and supplemented results generated under NRC research programs. Recently, the HRP has been an essential partner in evaluating the irradiation-assisted stress-corrosion cracking (IASCC) of light-water reactor (LWR) materials. The HRP has irradiated materials that were later tested under the NRC's research program at Argonne National Laboratory to measure crack initiation, fracture toughness, and crack growth rate under representative LWR conditions. The HRP's ongoing work on IASCC and other areas (e.g., irradiation-induced stress relaxation) supplements NRC-sponsored research and addresses existing knowledge gaps. The NRC staff is using this information to inform reviews of licensee aging management programs.

Man-Technology-Organization Laboratory

IFE's Halden facility also includes the IFE Man-Technology-Organization (MTO) Laboratory. The Halden Man-Machine Laboratory (HAMMLAB) (see Figure 12.2) is one of the principal experimental facilities in this laboratory. HAMMLAB uses a reconfigurable simulator control room that facilitates research into instrumentation and control (I&C), human factors, and human reliability analysis (HRA). Currently, HAMMLAB has hardware and software enabling it to simulate the Fessenheim pressurized-water reactor (PWR) plant in France, the Forsmark-3 boiling-water reactor plant in Sweden, and the Ringhals-3 PWR plant in Sweden.

Figure 12.2 HAMMLAB control room simulator

Many of the HAMMLAB experiments are performed with the control room configured as a prototype advanced control room with an integrated surveillance and control system. This setup is used to explore the impacts of automation and advanced human-system interfaces on operator performance. HAMMLAB has extensive data collection capabilities and typically uses qualified nuclear power plant operators (who are familiar with the plants being simulated) as test subjects.

Recently, HRP-designed and executed HAMMLAB experiments provided the foundation for the International Empirical HRA Study, a multinational study aimed at developing an empirically based understanding of the performance, strengths, and weaknesses of HRA methods used in risk-informed regulatory applications. The NRC will be using the study's results to address outstanding HRA technical issues, including those related to HRA model differences identified in a November 8, 2006, Staff Requirements Memorandum (SRM)-M061020. Currently, ongoing HRP experiments are addressing a number of topics of interest to the NRC, including control room staffing strategies, the role and effects of automation in advanced control room designs, and aids to improve control room teamwork. The NRC expects that this research will contribute to the technical basis for human factors guidance, especially for new reactor designs.

The IFE MTO laboratory also includes a virtual environment center and an integrated operations laboratory. The former is used to perform research involving mixed reality applications (e.g., training), and the latter is used to address issues associated with remote operations.

Finally, the MTO laboratory also conducts research on I&C systems. Past efforts include work in the area of instrumentation surveillance and monitoring techniques based on advanced decision algorithms. A number of HRP-developed systems have been evaluated for use by U.S. plants.

The current HRP digital systems research activities contribute to three phases of a system lifecycle:

- Development, assurance, and deployment of high integrity software important to nuclear power plant safety.

- Condition monitoring and maintenance support, where engineering and technical support teams are the intended beneficiaries of the research results. This research will improve accuracy and usability of current methods and develop novel techniques to improve diagnostics and condition-based maintenance.

- Development and application of software systems for operational support, where plant operators are the intended beneficiaries of the research results. The research program includes interaction of advanced control systems with human operators and issues related to the implementation and use of operational procedures.

Summary

The HRP has provided and continues to provide valuable information to the NRC. Much of this information addresses gaps that are otherwise not being addressed by current NRC research activities, and some of this information is foundational to the NRC's efforts to improve the technical basis of key models, methods, and tools. Furthermore, because the NRC is one of several contributors to the HRP budget, the HRP enables the NRC staff to significantly leverage its resources. More information regarding the NRC's participation in the OECD Halden Reactor Project can be found in SECY-11-0148 at http://www.nrc.gov/reading-rm/doc-collections/commission/secys/2011/2011-0148scy.pdf.

For More Information
Contact Matthew Hiser, RES/DE, at Matthew.Hiser@nrc.gov.

The Organisation for Economic Co-operation and Development/ Nuclear Energy Agency PKL2 Project

Background

Since 2001, the U.S. Nuclear Regulatory Commission (NRC) has been involved in a series of experimental programs fostered by the Organisation for Economic Co-operation and Development (OECD) that use the Primärkreislaufe-Versuchsanlage (PKL, primary coolant loop facility) to investigate safety-related issues relevant to current and new pressurized-water reactor (PWR) designs. In April 2008, the NRC became involved with the OECD/Nuclear Energy Agency (NEA) PKL2 Project (PKL2), a 3.5-year program that investigated a series of topics, such as complex heat transfer mechanisms in steam generators (SGs), controlled cooldowns under natural-circulation conditions, and boron precipitation processes following a large-break loss-of-coolant accident. Having achieved its objective, PKL2 concluded in September 2011. A few months later, a new experimental program, OECD/NEA-PKL3, was developed to address a new series of topics, to include beyond-design-basis accidents with significant core heatup, accidents during shutdown conditions, and a followup to the boron-precipitation test conducted under PKL2. PKL3 began in April 2012 and will conclude in December 2015.

Designed and built in the 1970s by AREVA NP GmbH (formally Siemens/KWU), the PKL facility is a full-height, 1:145 volume-scaled replica of a German PWR. Configured for enhanced realism, the facility has four identical reactor coolant loops arranged symmetrically around a reactor pressure vessel (RPV) that contains a simulated core. Each of the four loops is equipped with a fully functional SG and a reactor coolant pump, and the core is simulated using 314 electrically heated rods. Each SG contains 30 U-tubes of original size and material, and each reactor coolant pump is equipped with an active speed controller to enable the simulation of different pump characteristics. The bundle of rods representing the core are capable of generating 2.5 megawatts (MW) of core power, which is equivalent to 10 percent of the scaled nominal power rating of the 3,600-MW PWR used as the basis for the facility's design.

Objective

The OECD/NEA PKL programs contribute to the understanding of complex thermal-hydraulic processes involved in accident scenarios and to a better assessment of the countermeasures implemented for accident control. In addition, these programs supply valuable information regarding plant safety margins and provide an extensive database for use in the further development and validation of thermal-hydraulic computer codes.

Approach

Before experiments are conducted, all participants agree on the subject matter and scope of the topics to be explored. A schedule then is finalized, and the experiments are conducted systematically. At the completion of each experiment, a preliminary data report is disseminated to each member organization for evaluation.

The following is the program of experimentation agreed upon for the PKL2 program:

- G1: heat transfer in the SGs (failure of residual heat removal system under ¾ loop operation).

- G2: cooldown under asymmetric boundary conditions (i.e., isolated SGs).

- G3: cold water transients following a main steam line break.

- G4: influence of secondary-side parameters on heat transfer under reflux conditions.

- G5: boron precipitation processes after a large-break loss-of-coolant accident.

- G6: formation and behavior of upperhead void during cooldown.

- G7: effectiveness of secondary-side depressurization and the performance of core-exit thermocouples.

Many significant findings came out of the PKL2 program, most notably the findings from Test G2 and G6. One of the findings from Test G2 was that natural circulation should be expected to stagnate if the SGs are continuously cooled down at 50 kelvin per hour; however, natural circulation can be maintained if the cooldown is done in a stepwise manner. Test G6 showed that coolant volumes in the RPV, up to around 0.5 meters above the coolant-loop piping, participate in the cooldown process, which represents the limiting factor for void growth. Also, after activation of two reactor coolant pumps, the upperhead void does not collapse completely. Instead, it collapses up to the point where the coolant flow into the RPV dome through the upperhead bypass is no longer injected into steam but into the now elevated coolant level.

Because of scaling considerations, some of the results from the PKL programs are directly transferable to a reactor, while others are only qualitatively transferrable. The more directly transferable results are related to natural circulation, the development of level swell, and SG heat-transfer mechanisms. Results that may be distorted as a result of geometrical differences in the SG plena and the upper-plenum of the RPV, such as two-phase flow distributions and the dynamics of void propagation, are less transferrable.

Computer-code validation is one of the objectives of the PKL programs, and participants are encouraged to model completed experiments using their computer codes of choice and to compare the results to the data. Using the NRC's TRAC/

RELAP Advanced Computational Engine (TRACE), a post-test calculation was performed for PKL2 Test G3. To perform the calculation, a TRACE model representing each of the major components and control systems present in the facility was developed (Figure 12.3). The TRACE results showed that the code was capable of predicting all of the key phenomena reasonably well. The results also made apparent the uniqueness of the four-loop data in illuminating the asymmetric effects of the test, which proved to be a challenge for the code to simulate.

As PKL3 tests are completed, similar calculations will be performed and analyzed to assess the applicability of TRACE and provide further insight into safety-related issues.

For More Information
Contact Shawn O. Marshall, RES/DSA, at
Shawn.Marshall@nrc.gov.

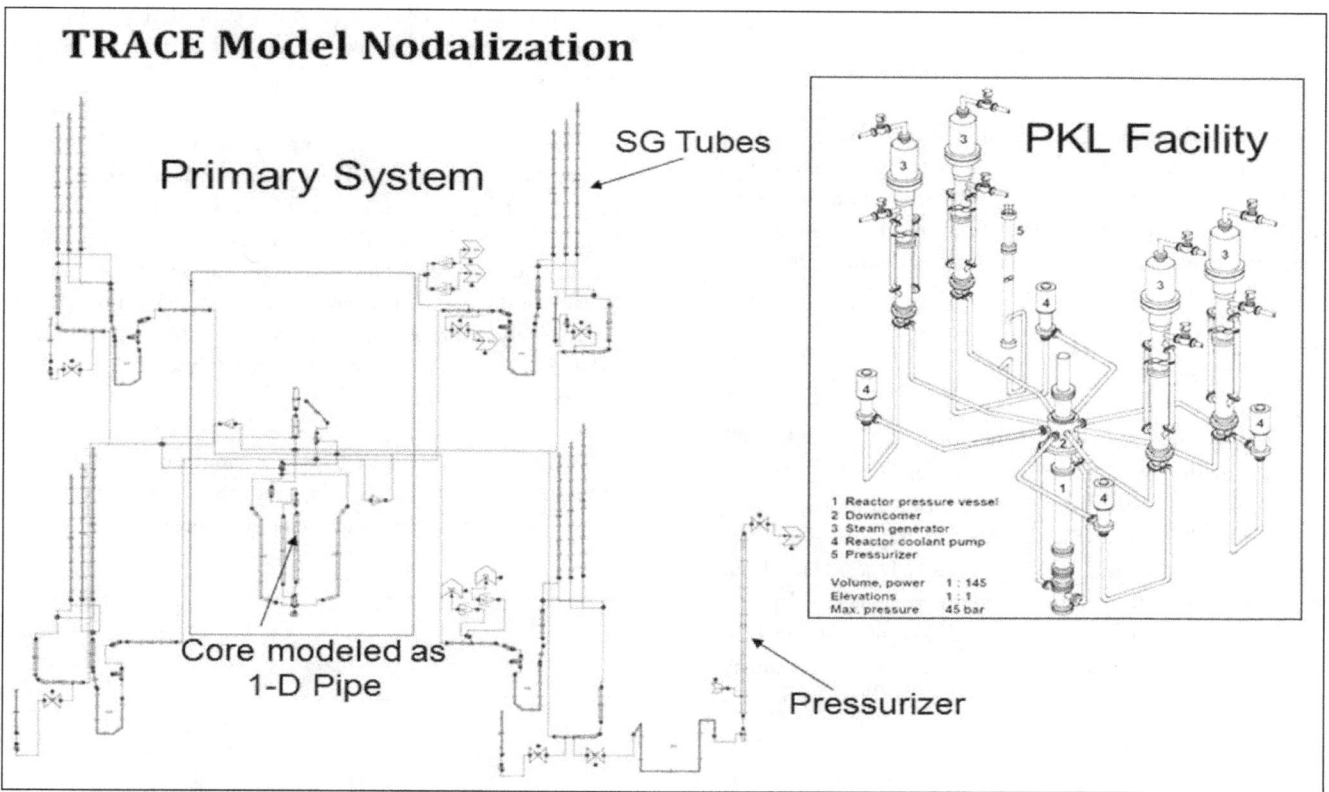

Figure 12.3 TRACE nodalization and schematic of PKL facility

The Organisation for Economic Co-operation and Development ROSA-2 Program

Background

The U.S. Nuclear Regulatory Commission (NRC) has been participating in the Rig of Safety Assessment (ROSA) program for many years under the Organisation for Economic Co-operation and Development/Nuclear Energy Agency. The ROSA-2 program is the latest phase of the program to conduct thermal-hydraulic (T/H) accident experiments in pressurized-water reactors (PWRs). The ROSA-2 program started in 2009 and completed in 2012.

Objective

The NRC staff members participating in this international project investigate potential safety issues relevant to current PWRs and new PWR designs. The ROSA-2 test data is being used to validate the TRAC/RELAP Advanced Computational Engine (TRACE) computer code and expand the usefulness of the code as an audit tool.

Approach

The ROSA programs use the Large Scale Test Facility (LSTF) operated by the Japanese Atomic Energy Agency (JAEA) to conduct T/H accident experiments (see Figures 12.4 and 12.5). The LSTF, which has been in use since 1985, is an instrumented full-height, 1/48 volumetrically scaled test facility intended to perform system integral experiments simulating the T/H response at full-pressure conditions of existing and next-generation PWR designs during loss-of-coolant accidents (LOCAs) and other operational and abnormal transients.

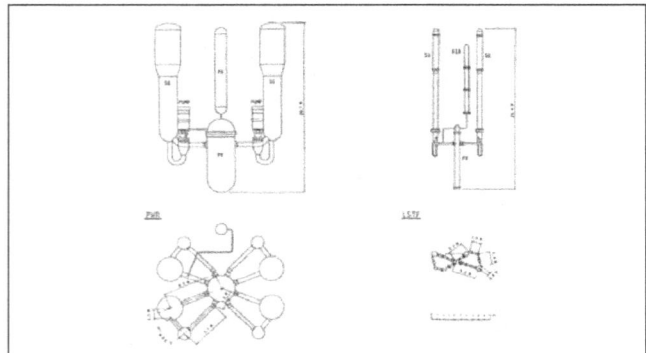

Figure 12.4 Size comparison of ROSA/LSTF to a four-loop PWR

All seven tests proposed for the ROSA-2 program have been performed. As part of the ROSA-2 program, testing at the LSTF facility investigated the following safety issues:

- Three intermediate-break LOCA tests, which address risk-informed break size definition and verification of safety analysis codes, were performed.

- Improvements and new proposals for accident management mitigation and emergency operation were investigated. Two completed tests focused on the recovery from a steam generator tube rupture—one with and one without—a main steam line break.

A counterpart test with the Primärkreislauf-Versuchsanlage (primary coolant loop test facility) PKL test facility was performed. The PKL facility in Erlengen, Germany, is operated by AREVA NP. Counterpart testing at the ROSA-2/LSTF and PKL facilities provides test data that reflect the design scaling of the two facilities and produced two sets of test data for computer code validation.

The ROSA-2 test program has completed testing of a 17-percent intermediate hot-leg break, a 17-percent intermediate cold-leg break, and a 13-percent intermediate cold-leg break. The NRC staff has developed a TRACE model of the primary and secondary sides of the LSTF test facility to analyze these tests. Test data for the two 17-percent break tests have been compared to TRACE blind and post-test predictions. TRACE predictions for the 13-percent cold-leg break test currently are being compared to preliminary test data.

The NRC staff also is updating TRACE models of the ROSA-2 LSTF and PKL facilities to develop test predictions that can be compared to the counterpart test data from the two facilities.

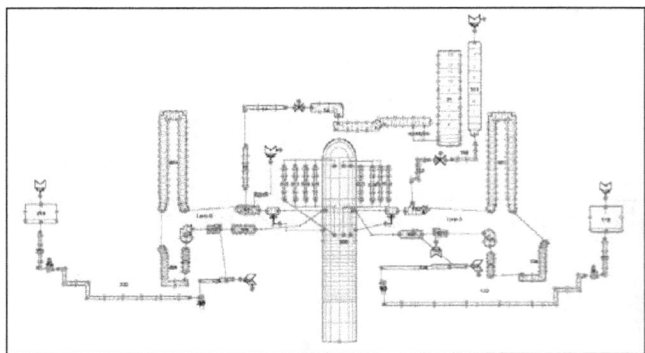

Figure 12.5 LSTF primary system TRACE model used for 17-percent intermediate cold-leg break

For More Information
Contact William Krotiuk, RES/DSA, at
William.Krotiuk@nrc.gov.

Studsvik Cladding Integrity Project

Background

The Studsvik Cladding Integrity Project (SCIP) is an international program supported by the Organisation for Economic Co-operation and Development/Nuclear Energy Agency and launched in 2004. The program has now been extended to 2014, with participants from Europe, Japan, the United States, Russia, and Korea. The participants represent four categories: those that supply and manufacture the fuel, the power companies themselves, regulators, and laboratories with similar assignments to Studsvik's.

Objective

SCIP is focused on improving the ability to predict mechanisms that can cause damage to cladding under normal operation and during transients. The program is conducted in the form of experiments, studies of fundamental mechanisms, development of suitable testing methods, and knowledge transfer.

The SCIP experiments and studies of fundamental mechanisms enable the understanding and quantification of key parameters important to hydrogen-induced failures, stress-corrosion cracking failures, and pellet-cladding mechanical interaction failures. This work provides valuable information for the development of operating restrictions.

The development of testing methods includes in-cell and out-of-cell mechanical testing techniques, as well as postirradiation analysis methods. This work enables the characterization of changes in cladding and pellets that take place with irradiation and provides valuable and unique characterization of advanced cladding and fuel pellet designs.

Approach

Multiple laboratories are performing the technical work in SCIP II. Power transient testing is conducted in the Halden boiling-water reactor. Studies of the irradiated rods are then made at the Studsvik Hot Cell Laboratory, leading to a series of mechanical tests in other laboratories at Studsvik.

Use of SCIP Data in the Integral Assessment of Fuel Rod Computer Codes

As part of the U.S. Nuclear Regulatory Commission's (NRC's) fuel performance code development effort, new code versions are exercised to assess the integral code predictions to measured data for various performance parameters. The documentation of the integral assessment is publicly available and serves to demonstrate the code's ability to accurately predict integral fuel response under normal and off-normal conditions. As new data are generated, new assessment cases are added to the integral assessment suite.

The latest integral assessment added 10 SCIP ramps to the assessment suite, with 12 more being considered. The ramps were modeled to assess the ability of FRAPCON 3.4 to predict cladding hoop strain during power ramps. Peak node plastic strain values from SCIP ramp data were compared to predicted values. Measured versus predicted values of plastic strain were compared as a function of burnup and ramp terminal level. These ramp tests were the first ramp tests that FRAPCON 3.4 was compared to with burnup greater than 45 gigawatt day per metric ton of uranium.

The comparison of predicted to the measured values in these ramp tests provided valuable insight into FRAPCON's ability to predict fuel and cladding response during power ramps. In this comparison effort, it was noted that FRAPCON 3.4 under-predicted the measured hoop strain in high burnup rods. The under-prediction was most severe for those ramp tests with long hold times, as can be seen in Figure 12.6. The NRC is now revisiting the FRAPCON 3.4 strain model to investigate the source of this under-prediction, and, if possible, to improve the modeling capabilities of FRAPCON. Furthermore, the NRC has plans to use the SCIP ramp data and the results from the FRAPCON 3.4 benchmarks to develop a pellet-cladding interaction failure model for FRAPCON 3.4.

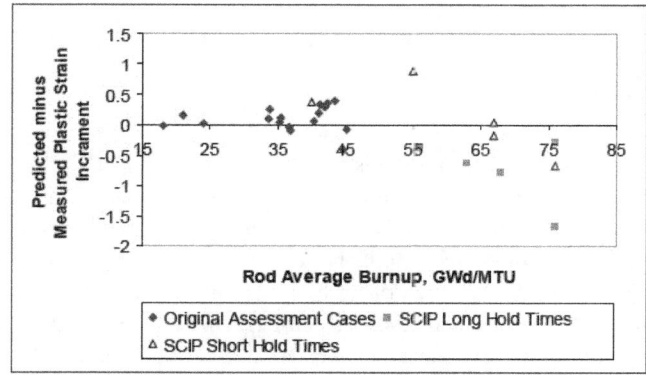

Figure 12.6 FRAPCON 3.4 predicted minus measured permanent hoop strain as a function of burnup, indicating an underprediction at high burnups

For More Information
Contact Patrick Raynaud, RES/DSA, at
Patrick.Raynaud@nrc.gov.

Zirconium Fire Research

Background

In 2001, the U.S. Nuclear Regulatory Commission (NRC) staff performed an evaluation of the potential accident risk in a spent fuel pool (SFP) at decommissioning plants in the United States. NUREG-1738, "Technical Study of Spent Fuel Pool Accident Risk at Decommissioning Nuclear Power Plants," describes a modeling approach for a typical decommissioning plant with design assumptions and industry commitments, the thermal-hydraulic (T/H) analyses performed to evaluate spent fuel stored in the SFP at decommissioning plants, the risk assessment of SFP accidents, the consequence calculations, and the implications for decommissioning regulatory requirements. Some of the assumptions in the accident progression in NUREG-1738 were known to be conservative, especially the estimation of the fuel damage. The NRC continued SFP accident research by applying best-estimate computer codes to predict the severe accident progression following various postulated accident initiators. The best-estimate computer code studies identified various modeling and phenomenological uncertainties that prompted a need for experimental confirmation. The NRC conducted the present experimental program to address T/H issues associated with complete loss-of-coolant accidents (LOCA) in pressurized-water reactor (PWR) SFPs. This experimental program is part of an international effort established with the Organisation for Economic Co-operation and Development (OECD) and includes the following 13 countries: Czech Republic, France, Germany, Hungary, Italy, Japan, Norway, Republic of Korea, Spain, Sweden, Switzerland, United Kingdom, and the United States (with the U.S. Nuclear Regulatory Commission as the operating agency).

Objective

The objective of this project is to provide basic T/H data associated with an SFP complete LOCA. The accident conditions of interest for the SFP were simulated in a full-scale prototypic fashion (electrically heated, prototypic assemblies in a prototypic SFP rack) so that the experimental results closely represent actual fuel assembly responses. A major impetus for this work is to facilitate accident code validation (primarily MELCOR) and reduce modeling uncertainties within the code.

The experimental program was conducted at Sandia National Laboratories. The first phase of the program focused on axial heating and burn propagation in a single PWR 17x17 assembly, and the second phase on radial and axial heating and zirconium fire propagation, including effects of fuel rod ballooning in a 1x4 assembly configuration. The first two sections of this article summarize the background and objectives of the experiments. The subsequent sections describe the testing approach and results of the first phase of the experimental program.

Testing Approach

The study was conducted in two-phases. Phase 1 focused on axial heating and burn propagation. A single full-length test assembly was constructed with a prototypic fuel skeleton (see Figure 12.7) and zirconium-alloy clad heater rods. As demonstrated in the previous study for boiling-water reactors (BWRs), the thermal mass of the compacted magnesium oxide (MgO) powder used to make the electric heater is an excellent match to spent fuel. The assembly was characterized in two different-sized storage cells. Phase 1 started with separate effect tests where the assembly hydraulic and T/H response was investigated. It concluded with an ignition test to determine where in the assembly ignition first occurs and the nature of the burn in the axial direction of the assembly. The pool cell was completely insulated to model boundary conditions representing a "hot neighbor," which is a typical bounding scenario.

Figure 12.7 Single fuel assembly for Phase 1 testing in construction stage

Phase 2 addressed axial and radial heating and burn propagation, including effects of fuel rod ballooning. Five full-length assemblies were constructed. The center assembly was of the same heated design as used in Phase 1. The four peripheral assemblies were unheated but highly prototypic, incorporating prototypic fuel tubes and end plugs. These boundary conditions experimentally represent a "cold neighbor" situation, which complements the bounding scenario covered by Phase 1. The peripheral fuel rods were filled with high density MgO ceramic pellets, sized to precisely match the thermal mass of spent fuel. Similarly, this phase started with separate effect tests, including hydraulic and T/H measurements. Studies using this test assembly concluded with a fire test in which the center assembly was heated until ignition occurred, which eventually propagated axially and radially to the peripheral assemblies, as predicted. Fuel rods in two of the four peripheral assemblies were pressurized with argon, and the fuel rods ballooned when the zirconium-alloy cladding reached a high enough temperature. The two peripheral assemblies without pressurized rods were compared to evaluate the effect of ballooning.

Analysis Support and Status

As in the previous BWR study, all stages of testing used MELCOR modeling results. Pretest MELCOR modeling results were used to guide the experimental test assembly design and instrumentation. MELCOR modeling results also were used to choose experimental operating parameters, such as the applied assembly power. At each step in the testing, improvements were made to the MELCOR model to continually increase confidence in the modeling validity. Experiments are complete as of June 2012 and reports for Phase 1 and 2 are in process and will be transmitted to OECD in 2013.

For More Information:
Contact Ghani Zigh, RES/DSA, at Ghani.Zigh@nrc.gov.

International Nondestructive Examination Round Robin Testing

Background

Primary pressure boundary components made of nickel-based alloys are susceptible to primary water stress-corrosion cracking (PWSCC). Between November 2000 and March 2001, leaks were discovered in Alloy 600 control rod drive mechanism (CRDM) nozzles and associated Alloy 182 J-groove attachment welds in several pressurized-water reactors. Destructive examination of several CRDMs showed that the leaks resulted from PWSCC. By mid-2002, over 30 leaking CRDM nozzles had been reported domestically. The cracking resulted in leaks in the primary pressure boundary and caused coolant leakage in several dissimilar metal welds (DMWs), CRDM weldments, and bottom-mounted instrumentation (BMI) nozzles. Such events, both domestic and international, focused additional research to address PWSCC in DMWs.

The U.S. Nuclear Regulatory Commission (NRC) established the Program To Assess the Reliability of Emerging Nondestructive Techniques (PARENT) as the follow-on to the international cooperative Program for the Inspection of Nickel Alloy Components (PINC). The goal of PINC was to evaluate the capabilities of various nondestructive evaluation (NDE) techniques to detect and characterize surface-breaking PWSCC in DMW in BMI penetrations and small-bore (approximately 400 millimeters (mm) in diameter) piping components.

Commercial and university inspection teams conducted a series of international blind round robin (RR) tests. Results from these tests show that a combination of conventional and phased-array ultrasound techniques provided the highest performance for flaw detection and depth sizing in DMWs in piping. The effective detection of flaws in BMI by eddy current and ultrasound shows that it may be possible to reliably inspect these components in the field. The results of the PINC program with respect to the probability of detection of the NDE techniques have been published in NUREG/CR-7019, "Results of the Program for the Inspection of Nickel Alloy Components," issued August 2010, which is available on the NRC's Web site at www.nrc.gov.

Objective

The purpose of PARENT is to compile (1) a knowledge base on cracking in Alloy 600 and similar nickel-based alloy welds in nuclear power plants and (2) RR test results, including the crack morphology and NDE responses from emerging techniques. PARENT will include both open and blind testing because the blind testing approach puts restrictions on teams using experimental NDE techniques, which hinder their ability to maximally demonstrate the technique's strengths and weaknesses. The objective of the blind tests is to evaluate commercially available NDE inspection techniques to determine which are the most effective for detecting and sizing PWSCC with the requirement that only qualified inspectors and procedures are used. The objective of the open tests is to evaluate novel and emerging NDE inspection techniques that industry and universities are developing to determine which ones show the most promise in regard to responses to (e.g., signal/noise) and sizing of realistically simulated PWSCC (especially small flaws) in components that have realistic geometries.

Approach

The NRC implemented international agreements with organizations from Finland, Japan, the Republic of Korea, Sweden, and Switzerland to establish PARENT. Pacific Northwest National Laboratory is assisting the NRC with the RR tests and coordination of the program. As part of their international collaboration, PARENT participants identified, ranked, and determined which component configurations should be considered for the program. The program used a series of test blocks with cracks available from different programs or fabricated by different contributors (Figures 12.8, 12.9, and 12.10) to simulate the selected component configurations.

Figure 12.8 LBDMW from Sweden

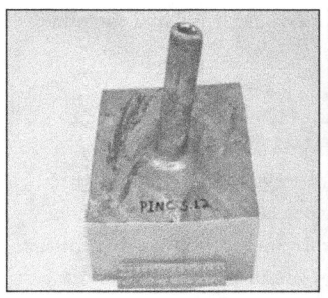

Figure 12 9 BMI penetrations

Figure 12 10 SBDMW from PNNL

The goal of PARENT is to continue the work that had begun in PINC and to apply the lessons learned to a series of open and blind international RR tests that will be conducted on a new set of piping components, including BMI, large-bore DMWs (LBDMWs), and small-bore DMWs (SBDMWs).

The surface conditions, access to both sides of the weld, and inspection conditions for the PARENT specimens provided the inspectors with less challenging conditions than those expected in field inspections of components in pressurized-water reactors. This finding supports the continuation of performance demonstration efforts in the nuclear industry to ensure adequate qualification of inspectors. The variability in team performance should be factored into the decisionmaking process when applying the results of this study.

The program developed and coordinated testing protocol and international testing and shipping schedules with the open and blind RR PARENT testing teams. Blind testing began in 2011, and open testing began in early 2012. The overview below describes the test blocks, test teams, NDE techniques, test protocol, and test schedule that will be used in the open and blind RR testing.

PARENT Teams and Blind RR Test Blocks

For PARENT blind RR testing, 14 teams will apply 6 unique NDE techniques to 14 test blocks, 5 BMIs (7 teams), 2 SBDMWs (approximately 300mm-diameter, 30 to 40mm thick walls/welds [7 teams]), 6 LBDMWs (approximately 900mm-diameter, 68 to 80mm-thick walls/welds [8 teams]), and 1 weld overlay. The flaw types in these test blocks include an electronic discharge machined notch, lack of fusion, laboratory-grown stress-corrosion cracking, weld solidification cracking, mechanical fatigue, and "reference reflectors" for each of the three categories of test blocks. During data analysis, the data taken on the reference reflectors may allow a comparison of results among different test teams that have used the same NDE technique.

PARENT Open RR Test Blocks and Teams

For PARENT open RR testing, 22 teams will apply 11 unique NDE techniques to 18 test blocks, 4 BMIs (8 teams), 10 SBDMWs (approximately 400mm-diameter, 30 to 40mm-thick walls/welds [19 teams]), and 4 LBDMWs (approximately 900mm-diameter, 85 to 90mm-thick walls/welds [4 teams]). In each of the three categories of test blocks, a variety of flaw types offer a wide range of flaw characteristics.

PARENT RR Test Results and Schedule

Two NRC NUREG reports will document the results of this research in 2015, one report for open testing and the other for blind testing. This research will provide the NRC with information that will aid in its independent evaluations of inservice inspection programs of licensed nuclear energy utilities that make assessments of the integrity of components that have DMWs in operating U.S. nuclear power plants. RR tests began in 2011 and are scheduled for completion during summer 2014.

For More Information
Contact Dr. Iouri Prokofiev, RES/DE, at
Iouri.Prokofiev@nrc.gov.

International Cooperative Research on Impact Testing

Background and Objectives

The U.S. Nuclear Regulatory Commission (NRC) believes that it is prudent for nuclear power plant designers to take into account the potential effects of the impact of a large, commercial aircraft on nuclear facilities. The agency's Office of Nuclear Regulatory Research (RES) has been conducting research in the area of impact loads on nuclear power plant structures that contributes to maintaining and developing critical skills needed to carry out the agency's mission of ensuring the safety of nuclear installations. Currently, the NRC participates in two international collaborative research programs in this area—one with the Technical Research Center of Finland (VTT) and one with the Nuclear Energy Agency (NEA) Committee on the Safety of Nuclear Installations (CSNI's) Working Group on Integrity and Aging of Components and Structures (IAGE) Concrete Subgroup.

Primary objectives of these programs are (1) to benchmark the various computer codes that the NRC staff and its contractors use in impact assessments against experiments, and (2) to synthesize the results of benchmarking into recommendations for good practices. These collaborative programs also provide opportunities to interact and exchange information with nuclear regulators abroad and with international nuclear safety organizations, ensuring NRC cognizance of ongoing impact research in various countries.

Anticipated benefits to the NRC from its participation in these programs include (1) reducing uncertainty associated with assessments of impact loads on nuclear installations, and (2) ensuring that the assessments performed for U.S. reactors represent the state-of-the-art in ensuring the safety of the public and protection of the environment.

Approach

Impact Test Agreement with the Technical Research Center of Finland

The NRC, the VTT, and nuclear regulators and nuclear safety research organizations in other countries participate in a multiyear international experimental research program, called IMPACT, to collect and analyze new data on the performance of reinforced and prestressed concrete walls subject to impact loads. All testing data under this program are provided by VTT using unique testing facilities not readily available elsewhere in the world, while the technical work of the NRC and the other participants focuses on analytical efforts.

Specific aims of the project include (1) obtaining new data on the time-varying shockloads from the impact of empty tanks, liquid-filled tanks, and hard missiles on rigid structures, (2) collecting new data on the response of reinforced concrete walls (e.g., displacements, strains) to impacts from those missiles (see Figures 12.11 and 12.12), (3) use of the new data to develop insights into the behavior of structures under impact conditions (e.g., perforation speeds); and (4) use of the new data to benchmark computer simulation codes.

Figure 12.11 Impact of hard missile on prestressed concrete wall

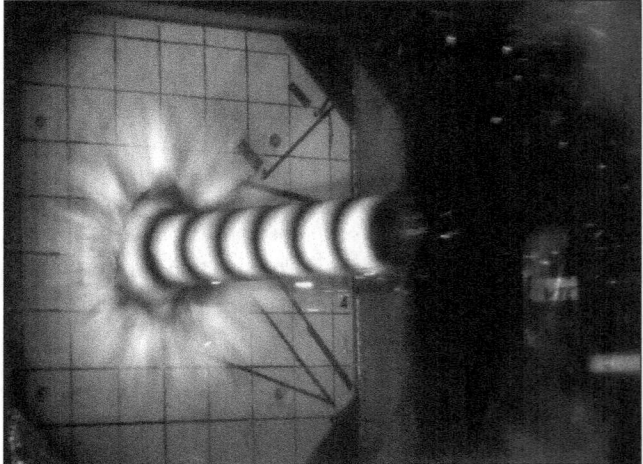

Figure 12.12 Impact of liquid-filled tank on reinforced concrete wall

VTT tests for the IMPACT program assess various reinforcement conditions, including prestressing, support conditions, slab thickness, impact speeds, and missile hardness. Each of the first two-phases of the program tested over 20 impacts on concrete slabs, and a similar number of those tests is planned for the third phase of the program already underway.

The IMPACT program includes regular workshops in which the participants exchange information on benchmarking, including benchmarking being done by RES staff (see Figure 12.13).

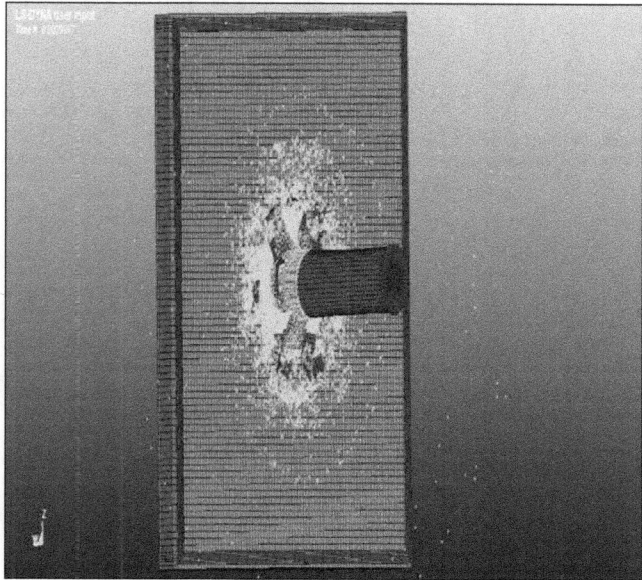

Figure 12.13 Modeling of liquid-filled soft missile impact on reinforced concrete slab

Research to Support the CSNI Project on Impact Assessment

The Concrete Subgroup of the IAGE, developed a round robin benchmark exercise entitled, "Improving Robustness Assessment Methodologies for Structures Impacted by Missiles." The purpose of this project is to develop guidance that outlines effective methods of evaluating the integrity of structures impacted by missiles and to compare various methods in a round robin study of impact data. The project uses publicly available data from simple, reduced-scale tests and will reinterpret previous tests with newly available data, modeling capabilities, and results. The exercise considers several types of structures, ranging from structural components and box-shaped structures of reduced size to reactor building-like structures of reduced size. The project is expected to produce state-of-the-art reports that collect the contributions, synthesize them, and propose recommendations for good practices. The first report was issued in December 2011 as Nuclear Safety Report NEA/CSNI/R(2011)8.

To support its participation in this program, the NRC contracted Sandia National Laboratories to benchmark different types of numerical simulation tools and to develop improved insights on modeling and damage criteria aimed at increasing confidence in numerical simulations for the assessment of existing and planned facilities.

For More Information:
Contact Dr. Jose Pires, RES/DE, at Jose.Pires@nrc.gov.

Round Robin Analysis of Containment Performance Under Severe Accidents

Background

The U.S. Nuclear Regulatory Commission (NRC) and the Atomic Energy Regulatory Board (AERB) of India are working together through the USNRC–AERB Nuclear Safety Co-Operation Program administered by Office of International Programs (OIP) as part of the Indo–U.S. Civilian Nuclear Agreement.

Objective

Through this program, the NRC and AERB agreed to organize and participate in the Standard Problem Exercise #3 (SPE#3) round robin analyses. The SPE#3 was built on the previous round robin analysis of the NRC and the Nuclear Power Engineering Corporation of Japan 1:4–Scale Prestressed Concrete Containment Vessel (PCCV) model tests conducted at the Sandia National Laboratories (SNL). Following the 1:4 scale PCCV, shown in Figure 12.14, and the International Standard Problem #48 (ISP 48) efforts, there was interest in investigating local effects and questions that had been unanswered previously because of modeling and computational limitations at the time and scope limitations of the previous efforts. One question that remained from the ISP 48 effort was how to reduce uncertainties in predicting the leak tightness and structural integrity of a PCCV under severe accident pressure and temperature conditions.

Approach

At the kickoff meeting of the SPE#3, held in Mumbai, India, in June 2010 the scope of the first phase of the SPE#3 was agreed upon. There was an interest in investigating the effects of containment dilation or ovalling on prestressing force, slippage of prestressing cables, steelconcrete interface failure mechanisms, and the use of nominal concrete strength properties versus in-situ conditions. Using the 1:4 Scale PCCV model as a starting point, seven international organizations (AERB, Nuclear Power Corporation of India Limited, Electricité de France, Fortum (Finland), Gesellschaft fur Anlagenund Reaktorsicherheit (German Agency for Reactor Safety), the NRC, and Scanscot Technology (Sweden) have participated in the round robin SPE#3. The participants agreed to create two local models and a full threedimensional model to investigate the local and global effects mentioned above.

Figure 12.14 Completed pre-stressed concrete containment vessel 1:4 scale model at SNL (Source: Figure 52, NUREG/CR-6906 (SAND2006-2274P)

The second workshop took place in Washington, DC, in April 2011 at NRC headquarters. In this workshop, participants discussed outcome of their investigations on tendon force as a function of containment dilation and tendon slippage, steelconcrete interface and failure mechanisms in the liner, various local effects, and the use of nominal versus in-situ conditions in the previous round robin analyses.

The third and final workshop took place in March 2012 at NRC headquarters to discuss phase two of the SPE#3 analyses. This phase of work has two distinct parts. In the first part, the participants examined methods to estimate leakage rate as a function of temperature and pressure. The second part consisted of enumeration of methods for predicting leakage of prestressed concrete containment vessels as a function of pressure and temperature; application of these methods to characterize the performance, in terms of leakage rate, under pressure and temperature; and the transition to probabilistic space.

For More Information
Contact Madhumita Sircar, RES/DE, at Madhumita.Sircar@nrc.gov.

Collaborative Research with Japan on Seismic Issues

Overview

The U.S. Nuclear Regulatory Commission (NRC) has a cooperative agreement with Japan Nuclear Energy Safety Organization (JNES) in the area of seismic engineering research. The intent and purpose for the collaboration is to maximize the overall benefits of each party's individual programs in the area of seismic safety research. The research performed and information exchanged under the collaborative program has expanded the data and knowledge base in the area of seismic testing and analysis. The continued largescale testing, as well as extensive development in the seismic risk assessment methods, provides valuable insights to the nuclear engineering community. This research program also provides the opportunity to interact with and exchange information with other Japanese organizations, ensuring NRC cognizance of all ongoing seismic research in Japan.

In particular, the 2011 Fukushima nuclear accident, the only nuclear accident caused by a natural disaster, shows the importance of seismic engineering research in enhancing the understanding of how nuclear power plants perform during rare but very large earthquakes and in improving the safety of nuclear power plants. This collaboration program provides an avenue to access the actual data for post-Fukushima research.

Objectives

The goal of this project is to better understand the seismic behavior of nuclear power plant (NPP) structures and components, obtain largescale seismic test data to benchmark analytical techniques, assess equipment fragility test data to reduce uncertainty associated with seismic probabilistic risk assessment (PRA) and seismic margin assessments, confirm and advance current seismic design and analysis methods, and provide the basis for regulatory positions for use in the evaluation of new reactor applications. The exchange of seismic information with Japan is very beneficial for the NRC in supplementing the agency's knowledge and in obtaining technically sound earthquake impact data.

Current Research Activities

The scope of the program includes analyses of various structures, systems, and components (SSC) for which JNES performs seismic tests and provides test results data to the NRC. These SSC include equipment, such as pumps, valves, fans, tanks, and

electric panels; degraded piping; concrete-filled steel members; and base-isolated structures and components. The project also includes the study of available earthquake response data recorded on structures at the Kashiwazaki nuclear power plant to assess the need to upgrade seismic response analysis methods for structures.

Application of JNES Equipment Fragility Tests

The goal is to assess the impact that the new JNES fragility test results may have on current U.S. PRAs and ways the data can be used for future U.S. PRAs. The collaboration uses both Phase I and Phase II JNES Fragility Test data. Phase I includes horizontal and vertical shaft pumps, electric panels, and control rod insertion capability, and Phase II includes valves, tanks, fans, support structures anchored to concrete, and overhead cranes. Figure 12.15 shows a JNES seismic fragility test of large vertical shaft pump. Figure 12.16 shows the JNES ultimate strength test of cylindrical liquid storage tanks subjected to simulated earthquake loading. Figure 12.17 shows the peak acceleration recorded at Kashiwazaki-Kariwa NPP (gal=cm/s^2, design values in parenthesis).

Figure 12.15 JNES seismic fragility test of large vertical shaft pump

Figure 12.16 JNES Ultimate Strength Test of Cylindrical Liquid Storage Tanks Subjected to an Earthquake

Assessment of Concrete-Filled Steel Components

The current standard and available test data from Japan will be used to analyze and assess the structural performance of concrete-filled steel members under seismic loading for a range of design configurations.

Assessment of Base-Isolated NPP SSC

The performance of base-isolated NPP SSC will be assessed, and recommendations will be provided on the application of base-isolation technology to existing and new NPP structures and components. JNES will provide the Japanese concept to apply base-isolation for NPP structures and components. The research will include collection of available test data and experience.

Study of the Effects of Floor Flexibility on Structural Response

To further apply the lessons learned from the Kashiwazaki earthquake regarding the effect of floor flexibility on building response, the available Kashiwazaki recorded response data are being used to investigate the effectiveness of the use of lumped mass and finite element models for determining the seismic response of structures with flexible floors. Information and earthquake recorded response data provided by JNES are being reviewed and evaluated. Seismic analysis methods and structural modeling approaches are being investigated to assess their suitability to match the recorded earthquake response for that structure. Insights gained will be used in upgrading NRC staff guidance regarding acceptable methods for determining the seismic response of buildings considering three-dimensional effects.

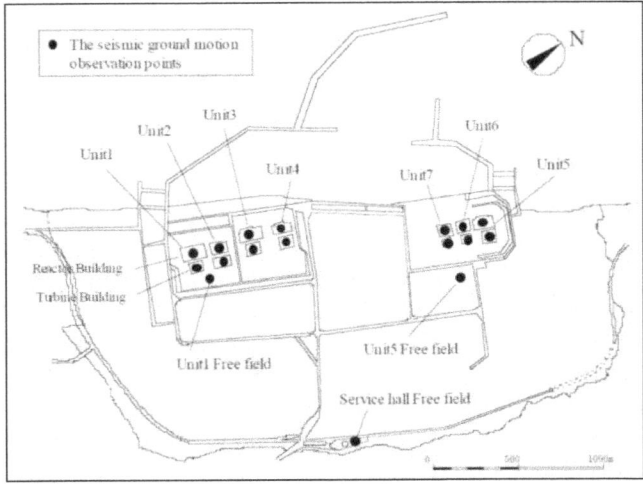

	North-South Component	East-West Component	Vertical Component
Unit 1	311 (274)	680 (273)	408 (235)
Unit 2	304 (167)	606 (167)	282 (235)
Unit 3	308 (192)	384 (193)	311 (235)
Unit 4	310 (193)	492 (194)	337 (235)
Unit 5	277 (279)	442 (274)	205 (235)
Unit 6	271 (263)	322 (263)	488 (235)
Unit 7	267 (263)	356 (263)	355 (235)

Figure 12.17 Peak Acceleration Recorded at Kashiwazaki-Kariwa NPP (gal=cm/s², design values in parenthesis)

Information Exchange Meetings

On a periodic basis, information exchange meetings are held in the United States and Japan to discuss the findings related to the above collaboration activities, as well as other information that each side may have developed related to the seismic safety research being performed in either country. At least one information exchange meeting is held each year in either the United States or Japan.

For More Information
Contact Dr. Scott Stovall, RES/DE, at Scott.Stovall@nrc.gov.

Agency Forward-Looking and Long-Term Research

Background

Forward-Looking Research

The U.S. Nuclear Regulatory Commission (NRC) currently identifies, as a matter of routine planning, Forward-Looking research activities that support potential future regulatory needs. The agency identifies and pursues these Forward-Looking research activities during the normal course of the planning and budgeting processes.

Long-Term Research

Each year since 2007, the staff has prepared Commission papers on Long-Term research activities. The papers discuss candidate Long-Term research topics and estimate resource needs for use in budget preparation. For the purposes of the annual Commission papers, Long-Term research is defined as research that is not already funded or otherwise being conducted that will provide the fundamental insights and technical information needed to address potential technical issues or knowledge gaps to support anticipated NRC needs in the future (i.e., more than 5 years). Long-Term Research Program (LTRP) projects generally last 1 to 2 years and are feasibility or scoping studies that assess if future research on the topic should be pursued with additional research.

Approach

The NRC performs regulatory research to support the achievement of the goals identified in its Strategic Plan. These goals are established to provide protection of public health and safety and the environment; provide for the secure use and management of radioactive materials; promote openness in the NRC's regulatory processes; ensure that NRC actions are effective, efficient, realistic, and timely; and promote excellence in agency management.

Both Forward-Looking and Long-Term research could support possible new program areas, support development of technical bases for a range of anticipated regulatory decisions, address emerging technologies that could have future regulatory applications, or could be used to develop plans to implement needed research.

Process for Determining Long-Term Research Projects

The process for determining the research activities that should be funded under the Long-Term research plan starts with a candidate list of potential projects submitted by the research, regulatory, and regional offices. In addition, previously suggested projects that were not funded are included in the candidate list. A committee composed of eight senior level staff members from the Office of Nuclear Regulatory Research (RES) and the regulatory offices reviews, evaluates, and rates these potential projects. The committee uses five evaluation criteria to rate each project. The five criteria address leveraging resources, advancing the state-of-the-art, providing an independent assessment tool to the NRC, applying to more than one program area, and addressing technical or regulatory gaps created by technology.

The committee forwards the results of the review to the RES Office Director, who, in coordination with the regulatory offices, submits an annual information paper to the Commission that describes the proposals identified for funding, projects in progress, and the status of the overall program. During the planning, budgeting, and performance management process, the RES Office Director, along with the directors of the agency's regulatory offices, agrees on those Long-Term research proposals that should receive a "high" priority and be recommended for funding.

Long-Term Research Projects

The LTRP began funding projects in fiscal year (FY) 2009. Since then, numerous projects have been funded and completed. These include the following:

Advanced Level 2/3 Probabilistic Risk Assessment (PRA) Modeling Techniques. An internal scoping study was completed in May 2009. Based on this LTRP project, it was recommended that continuation of the work occur following the budget planning of research and development of a dynamic event tree methodology. That development work is ongoing at Sandia National Laboratories and the University of Maryland. The work at this point focuses on the implementation of a tool development plan that was finalized in early January 2011. Specifically, the contractors are developing an ADS-IDAC-MELCOR model to dynamically simulate station blackout at the Surry plant.

Integral Effects Test Facilities for Advanced Non-Light-Water Reactors. A literature review was conducted to determine the state-of-the-art in gas reactor experiments and analysis tools. Based on the literature review, a list of key thermal hydraulic and reactor physics phenomena was developed and examined to determine which phenomena require additional investigation through analysis or experiment. As a result of this LTRP project, a decision to construct a test facility was made and supported.

The facility is being constructed with the experimental test program, and code validation efforts are expected to be complete in 2013.

Fire Safety of Digital Instrumentation and Control and Electrical Systems (see Figure 12.18). NUREG/CR-7123, "A Literature Review of the Effects of Fire Smoke on Electrical Equipment," was published to document the current state of knowledge of smoke damage to control circuits.

Figure 12.18 Cone calorimeter for analyzing the impact of smoke on electrical equipment

Sensors and Monitoring to Assess Grout and Vault Behavior for Performance Assessments. To support performance assessment of large concrete vaults and cementitious grout monoliths containing radioactive waste (waste incidental to reprocessing), the National Institute of Standards and Technology has performed a preliminary evaluation of the state-of-the-art of sensors, nondestructive evaluation methods, and relevant geophysical techniques that may be used to quantify changes in the chemical (e.g., redox state) and structural properties (e.g., crack initiation, development, and propagation) of large engineered waste isolation systems. Nondestructive methods, such as acoustic emission sensing (Figure 12.19), measurement of electrical properties, and ultrasonic pulse velocity were assessed in the draft report entitled, "Sensors and Monitoring To Assess Grout and Vault Behavior for Performance Assessments.

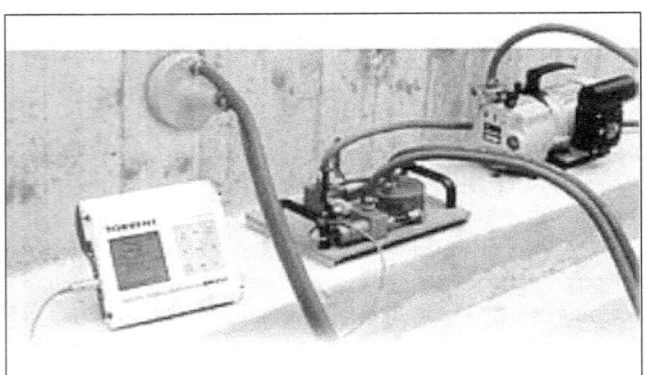

Figure 12.19 In-situ surface air concrete permeability test apparatus

Other project examples presently included in the LTRP are:

- Advanced fabrication techniques (FY 2010).

- Boiling-water reactor burnup credit and evaluation of newly available isotopic and criticality data (FY 2011).

- Advanced PRA (FY 2011).

- Nondestructive evaluation and surveillance of civil structures (FY 2011).

- Uncertainty methods for PRA (FY 2011).

- Smoke effects and transport (FY 2011).

- Extended in situ real time monitoring (FY 2011).

- Advanced light water reactor fuels (FY 2011).

- Smart grid impacts on nuclear power plants (FY 2012).

- Advanced reprocessing: identification of regulatory issues associated with electrochemical processing of spent nuclear fuel (FY 2012).

- Safety and regulatory issues of the thorium cycle (FY 2012).

- Evaluating remaining service life of nuclear power plant concrete structures (FY 2013).

- Using paleoflood information to assess climate variability contribution to flooding risks at nuclear power plant sites (FY 2013).

- Spectroscopy for early detection of concrete degradation (FY 2014).

- Reducing uncertainty in dam risk analysis (FY 2014).

- Advanced knowledge engineering tools to support risk-informed decisionmaking (FY 2014).

For More Information
See "The Office of Nuclear Regulatory Research Long-Term Research Program," August 2012, NUREG/BR-0506.

Contact Sergio Gonzalez, RES/DSA, at Sergio.Gonzalez@nrc.gov.

NRC FORM 335
(12-2010)
NRCMD 3.7

U.S. NUCLEAR REGULATORY COMMISSION

BIBLIOGRAPHIC DATA SHEET

(See instructions on the reverse)

1. REPORT NUMBER
(Assigned by NRC, Add Vol., Supp., Rev., and Addendum Numbers, if any.)

NUREG-1925, Rev. 2

2. TITLE AND SUBTITLE

Research Activities FY 2012-FY 2014
U.S. Nuclear Regulatory Commission

3. DATE REPORT PUBLISHED

MONTH	YEAR
August	2013

4. FIN OR GRANT NUMBER

5. AUTHOR(S)

Office of Nuclear Regulatory Research (RES) Staff

6. TYPE OF REPORT

Technical

7. PERIOD COVERED (Inclusive Dates)

2012-2014

8. PERFORMING ORGANIZATION - NAME AND ADDRESS (If NRC, provide Division, Office or Region, U. S. Nuclear Regulatory Commission, and mailing address; if contractor, provide name and mailing address.)

Office of Nuclear Regulatory Research
U.S. Nuclear Regulatory Commission
Washington, DC 20555-0001

9. SPONSORING ORGANIZATION - NAME AND ADDRESS (If NRC, type "Same as above", if contractor, provide NRC Division, Office or Region, U. S. Nuclear Regulatory Commission, and mailing address)

same as above

10. SUPPLEMENTARY NOTES

11. ABSTRACT (200 words or less)

The Office of Nuclear Regulatory Research (RES) supports the regulatory mission of the U.S. NRC by providing technical advice, tools, and information to identify and resolve safety issues, make regulatory decisions, and issue regulations and guidance. This includes conducting confirmatory experiments and analyses; developing technical bases that support the NRC's safety decisions; and preparing the agency for the future by evaluating the safety aspects of new technologies.

The NRC focuses its research primarily on near-term needs related to the oversight of operating reactors, as well as to new and advanced reactor designs. RES develops technical tools, analytical models, and experimental data to allow the agency to assess safety and regulatory issues. The RES staff uses its expertise to develop these tools, models, and data or uses contracts with commercial entities, national laboratories, and universities or in collaboration with international organizations.

This NUREG presents research conducted across a wide variety of disciplines, ranging from fuel behavior under accident conditions to seismology to health physics. This research provides the technical bases for regulatory decisions and confirms licensee analyses. RES works closely with the NRC's licensing offices in the review and analysis of high-risk events and provides its expertise to support licensing. RES has organized this collection of information sheets by topical areas that summarize projects currently in progress. Each sheet provides the names of the RES technical staff who can be contacted for additional information.

12. KEY WORDS/DESCRIPTORS (List words or phrases that will assist researchers in locating the report.)

research, reactor safety codes and analysis, severe accident research and consequence analysis, radiation protection and health effects, risk analysis, human factors and human reliability, fire safety research, seismic and structural research, materials performance research, digital instrumentation and control research, Fukushima research, international cooperative and long-term research,

13. AVAILABILITY STATEMENT

unlimited

14. SECURITY CLASSIFICATION

(This Page)

unclassified

(This Report)

unclassified

15. NUMBER OF PAGES

16. PRICE

www.nrc.gov
NUREG-1925, Rev. 2
August 2013

www.ingramcontent.com/pod-product-compliance
Lightning Source LLC
Chambersburg PA
CBHW080238180526
45167CB00006B/2320